A Greener Vision of Home

Social History, Popular Culture, and Politics in Germany
Geoff Eley, Series Editor

A History of Foreign Labor in Germany, 1880–1980: Seasonal Workers/Forced Laborers/Guest Workers
Ulrich Herbert, translated by William Templer
Reshaping the German Right: Radical Nationalism and Political Change after Bismarck
Geoff Eley
The Stigma of Names: Antisemitism in German Daily Life, 1812–1933
Dietz Bering
Forbidden Laughter: Popular Humor and the Limits of Repression in Nineteenth-Century Prussia
Mary Lee Townsend
From Bundesrepublik *to* Deutschland: *German Politics after Unification*
Michael G. Huelshoff, Andrei S. Markovits, and Simon Reich, editors
The People Speak! Anti-Semitism and Emancipation in Nineteenth-Century Bavaria
James F. Harris
The Origins of the Authoritarian Welfare State in Prussia: Conservatives, Bureaucracy, and the Social Question, 1815–70
Hermann Beck
Technological Democracy: Bureaucracy and Citizenry in the German Energy Debate
Carol J. Hager
Society, Culture, and the State in Germany, 1870–1930
Geoff Eley, editor
Paradoxes of Peace: German Peace Movements since 1945
Alice Holmes Cooper
Jews, Germans, Memory: Reconstruction of Jewish Life in Germany
Y. Michal Bodemann, editor
Exclusive Revolutionaries: Liberal Politics, Social Experience, and National Identity in the Austrian Empire, 1848–1914
Pieter M. Judson
Feminine Frequencies: Gender, German Radio, and the Public Sphere, 1923–1945
Kate Lacey
How German Is She? Postwar West German Reconstruction and the Consuming Woman
Erica Carter
West Germany under Construction: Politics, Society, and Culture in Germany in the Adenauer Era
Robert G. Moeller, editor
A Greener Vision of Home: Cultural Politics and Environmental Reform in the German Heimatschutz *Movement, 1904–1918*
William H. Rollins
A User's Guide to German Cultural Studies
Scott Denham, Irene Kacandes, and Jonathan Petropoulos, editors

A Greener Vision of Home

Cultural Politics and Environmental Reform in the German *Heimatschutz* Movement, 1904–1918

William H. Rollins

Ann Arbor

The University of Michigan Press

Published in the United States of America by
The University of Michigan Press
Manufactured in the United States of America
⊗ Printed on acid-free paper

2000 1999 1998 1997 4 3 2 1

A CIP catalog record for this book is available from the British Library

Library of Congress Cataloging-in-Publication Data

Rollins, William H., 1962–
A greener vision of home : cultural politics and environmental reform in the German Heimatschutz movement, 1904–1918 / William H. Rollins.
p. cm. — (Social history, popular culture, and politics in Germany)
Includes bibliographical references (p.) and index.
ISBN 0-472-10809-3 (acid-free paper)
1. Green movement—Germany. 2. Germany—Environmental conditions—History. 3. Environmental protection—Germany—History—18th century. 4. Environmental protection—Germany—History—19th century. I. Title. II. Series.
GE199.G3R65 1997
363.7'0576'0943—dc21 97-4705
CIP

Contents

Preface

One of the most striking aspects of today's ecological crisis is the mental paralysis that has accompanied it. Never, one could say, have so many known so much and yet done so little. We know that fossil fuels contribute to global warming—and yet we in the industrialized world go on burning them as if there were no tomorrow. We know that many of our agribusiness techniques are exhausting soils, poisoning waters, and destroying entire regions—nevertheless we continue to shut our eyes to these costs. We know that atomic fuels, heavy metals, and PCBs have the potential to contaminate entire food chains—and still we tolerate them in our industrial processes. In short, we know of hundreds, if not thousands, of measures that should be taken if life as we know it is to continue on this planet. In spite of all this knowledge real change has been slow in coming. There can be no doubt that strong material interests are responsible for much of the inertia: automakers, for example, will continue to fight for more highways, more parking places, absurd guidance systems, and more pollution as long as they live. But why do the rest of us follow their narrowly focused agenda? We know categorically that they are wrong, and yet all too often we succumb to the kind of comfortable status-quo solutions they offer. This curious state of affairs indicates that there is another dimension to our environmental mess, one perhaps more intractable than the political power of industry lobbies. If we are unable to act on even the clearest information, it is obvious that "the environmental crisis lies equally in our consciousness" (Meyer-Abich 1990, 11).

This crisis of consciousness is rooted, moreover, at a very basic level—one could perhaps better call it a crisis of unconsciousness, for we tend to disregard ecological priorities most consistently in our unreflected behavior, our mental habits, our everyday perceptions and values. Scientific predictions and theoretical analyses are often powerless to overcome this kind of overall cultural inertia. A mundane example is the need in the United States for a rail-based system of local and regional mass transportation. In spite of the fact that study after study has shown that a massive switchover to this kind of post-automobile system is necessary, very few Americans are willing to devote large amounts of tax money to rebuilding the passenger-rail infrastructure. It is not that they are necessarily *against* rail travel per se, they are simply comfortable with their automobiles and unfamiliar with the different rhythms and spatial patterns of a life structured around rail transportation. Here, as in other aspects of our ecological problems, change will be dependent on efforts to approach people within the context of their day-to-day existence. Only by showing how the most ordinary things can and ought to change will we be able to reorient modern industrial culture as a whole. We must therefore challenge the assumptions that we take every day for fact, we must seek to construct new ideals for the future, and we must also be prepared to lay new accents on history.

This book is aimed chiefly at fulfilling the latter requirement by examining a small part of German environmental history. I have concentrated here not on a factual account of ecological destruction, but have rather retained the humanist's concentration on an analysis of societal values. It is a methodological emphasis that Donald Worster has defended with admirable succinctness: "We are facing a global crisis today, not because of how ecosystems function but rather because of how our ethical systems function" (Worster 1993b, 27). But as Worster himself has shown, the ethical systems that really matter are not the ones discussed in textbooks of philosophy: instead, what happens to the environment is determined by widespread economic and cultural priorities, by people's "common sense" attitudes toward nature. This foundation in the most quotidian forms of practical consciousness gives a study of German environmental history at the turn of the century an unfortunate topicality. The problems faced then and in that country are, to a great extent, the problems we are still facing today, and for which no one has yet found a reliable cure. Nevertheless I will suggest in the conclusion that we can learn from the Heimatschutz movement. Precisely because these Germans were unable to rely on ecological data to make their arguments for them, they were able to grasp the need for a culturally active brand of environmentalism, one that gave people something positive to aim for: a greener vision of home.

Acknowledgments

The initial phase of research for this book was carried out in 1990 and 1991 and was made possible by a generous grant from the Berlin Program for Advanced German and European Studies of the Free University of Berlin and the Social Science Research Council with funds provided by the Volkswagen Foundation and the German Marshall Fund of the United States; the work profited at this stage from Jürgen Kocka's comments and the cross-disciplinary enthusiasm of my fellow researchers. For a second opportunity to do necessary on-site research I am indebted to the German Academic Exchange Service, which provided a Research Grant for Recent PhDs in the fall of 1995. For a chance to discuss some of the thoughts in my conclusion I thank the Max Kade Institute in Madison, Wisconsin, which sponsored a conference on "Aldo Leopold: His Land Ethic and Influence in Germany and the U.S." in October 1994; I would further like to express my sincere appreciation for the support of Gerhart Trommer and his colleagues at the Institut für Biologie-Didaktik of the J. W. Goethe-Universität in Frankfurt, who heard an expanded version of the United States/Germany comparison in November 1995. For their kind assistance I would like finally to thank the personnel at all of the archives listed in the bibliography; my thanks apply equally to all of the hardworking interlibrary loan librarians who have helped obtain indispensable materials at various stages of the project, starting at the University of Wisconsin-Madison and continuing on to Gustavus Adolphus College and the University of Kentucky in Lexington.

A complete list of acknowledgments for any work would probably read something like a biography; in the case of a study on a landscape protection organization, it seems allowable to link this with a topography of influence. I was probably predisposed to look at the Heimatschutz more closely by my mother, Jane Ann Muehlke, and my father, James Alfred Rollins. I would like to think that the book represents a kind of delayed expression of thanks not only for their parenting, but more specifically for their lessons in naturalism in the burned-over scrub forests and abandoned fields of southern Maine. My studies at Dartmouth College and at the university in Freiburg were also propitiously situated, taking place not only in areas of great natural beauty, but more importantly in communities where a kind of environmental culture is increasingly palpable. As I gaze out my window at the volcanic tufts of the Port Hills, though, it becomes clear to me that the book would have been impossible without the unique array of people and experiences that are, for me, connected to Madison, Wisconsin. Klaus Berghahn, Jim Steakley, and Mark Bassin all helped with input to the dissertation; an even greater debt is owed to my thesis adviser, Jost Hermand, whose engaged readings of German culture have pointed the way for me. My hope is that, in its better parts, the book may reflect something of the spirit of these and other fellow Madisonians, including Kathrin Bower, Antje Harnisch, Peter Höyng, Mark Kreitzer, Sherri Sokeland, Amy Stapleton, Friedemann Weidauer—and above all my partner, Lisa Bilsky.

William H. Rollins
Christchurch, New Zealand
May 1997

Introduction

The turn of the century was a time of unprecedented economic expansion in Germany. The population soared, cities sprawled and fused into urban conglomerations, and massive infrastructure projects—new roads, canals, telegraph and electrical wires, more rail corridors—sliced up previously intact areas of the landscape. German industry boomed, but at the same time it dumped more and more pollutants into the waters, soil, and sky. Cities and municipalities were often no better. Their recently installed drain systems only served to guide untreated wastes quickly and efficiently into German rivers, transforming once pleasant streams into reeking open-air sewers. As a result of all this, local air and water quality deteriorated to almost unbelievable levels, and large stretches of the wounded German landscape began to resemble a raw construction site. A sketch from 1910 (fig. 1) gives us an idea of how the countryside around Essen disappeared under the onslaught of industrial development. Farther away from the cities, extractive industries gashed their way through the environment in search of building stone, coal, metal ores, and potassium salts. There was an overdose of progress even in the agricultural hinterlands, where farmers and foresters laid out vast monocultures in their desperate efforts to increase efficiency and boost production. Only one lifespan earlier, in the time of the Romantics, the German countryside had seemed to abound in leafy hiding places for the elusive "blue flower" of spiritual redemption. Now the wanderer was more likely to encounter dreary plains of sugar beets and plantations of spindly young pines, barren expanses punctuated only by stone-lined gutters that had once been lush streambeds. To those with eyes to see, the land bore all of the signs of man-made ecological disaster.

The distinctive feature of life at the turn at the century, however, was that hardly anyone was able to see and to react to the mounting crisis. However divided they might have been by class and confession, Germans stood firmly united by their belief in technological and industrial progress,

Fig. 1. The wounded countryside of Germany's urban-industrial expansion. (From Th. Goldschmidt AG archive. Used by permission.)

a process whose future benefits seemed so certain and so clear to the conventional wisdom of the time that evidence to the contrary, and any perception of environmental problems in particular, was rejected almost before it had been consciously registered. To have finally activated this area of public perception and concern was the great achievement of the *Bund Heimatschutz,* a national citizens' group that was founded in 1904 and grew to about 30,000 members by 1914. Long before the "Limits to Growth" and Greenpeace, this "League for Homeland Protection" was centrally concerned to criticize the denuded and degraded modern landscape. In doing so it opened people's eyes to the world around them and helped initiate a public debate about man's use of the environment that continues to this day.

The purpose of the new organization was defined in its charter as "protecting the natural and historically developed uniqueness of the German homeland" (*Mitteilungen des Bundes Heimatschutz* 1904, 7). In principle this meant that the Heimatschutz was interested in thousands of things that might contribute to a given locale's character, from large mountains right down to the smallest bit of native craftsmanship. What the Heimatschutz did in actual practice was to concentrate on the kind of perspective shown in the simple woodcut in figure 2. Adopted by the *Bund Heimatschutz* for the cover of its journal, this picture reveals that the Heimatschützer's primary interest was in large-scale ensembles. This is not to say that they ignored smaller items: there was much reflection among Heimatschutz members about street signs, for instance, and even the most modest piece of advertising or door hardware was likely to come under their scrutiny. Nevertheless it was the broad sweep of one's vision that counted most for them, for it alone allowed the synthesis of buildings, roads, fences, billboards, fields, hedges, woods, riverbanks, bridges, and countless other disparate components of the landscape into that overall physical–aesthetic entity to which they gave the name *Heimat.*

The Heimatschützer were driven by a conviction that the cumulative effect derived from the community surroundings constituted a key element in one's quality of life. But Germany's capitalist-industrial society, they found, programmatically ignored this aspect of collective existence in its favoring of private rights to the environment. Heimatschützer therefore carried their case to the public in a series of individual battles and campaigns for legislation. Perhaps the most visible result of their pressure was a law passed in Prussia in 1907 that regulated construction or advertising in particularly noteworthy public vistas. The idea at the heart of Heimatschutz went much further, though. The Heimatschützer insisted on maintaining the balance and beauty of the entire ordinary working landscape, not just individual scenic areas. This was environmentally

Fig. 2. *Bund Heimatschutz* icon by the artist Otto Ubbelohde. (From *Heimatschutz* 1908, 1)

significant in that the aesthetic ideal of Heimat led to a demand for a thoroughgoing reform of German land-use practices, including such things as the retention of hedgerows, a commitment to mixed forests, and a halt to excessive stream regulation. This new vision of home was socially progressive as well: its conservationist agenda, unlike strictly preservationist approaches, did not shy away from the difficult problems of industrial development and social equity that are implicit in true environmental reform. As the first chairman of the *Bund Heimatschutz* put it, the goal was an ecologically tenable balance of all these things, a "harmonious culture, one that combines respect for the earth with its use" (Schultze-Naumburg 1917, 330).

As important as the objectively ecological aspect of Heimat is, the subjective dimensions are no less noteworthy. *Heimat* as a description of a place went beyond the immediately sensible elements of a landscape; even today the word resonates powerfully at more emotional-affective levels of meaning, connoting psychological comfort as well as a wider social cohesiveness and rootedness. In this double layer of meaning lies not only the key to the movement's achievements, but also its claim to continuing significance. For with the word *Heimat* it became possible to express the important new view that each and every one of us has an emotional and ethical stake in our local environment, as a commonly held home. This is not to say that the movement can be explained through lexical analysis, however. On the contrary, it is necessary to go in quite the opposite direction. As we shall see, it required the active intervention of the Heimatschutz movement in order to *produce* this specific meaning of *Heimat* and to make it available as a paradigm of value. As we investigate its use of the word, we will find that its meaning—not in a dictionary, but rather in the sphere of real social relations—was essentially located in a program of applied environmental reform.

This study is thus an investigation of environmental culture. People did not always worry about the health of the land around them; if they did, they often worried about completely different aspects than those we now focus on. It is important to see that this historical change in priorities did not come about overnight, nor even primarily as the result of explicit scientific teachings. Instead it seems clear that the range of ways in which we think about our world is and always has been partly determined by the reinforcing, innovating, and sanctioning effects of what are often called cultural *norms.* An example of how this cultural element can influence environmental behavior is given by the restrictions placed on miners in medieval Europe. Historians have remarked that these miners were encouraged to think about their activities as a violation of the body of Mother Earth; although ore was obviously still removed from hillsides, the miners were clearly conditioned to avoid all-too-zealous production. This conditioning went so far as to require elaborate ceremonies of apology for and ritual cleansing of what was, for all intents and purposes, an environmental sin (cf. Merchant 1980, 29ff.).

No one is arguing that this is the only kind of pro-environment culture possible, though. Like the more explicitly enforced behavioral norms we call laws, cultural norms are the products of flesh-and-blood human beings. Changing material interests, new perceptions, and new philosophies often require entirely new definitions of what is acceptable, important, good, or pleasing. In the case of the miners' sanctions we can say that their religious cast was appropriate to the images, philosophy, and kind of education—in short, the culture—in which they existed. In our secularized age the generally available systems of thought and value are quite different. Few of us are worried anymore about offending God or a personified Mother Earth with our actions. If we are nonetheless beginning to share with our medieval forebears an affective revulsion against "environmental sin," it is because something is starting to take root in our emotional life without, however, transgressing against our secular logic.

This "something" that is still in the making—and to which the Heimatschutz made a significant early contribution—could be labeled an "environmental consciousness," or alternatively an "environmental culture." The latter term is perhaps to be preferred, as the word "consciousness" has specific connotations of rationality and the Rationalist tradition of Descartes. For a study of environmentalism, this heritage brings with it two unwanted dimensions: first, its association with radically anthropocentric forms of philosophy, and second, its tendency to privilege fully articulated phenomena, distinct "ideas" as opposed to diffuse "feelings" and the evidence of the senses. The term *consciousness* is also problematic in that it conveys no sense of a social or material basis. It simply "is," somewhere, and for no apparent reason. By contrast this investigation of the Heimatschutz proceeds from the assumption that environmental ideas are made and fought for by men and women under historically specific conditions, in relation to the whole range of their social and material existence. It is in this sense that the term *environmental culture* is preferred, since we are interested not only in an articulated form of ecological knowledge, but also in the processes by which a wider form of environmental behavior was developed, propagated, and brought to bear in the realm of social practice.

After a cursory examination of the Heimatschutz movement one cannot escape the impression that its ideas about controlling growth and maintaining landscapes anticipated key concerns of today's "Green" and ecology movements. Given the widespread interest in and support for the aims of these modern movements, one would expect that a predecessor such as the Heimatschutz would be concomitantly well researched, well understood, and perhaps even appreciated. Surprisingly, this is not the case. In part it has to be attributed to the enormity and urgency of the tasks facing all of today's environmentalists, for whom practical problem-solving and scientific research are naturally a priority. The first wave of historical research into environmentalism tended to mirror this set of pragmatic concerns, focusing either on ecological science and the history of its ideas or on the development of well-defined governmental institutions. Only in the last fifteen years or so have scholars begun to tackle a much wider brand of environmental history, one that tries to link a given society's regime of ecological interactions with the priorities and values that are propagated at the level of everyday popular culture. The amount of progress being made on this front is moreover different in relation to different nations: research on popular environmental history in the United States, for example, began earlier and arguably dwarfs the comparatively small body of scholarly work that has been done on Germany and other European countries.

For Germany in particular the legacy of National Socialism has had an additional powerful restraining effect. To all who would listen, Hitler's National Socialist regime advertised itself as environmentally friendly; many conservationists of the time were taken in by the Nazis' finely tuned rhetorical support for the popular environmental goals of "Heimat," "landscape," and "nature." Since a number of Germany's leading conservationists also actively embraced the entire racial philosophy of the Nazis, the melding of the two phenomena in the public mind was nearly complete. This difficult historical background helps explain why Heimat remains a controversial ideal even under the renewed attention being given to regionalism, with today's proponents summarily rejecting all earlier forms of Heimat enthusiasm as "pure resentments, ideologies and utopias" (Lipp 1986, 338). Such blanket generalizations help explain why the Heimatschutz movement until recently was never anything more than a negatively tinged footnote in cultural, historical, and even environmental studies of Wilhelmine Germany.

Scholarly discourse has been further influenced by the more general political connotations of *Heimat.* It is a word like *nation* or *community* in that it has always connoted a feeling of belonging to a social collective. Critics have long observed that such concepts of group identity can be dangerous. Some focus on "positive integration" and worry that a monolithic identity will obfuscate real differences within a group, thereby contributing to a dominant group's hegemony. The best example of this is the nineteenth-century ideology of nationalism, which is widely interpreted as the attempt by elites to cover up the divergence of interests in class society. Heimat has been seen in this manner as a kind of sub-discourse of nationalism, an "attractive phrase of State ideologies of integration" (Greverus 1972, 333).

Negative integration is the focus of a second line of critique, which starts from the fact that words such as *nation* and *community* function to define the boundaries of a collective. While this bounding includes some aspects, individuals, or territories, critics observe, it necessarily excludes others—a rhetorical distinction that can provide the basis for more violent, physical forms of social exclusion. This destructive potential has repeatedly been identified in examinations of national and religious communities, and it comes as little surprise to see that the critique has been applied to *Heimat* as well. According to a recent article this concept is fundamentally "about sustaining cultural boundaries and boundedness" and is necessarily "defined by contrast to all that is foreign or distant." The two authors conclude that Heimat is therefore "an ominous utopia . . . rooted in that intolerance of difference, that fear of the 'other', which is at the heart of racism and xenophobia" (Morley and Robins 1990, 4, 10, 19). This sharp rejection clearly builds on the term's fateful association with National Socialism, but it goes even further with its supposition that *any* use of the concept is inherently flawed.

The circumstances would seem to demand that any reappropriation of the idea of Heimat must examine its political contours in some detail. This demand is a fair one, but at the same time it is only reasonable that we conduct our assessment based on a historically specific usage of the word. Like *nation* and *community, Heimat* ought to be judged by taking a close look at the actions undertaken in its name. A recent book-length study by Celia Applegate, subtitled "The German Idea of Heimat," provides a useful starting point because it links the term with the specific history of nineteenth-century Germany. In a nutshell, Applegate shows that the cultural value of Heimat was constructed as an important compensatory realm in order to deal with the dislocating, centralizing, and homogenizing tendencies of Germany's development. The strength of this work is that it takes Heimat sensibilities seriously and shows how this value entered into the real political process of a specific area, the *Pfalz.* If Heimat was, as Applegate writes, "an invention and a mythology," it was also a "potentially useful" one in which lots of citizens believed and upon which they acted.

One of the first applications of the new mythology came in the 1820s by the "progressive bureaucracy" in Bavaria. Here a group of largely bourgeois reformers was able to use Heimat to broaden citizenship rules. Their newly crafted "right to Heimat" (*Heimatrecht*) eliminated narrow communal restrictions and gave Bavarians the same right to vote and receive social welfare anywhere within the boundaries of the larger state. As Applegate argues, the subsequent history of Heimat largely follows this same trajectory of mediating between established paradigms of "belonging" and the modern realities of an industrializing society. As a realm of regional identity it "suggested a basic affinity between the new, abstract political units and one's home, thus endowing an entity like Germany with the emotional accessibility of a world known to one's own five senses" (8, 10). In the era of a new and unfamiliar nationalism, Heimat was the sensible face of the popular collective.

There is much explanatory power in this model, some of which can be applied directly to an examination of the Heimatschutz. It suggests, for instance, that the value of Heimat was loaded with a real subjective commitment to one's fellow citizens. But Applegate's analysis also goes further and shows that this solidarity had a specific embodiment in the physical realm. Her discussion of the Pfalz Forest Association—the closest analog to the Heimatschutz movement that existed in the *Pfalz,* and one that flowered during the same years before World War I—makes clear that the new interest in nature was strongly informed by an "inclusive and tolerant" spirit. Indeed, the countryside offered a unique forum for constructing a community, and it functioned as "a source of collective identity not to be found in the urban environment" (77).

From the perspective of an investigation of the Heimatschutz as an early environmental movement, this is the point at which the limitations of Applegate's study become apparent. She does not develop the full implications of the striking conjunction of political community and physical surroundings that we find in Heimat. Her focus remains fixed on political history. Even though she observes that the turn-of-the-century Heimat movement "transformed people's conception of nature," how this was done and what it meant in practical terms is left virtually unexplored; the countryside is interesting to her only as a "source of collective identity" and not as an environmental issue (59). Still, Applegate's study constitutes an important rehabilitation of the idea of Heimat, and as such it is a building block for a reassessment of the Heimatschutz. As we look at this movement it will become clear that Heimat was as much about an environmental community as it was about a political one.

Sociologists and anthropologists have taken an approach to Heimat that is altogether different from the political mythology approach of Applegate. Heimat, for them, truly exists—as a real, universal human need. Ina-Maria Greverus has contributed several articles and books to this discussion, starting with her 1972 book, *Der territoriale Mensch.* Her work is doubly important because it responds to a range of specific uses of the concept *Heimat* in Germany, both contemporary and past. Against the majority of these uses she has defended an anthropological definition of *Heimat* that sees it as a kind of home territory, one that guarantees orientation and identity along with providing opportunities for individual growth and activity. The key here is Greverus's emphasis on the subjective needs of the individual. *Heimat* in her sense of the word is an overall social/political framework that provides a meaningful sphere of action and self-realization for each respective individual:

> The political task of creating Heimat conditions aims higher than the quantitative and necessary improvement of environmental protection, social justice and equality. It aims beyond all that at supporting the individual in his own respective and thus alternative arrangement in a territory that he can appropriate for himself as "Heimat" in an active and self-determining manner. The life quality "Heimat" is neither inborn, nor can it be prescribed (*verordnet*). Rather, it is a production of the active subject who appropriates the environment for himself. (1979a, 17)

Greverus's individual-psychological definition is informed by a careful consideration of what she sees as the dual pitfalls of Heimat programs in nineteenth- and twentieth-century Germany. One is the tendency, described above, to construct a monolithic political identity that squelches real class or individual differences and teaches only "a readiness to conform" (1979a, 10). Greverus associates this strategy primarily with conservative and fascist ideologies. The other pitfall comes in related attempts to reify Heimat as a particular constellation of objects, houses, or landscapes, or to rely on a universal set of criteria in the sense of social engineering. Aiming her criticism now more directly at the functionalist program of total planning, Greverus argues that the "experts" engaged in Heimat engineering deny the essential human need to creatively participate in the construction of one's environment:

> Heimat experts have never sought for Heimat, nor hoped for it. Instead they pointed to where it was. These self-styled educators of the people were the instruments of the ones who knew. The environment as Heimat was proffered, prescribed, planned by the rulers over this environment. (1979a, 14)

Greverus's emphatic defense of space for the individual follows in the footsteps of leading postwar critics such as Alexander Mitscherlich, Henri Lefebvre, and the Frankfurt School, all of whom recognized totalitarian and bureaucratic structures of control as one of the cardinal evils of our times. *Heimat,* in her eyes, had unfortunately lent itself to misuse by such structures, since self-appointed "experts" in Heimat were able to attest to the psychological correctness of certain designs—thereby reinforcing the dangerous tendency of the system toward developing a "perfectly preplanned" world of objects (1979b, 194). This overriding fear of a "prescribed Heimat" strongly colors Greverus's understanding of the Heimatschutz movement as well, which she sees as having been group of biased and pushy "experts" who wanted to impose static, life-denying preservation orders on the world around them. In the stead of this and other "prescribed Heimats" Greverus calls for a much more flexible and participatory process to arise, one oriented toward the varying needs and desires of the individual citizen.

In many ways Greverus's anthropological approach to Heimat is more amenable to an understanding of the Heimatschutz as an environmental movement than is Applegate's concentration on political mythology. It has the great advantage that it accepts the need for Heimat not just as a construct of political discourse, but as essential in human nature. Because of every individual's inherent material *interest* in Heimat, Greverus is thus able to make a strong case for taking it into consideration in our everyday lives, in areas ranging from schooling to administrative planning procedures to architecture and agriculture. She thus rightly maintains that Heimat is a "political task" of the first order.

The problem with Greverus's antitotalitarian argument, though, is that it appears to leave little room for the practical politics of achieving Heimat. Heimat cannot be prescribed, she tells us; planning, too, is associated with the stultifying oppression of unfeeling rulers. Instead it appears from her writings that Heimat is largely a kind of "self-determination." Unfortunately these individualistic and negative criteria are not at all well suited to activating a mass movement in favor of Heimat. The reduction of Heimat to a question of self-determination in fact perfectly positions this value to be co-opted by laissez-faire ideology. It must therefore be remembered that Greverus's purely personal kind of Heimat fulfillment always has to be squarely based, as even Greverus is willing to admit, on fundamental "Heimat conditions" that include "environmental protection." Meaningful environmental protection, in turn, is a matter of public and social concern—which brings us back to our investigation of the Heimatschutz. As I read it, the original Heimatschutz was far removed from being the kind of exclusive, controlling, and totalitarian pressure group that Greverus is worried about; it was a comparatively small group, true, but one that tried as best it could to truly popularize its intuition that important pieces of the environment were being lost. Heimat was conceived here as a new set of values genuinely shared by the wider community. And if the Heimatschutz also sought the passage of laws protecting landscapes and buildings, it is going too far to see this as an inherently life-denying activity. Instead we should see it for what it was: namely, the pragmatically necessary enforcement and embodiment of community values.

These comments about the Heimatschutz can of course be applied to the environmental movement as a whole. We would probably all prefer to live in a world in which people did the ecologically right thing as a matter of course, but we also realize that this is not a likely prospect for the near future: too many concrete interests stand in the way, aided and abetted by a wide range of customs and habits. To some extent regulations will therefore have to be "prescribed," and it is also quite certain that a whole series of practices, structures, vehicles, and systems will eventually be quite intolerantly discriminated against. This could possibly present opportunities for totalitarian control tendencies, but it is a risk that must be taken.

The other danger that Greverus brings up, that of positive reification, will also always be present and needs to be guarded against. Nevertheless, no movement for overall societal change can afford to abandon the field of positive imagery; if it does renounce the right to paint its vision in glowing colors, its enemies will surely do the painting instead. Opponents of ecological practice have already learned this lesson, denouncing ecology with the provocative image of a new "Stone Age." Green politics will suffer as long as this negative picture goes unmatched by a more favorable, positive one. Whether we call it "Ecotopia," "Heimat," or some other name, the positive goal must be described and embellished upon if people are ever to be won for it. To quote Greverus herself, that is simply part of the political task of Heimat.

While Heimat in its overt sense is an appropriate object of political struggle, it is equally and perhaps more centrally an object of cultural contention, being linked, as in figure 2, with a new way of seeing and evaluating the environment. As such the Heimatschutz was seemingly ripe for investigation by cultural disciplines. But until now relatively little work has been done on the movement from this angle. One major reason seems to be that, as an object of study, the Heimatschutz simply did not fit well with the institutional structure and common assumptions of research into cultural topics. Until very recently such attention as was paid to the movement was directed strictly along disciplinary boundaries, resulting in a patchwork of various Heimat-related conceptions but little insight into the cultural-historical significance of the Heimatschutz as a whole.

Scholars interested in architecture, historical preservation, and landscape architecture have been among those most willing to address the Heimatschutz movement. Their interest can no doubt be traced to the fact that all three of these fields are concerned with the same relatively large-scale visual ensembles that inspired the Heimatschutz itself, and indeed, the movement is occasionally acknowledged for its pioneering fights against the wrecking ball and the land speculator. One common misconception, however, is that the Heimatschutz was simply an arm of the monuments-preservation movement (*Denkmalpflege*), one whose general aim was keeping alive "values of the past" (Pogacnik 1991).

Architectural historians have shown a markedly increased interest in the Heimatschutz just in the last several years; studies with particular insight into the wider movement are those by Otto (1983), Ringbeck (1991), and Frank (1992) (see also Barg 1992). There remains a considerable body of work to examine, too, as there were probably more than one thousand architects active in the movement. Their recommendations to planning councils and their own designs for new construction help prove that the Heimatschutz was never just about preservation, but rather included an attempt to mediate between traditional forms and modern technical requirements. All too often, however, the methodological emphasis in architectural criticism has remained on formal aspects, effectively reducing a wide-ranging movement to a simple "Heimat style" (cf. Petsch 1984; Hajós 1989; Lehne 1989; Achleitner 1989; Fehl 1995). In the field of garden design there is a similar tendency to study the Heimatschutz only for its contributions to a new, restrained, and functional style of garden (Richard 1984; Pogacnik 1991). Although one occasionally finds significant insights into the Heimatschutz in these studies, they all tend to be limited by their abstract and preconceived definition of the object of study—Heimatschutz as *Heimatstil,* or Heimatschutz as "national-nativist planning of leisure spaces" (Richard 1984, 294). Meanwhile, virtually everything else for which this movement stood is ignored, above all its concern for the natural environment, but also its critique of capitalism, its social base and its social ethics—in other words, its position within the wider cultural and political scene of Wilhelmine Germany. These disciplines have thus been good at defining a formal stylistic development, but have then often failed to tell us the full social meaning of it.

As a citizens' movement that encouraged concrete policy changes in areas such as planning, agriculture, forestry, construction, and advertising, the Heimatschutz was not the kind of movement that has traditionally interested students of art history or literature. The closest thing to Heimatschutz that we find in these fields is scholarship on *Heimatkunst* or the Heimat novel. The influence of this research on our view of the Heimatschutz has been great despite the fact that the artistic conception of Heimat was often only tenuously connected to the movement. The common terminology of Heimat helped mislead critics into equating the two phenomena, a confusion exacerbated by the fact that they arose at approximately the same time, appealed to portions of the same bourgeois audience, and, at least superficially, often pursued the same thematic interest in landscapes and nature. The result of these commonalities is that little distinction, if any, is made, and both artistic renderings of Heimat and Heimat-oriented practical reform movements are denounced with equal fervor. For the typical cultural critic, nothing less than a global ideological indictment of Heimat will do:

> The more that Wilhelmine society accustomed itself to the status of an urban-oriented industrial society, the more a virtually anti-historical (*gegengeschichtlich*) ideal of Heimat and the soil begins to flourish in parts of its literature and in the accompanying reform movements that felt called upon to save society from its decline. In these a preindustrial ideal of society and life is romantically idealized and the existing industrial world is virtually reduced to a demonized antipode. (Seidel 1967, 181)

This passage from 1967 is a significant one for an understanding of attitudes toward the idea of Heimat, for it reveals with exceptional clarity that an evaluative and selective scheme of history lies behind the rejection of Heimat. Neither Marxist nor capitalist, but subsuming both, this scheme of history valorizes the new, the urban, the industrial, and the technological as the constitutive forces of progress. In the cultural realm it can be called "modernism."

Because much of the best work done on *Heimatkunst* and Heimatschutz is informed by this progressivist evaluation of history, the result has been a series of gross misunderstandings. In the modernist critics' eyes any bit of nostalgia is suspect. Complaints about the unfamiliar and often barren world of capitalist industrialism are taken as a sign of "romantic" longings; any desire to escape William Morris's "age of Shoddy" is dismissed as having been wishful thinking. In a grotesque contortion of sympathies, the fact that a German might have felt something less than total enthusiasm for the direction of modern capitalism is denounced by modernists as *gegengeschichtlich*—a remarkable word for which others might say reactionary or unrealistic, but that is even more rigidly normative with its connotations of a one-way street in history. Given this arrogant sureness about which way the future is (and ought to be) going, it is not surprising to hear a like-minded modernist critic vilify the entire idea of Heimatschutz as a "defamation of human progress" (Schmidt 1965, 35).

This kind of aggressive modernist rejection of the Heimatschutz is complemented in the literature by a coolly dismissive stance. An analysis in this vein does not so much *criticize* the Heimatschutz movement as simply display a complete lack of understanding for its goals. The standard cultural history of the period, Hamann and Hermand's 1959 volume on *Stilkunst um 1900,* is representative. Although filled with incisive comments on other aspects of turn-of-the-century German culture, in the case of the Heimatschutz the book resorts to a mocking tone of disbelief:

> Suddenly people saw in every old oak tree a milestone of cultural tradition, and with volkish enthusiasm they advocated keeping the German soil free of international encroachments, all in order not to commit a sin against the holy ground in which the German Volk-Soul was rooted. (Hamann and Hermand 1977, 329)[1]

In this polemically condensed description of the Heimatschutz movement the authors rightly draw attention to some potentially problematic political vocabulary. Nevertheless their portrayal is perhaps more revealing for its dismissal of the organization's concern for trees and landscape. Characteristically, the tone here is not so much one of hot-blooded opposition as it is one of bemused astonishment. As a modernist, one simply didn't engage in tree-hugging—it was beyond *gauche,* it was unimaginable, foreign to one's entire cultural orientation. The Heimatschutz tendency to try to legitimate this behavior in terms of a "cultural tradition" is greeted with similar disbelief. All of this constitutes a very characteristic rejection and dismissal. For modernism can be defined precisely as the twofold attempt to "distance [society] from all previous cultural traditions and . . . to constitute society beyond nature" (Eder 1990, 26). With its attempt to save landscapes by playing on established aesthetic sensibilities, the Heimatschutz sinned on both counts.

This modernist attitude that has so deeply colored the reception of the Heimatschutz is directly linked with more specific readings of Germany's sociopolitical history. As alluded to above, an extremely important subtext of the scholarly interest in the Heimatschutz has been National Socialism. The teleological nature of the critics' thinking is evident in the very structure of their articles, which typically begin with or traverse the subject of Heimatschutz, and end with Nazism. Here the Heimatschutz has repeatedly been cited as a "Blut-und-Boden" or "crypto-fascist" ideology and likewise has gained the reputation of belonging to the "*völkisch*" Right and peddling an "obscure racist" agenda (e.g., Greverus 1979, 192; Hamann and Hermand 1977; Hartung 1991; Chickering 1984, 242). This is not to say that the critics are entirely wrong when they assert continuities between the Heimatschutz and the Nazi era; they are regrettably correct in noting that a number of Heimatschutz leaders such as Paul Schultze-

Naumburg became rabid National Socialists. But the key word here is perhaps *became.* There is comparatively little evidence to support the view that Heimatschutz ideas were inherently fascist or racist, or part of the much-discussed German *Sonderweg* in history. To the contrary, there is considerable evidence to the effect that the organization, during its formative Wilhelmine years, pursued a rather enlightened course of anticapitalist environmental reform and class reconciliation that had its counterparts in other European countries such as England. Only later did individual members turn away from or add to this set of convictions. It is accordingly the discontinuities between this cautiously progressive middle-class reform package and Nazism that deserve to be emphasized.

The ease with which critics have nonetheless connected these two phenomena has its roots in a specific reading of the Wilhelmine bourgeoisie as a whole. Those who see the Heimatschutz in a negative light generally see the entire middle class at the turn of the century as out of touch with industrial reality: behind the times both culturally and socially, inwardly weak, pessimistic, prone to grasp at illusions, and finally afraid of the proletariat and anxious to keep up its alliance with the old feudal castes at any cost. Let down by this ruling elite during World War I, this "weak" middle class supposedly continued its strategy during the years of the Weimar Republic, now grasping for the even more fateful straw of Nazism.

To date the single most influential interpretation of the Heimatschutz movement has been one in this vein provided by Klaus Bergmann in his 1970 book *Agrarromantik und Großstadtfeindschaft.* Much of the interpretation is based on the person of Ernst Rudorff, the Berlin music professor who originally proposed and helped found the *Bund Heimatschutz.* While some of Rudorff's statements are described as unusually conservative or "explicitly reactionary," it is clear that, for Bergmann, he is symbolically symptomatic not only of the movement he inspired, but indeed of the Wilhelmine bourgeoisie in general. This is evident even in Bergmann's denial:

> Rudorff was explicitly reactionary both in the content and the tenor of his publications, and only the fact that the Bund Heimatschutz did not take up all of the ideas of its initiator was able to prevent the movement from spending itself in senseless complaints against modern development . . . If the Heimatschutz league had blindly followed its founder Rudorff in all of his knee-jerk reactions against the social, economic, cultural and political developments, it would without a doubt have become a reactionary sectarian club, but not a neoromantic movement that appealed to large portions of an inwardly insecure bourgeoisie. (Bergmann 1970, 125)

What Bergmann actually ends up revealing with this halfhearted attempt to distinguish factions within the Heimatschutz is a telling portrait of his own historical model of the bourgeoisie. According to it the middle class of the Wilhelmine era consisted more or less of two different groups: genuine reactionaries such as Rudorff, and a larger class of crypto-reactionaries who, because of their class-based politico-psychological infirmity, followed the reactionary program and its illusions of turning back the clock.

Bergmann found his sociohistorical model fully confirmed by the Heimatschutz program of landscape protection. For, according to him, the landscape philosophy of the movement precisely mirrored the social values of the bourgeoisie that filled the ranks of the Heimatschutz: it was a "preservation of the *status quo*" in which there lay "a fundamental tendency that was, in the final analysis, ahistorical, and averse to any further development and change" (131, 133). An implied homology between environmental and social realms is a particularly pernicious feature of modernism. It holds, after all, that "modernizing" transformations of the environment are as necessary and as morally right as changes in unjust social conditions, and it often makes the latter dependent on the former. Given this view, any desire to protect landscapes is doubly wrong: it ignores the economic and technological imperative of change, and by extension it desires to put a stop to the transformative process that will, in the end, provide more social justice. Bergmann conveyed both of these points in his oft-cited accusation that the Heimatschutz landscape program "was rooted in a romantic aestheticism as well as in an imagined social idyll of peace and quiet" (128). This single phrase has deeply influenced our interpretation of the Heimatschutz movement until today. Implicitly or explicitly it can be found in Werner Hartung's 1990 dissertation, in Arne Andersen's 1987 article, in Ulrich Linse's 1986 book, and in Rolf-Peter Sieferle's 1985 book. What links all of these later studies with Bergmann is the conviction that the Heimatschützer were trying to turn back "inevitable structural transformations," that they were chasing something "romantic," something merely "aesthetic," something fundamentally not real (Bergmann 1970, 133).

The philosophical base of this argument is a rigid progress-vs.-reaction model that almost exclusively values technological innovation and increasing rationalization. These, Bergmann insists, are the "real" forces of history, and the only path to the future. Any subjective desire to oppose these forces, or even modify them, is "romantic" in a thoroughly negative sense—they are historically obsolete, unrealizable, "false." This is what Bergmann means when he criticizes the Heimatschutz project of protecting landscapes as "romantic aestheticism." This view does have a certain amount of internal coherence, but at the price of absurd conclusions. Having valorized and profoundly ontologized the entire process of technological development, it is forced to declare the desire for nonpolluted green countryside to be unreal. Such "reactionary" desires are then cleverly accounted for at the psychopolitical level as the "irrational longings" and the "irrational little games" of an "escapist" and disheartened Wilhelminian bourgeoisie (84, 87).[2]

In response to Bergmann it must be said that the concept of "escapism" does not apply to Heimatschutz efforts, at either level of its meaning. At the one level the charge of escapism says that this segment of the bourgeoisie was running away from a real and important situation, to wit, the big city, the problems of workers, the whole project of modernity. At the other level the charge of escapism implies that the goal toward which the bourgeoisie ran was illusory, devoid of anything but psychological reality. After a closer examination of the facts of the matter it will be seen that neither accusation is very convincing in the case of the Heimatschutz movement.

On the contrary, we shall see that the Heimatschutz was trying to engage the processes of modernity in their full breadth. The Heimatschützer did not labor under the illusion that they could somehow escape from urban industrialism—they knew all too well that wherever they might run, it would simply catch up to them. They therefore committed themselves to active reform from within, an informed, plugged-in, and enormously reflective process of guiding, limiting, and educating many different levels of society. All of these activities were aimed at an eminently *practical* vision of Heimat wherein human progress could peacefully coexist with historic and natural resources. In the years between 1904 and 1918 thousands of men and women spent untold tens of thousands of hours in meetings, at planning tables, and in the field in a valiant attempt to control booming development. If these were "irrational little games," it would appear that the standard of what is rational and what is not may need to be reconsidered. In the twenty or so years since Bergmann wrote his book this has actually begun to happen. Our ideas about what is reasonable have changed considerably in the wake of disappearing forests, oil spills, nuclear accidents, overconsumption, and undernutrition. "Progress," too, that watchword of modernism, appears more and more frequently in quotation marks, as a value that has lost its prima facie power to convince. It is perhaps not surprising, therefore, that critics have started to devote a little more thought to the positions staked out in the Heimatschutz discourse.

It should be remembered that the fact that until now very little has been known about the Heimatschutz is not because the movement's existence has been kept secret in inaccessible archives or indecipherable private notes (although Rudorff's handwriting in particular might be called that!). Rather, there has been a palpable lack of desire on the part of scholars to find out about this and other environmental reform movements. There can be no doubt that these academic and critical lacunae have in turn been directly influenced by the aims and understandings of Western society at large, for the silence on this matter complements our massive investments of economic and political capital in industrial growth. Scholarship is just one facet of the system, to be sure, but it is more than a coincidence that until recently even its "theory most critical of modern society share[d] with the competing systems of social and political thought . . . the basic assumption that nature is subject to man and that nature has to be conquered by man in order to further the progress of mankind" (Eder 1990, 24). The theory in question is, of course, Marxism, which for decades remained tied to various forms of economism and technological optimism. Raymond Williams's sober comment comes as an important reminder of the effort that must be put into rethinking the old categories: "It is a measure of how far we have to go," he said, "that socialists also still talk of the conquest of nature" (Williams 1980, 84).

As we saw above, the cultural disciplines can be especially revealing of scholars' biases. Art history and architectural criticism especially have long devoted themselves to constructing a teleology of the technological and urban development at the heart of Western and European societies. "Even today," a recent article critically observed, "modernism, the faith in progress and innovation, is the cultural standard that governs our conception of art and is still firmly entrenched in architectural practice" (Shinar 1992, 29). While the article cited focuses on Le Corbusier's work, in the German-speaking realm *Neues Bauen* can serve as an example. The outpouring of scholarship on this subject stands in marked contrast to its quantitatively small role within the total construction of the period. The obvious motive here was thus not so much a historical analysis of housing conditions, as rather the need on the part of subsequent scholars to define the tradition that could legitimate contemporary building practices. One could delineate many further developments in this and other applied arts, but the essential point is virtually self-evident: over most of the twentieth century, cultural value and scholarly approbation has tended to accrue to those styles, forms, and attitudes that are perceived as the most "modern," which here explicitly means also urban, industrial, and rationalized.

As part of its strategy ecological reform will have to challenge the modernist value of technological progress as a central part of a very real and very entrenched system. It is important to realize that this value is truly hegemonic, in the sense that it is neither arbitrary nor superficial, and is incapable of being changed through a simple act of will. It is rather only the cultural component of a worldwide system of domination by class, status, and financial interests. This is an opponent whose facticity stands in stark contrast to the fashionable whining of those who claim that "in a postmodern era, it is somewhat difficult to know what would count as 'dominant' against which culture could be 'oppositional.'" (Stauth and Turner 1988, 523). Such pessimism refuses to see our world for the definite selection that it is: a specific set of economic practices, a particular kind of society, and, most important here, a constituting and confirming set of cultural values. If we see this, we can begin to see, with Klaus Eder, "modernism as a specific cultural tradition"—one with specific biases, and to which there are alternatives, if one looks for them (Eder 1990, 26).

Modernism is in fact beginning to lose some of its once unchallenged attraction. In the wake of energy shortages, global warming, nuclear disasters, and the rampant destruction of wilderness, more and more people are coming to question the idea of unlimited technological progress that still dominates our Western societies' thinking. Since approximately 1980 these kinds of environmental concerns have led to a stronger critical interest in the Heimatschutz movement. In the past decade it has come under scrutiny in cultural-environmental histories of Germany by Sieferle (1984), Linse (1986), and Hermand (1991). Andersen's article of 1987 even takes the Heimatschutz as the central object of study. Less directly concerned with environmental aspects per se, but obviously embodying the same resurgence of interest in Heimatschutz, are Matthew Jefferies's article on architectural Heimatschutz (1992) and the thoroughly researched dissertation by Andreas Knaut (1993). With some exception for the last two authors, none have been willing to break with the tradition of a harsh assessment of the political and ideological thrust of the movement. Nevertheless, as a group they tend to take the environmental concern evidenced by the Heimatschutz more seriously.

Ironically, the same enormously increased ecological knowledge that inspired recent interest in the Heimatschutz seems to have simultaneously limited our understanding of it. This situation is illustrated most clearly in Arne Andersen's article. His study is obviously informed by today's rigorous scientific knowledge of pollution's effects on biological systems. Andersen sifts through the program of the Heimatschutz and argues, not without foundation, that it did not pay enough attention to some serous problems: water pollution, for example, as well as air emissions from industry, damage to human health, and the impact of development on biotopes. Why, Andersen asks, did the Heimatschutz not raise a bigger fuss over these important issues?

In answer to his own question Andersen identifies the chief drawback of the Heimatschutz as having been its visual-aesthetic approach to the environment, namely as landscape. For Andersen, this aesthetic method is what separates the Heimatschutz from "real," that is, strict scientific, ecological concern. Andersen points up this difference by emphasizing Rudorff's love for "the picturesque and poetic quality of the landscape" and goes on to criticize the Heimatschutz's "foundation of nature protection on aesthetic grounds." Discussing the group's proposal of an alternative to a 1905 dam project, Andersen implies a virtually absolute distinction between aesthetics and "real" ecology: "an ecological comparison of the two proposals was not submitted to debate; it was a matter of landscape beauty." The Heimatschutz was thus a far cry from today's "genuine" environmentalism: "Its conception of nature was not directed toward ecological connections" (Andersen 1987, 145f.).

On the one hand it is difficult to disagree with these findings. At the turn of the century Germany's water and air were indeed increasingly polluted, and it would have been better for all concerned if more individuals and organizations had gotten together to press for a thorough scientific understanding of the problems. The Heimatschutz—for whatever reasons—did not bring about this kind of policy discussion. On the other hand, though, it could be asked whether Andersen has posed the question in a fruitful and interesting manner. The negative result has not led to any greatly increased insight into the concrete movement known as the Heimatschutz. All we seem to know is that it wasn't Greenpeace. Did the Heimatschutz contribute anything of value to the contemporary discussions? Do we today share anything in common with the Heimatschützer? Or were this movement and its aesthetic method simply irrelevant to the future course of ecological awareness? Andersen can answer none of these questions, all of which would seem to be crucial for an evaluation of the environmental significance of the Heimatschutz movement.

As can be guessed from the preceding comments, I believe that it is wrong to draw too sharp a distinction between the science of ecology and the aesthetic-cultural program of Heimat. The two are by no means mutually exclusive and indeed should be seen as complementary. There are examinations of early German environmentalism that have recognized this simple fact, to which we now turn our attention.

If we disregard smaller reviews made by Heimat organizations largely for their own purposes, the first retrospective analysis of the Heimatschutz phenomenon of any stature was the 1954 study by Walter Schoenichen. As an examination of the Heimatschutz organization the study is unfortunately flawed, providing little concrete information on actual groups and concerned mainly with a distillation of wide range of phenomena into a somewhat artificial clash between two environmental agendas and two respective protagonists. Moreover Schoenichen displayed a distinct aversion toward the agenda most closely associated with the Heimatschutz. His prejudice was linked to his biography: he had worked for the Prussian State Agency for the Care of Natural Monuments (*Preußische Staatliche Stelle für Naturdenkmalpflege*), and as a professional nature-protector in the Weimar Republic and under the Nazis he had absorbed a considerable bias toward the "Conwentz principle" of *Naturschutz,* or scientific ecology; not surprisingly he tended to associate more negative terms with the "Rudorff principle" of Heimatschutz, which is seen as a merely "romantic" and "cultural" phenomenon. But in retrospect what is worth noting is that Schoenichen retained a sense of the "emphasis on emotion which is immanent in the idea of nature protection" and devoted many pages to an account of the preparatory role of poets and writers, above all those of the Romantic period (157, 5). Because of the study's approach we are not told precisely how this artistic or aesthetic tradition entered into the actual practice of the movement, but the connection is clearly implied.

For a long time afterward little attempt was made to rescue the idea that aesthetic impulses could be part of a true environmental movement. Following Schoenichen's example and radicalizing it, other studies "cleaned up" the history of what they saw as "true" environmentalism—Naturschutz—by assigning all "aesthetic" and "romantic" elements to a culture-mummifying Heimatschutz (Hellmich 1953, 9). Only recently have two studies appeared in which the conflation of aesthetics and ecology has become a topic of discussion again. Raymond Dominick's historical survey, *The Environmental Movement in Germany: Prophets and Pioneers, 1871–1971,* unfortunately perpetuates the tradition of separating out the Heimatschutz as a "historical preservation" movement, interested in "ruined castles and cathedrals, regional dialects, fashions of dress, and songs and dances," from the implicit mainstream of a more scientific Naturschutz (Dominick 1992, 21). But Dominick at least reminds us on five pages that aesthetic sensibilities, or "Romanticism," constituted one of a number of different rationales for environmental protection. The second study, Jost Hermand's 1991 essay on Heimatschutz and *Heimatkunst,* addresses the cultural production of the time more directly. Hermand pleads here for a new and more differentiated reading of both Heimatschutz aesthetics and *Heimatkunst* novels, one that would not simply dismiss the aesthetic presentation of their ideas as "ridiculous" or "sentimental," but would rather uncover their "green heart," that is, a new and positive ecological vision of the world (Hermand 1991, 132–33).

The studies by Dominick and Hermand are certainly on the right track. But in different ways both of these recent approaches suffer from the same basic problem as Schoenichen's earlier study. Neither gives us a sustained insight into the historical phenomenon of the Heimatschutz and the various factors that conditioned its aesthetic style of environmentalism. Dominick's study is largely descriptive; it dutifully documents the rise in membership of a far-flung consortium of conservation groups, and it includes a sampling of their various "rationales" for conservation. It is in essence a political science approach, concerned above all with the external mechanical aspects of a "constituency." Because of this focus Dominick is unable to answer some of the key questions that might be asked about the Heimatschutz: what its images meant, why they took the form they did, and how, if at all, they entered into the sociopolitical process.

This gap between aesthetic phenomenon on the one hand and political process on the other plagues Hermand as well. In spite of his sympathetic and, in my opinion, substantially correct reading of the environmentalist implications of the Heimat aesthetic, he does not demonstrate whether these implications—particularly in their less articulate and more aesthetic forms—were understood by contemporaries, how they were transmitted, and what groups or actions resulted from them. We are left with an aesthetic, but no political or historical resonance. The key to closing this gap lies not only in a more thorough examination of the Heimatschutz organizations, but also in a method that allows us to grasp the political potential of an aesthetic movement. For the two cannot be separated in the case of the Heimatschutz. An acute aesthetic sense saturated its language, motivated its arguments for reform, and surfaced again and again in its use of and respect for paintings. The highly aesthetic category of landscape was the central site of this movement's critique: a tool with which environmental damage could be not only perceived, but also discussed and, for the first time, turned into a public issue. If we look carefully at the Heimatschutz movement we will find that its aesthetic vision of Heimat was centrally related to the development of ecological knowledge and the building of an environmental culture.

It was stated above that much of the criticism dealing with the Heimatschutz has been informed by the modernist supposition that there is only one possible technological-industrial path for human history; we also saw that this was one area in which even oppositional Marxist theorists have tended to agree with the conventional wisdom of modernization theory. Both camps have further tended to agree that social phenomena are determined at the level of the economy and of technical organization. Vulgar materialist determinism, however, is belied by ordinary experience, which shows us that intellectual labor—or more broadly, the cultural sphere—has an important co-determining role to play in the day-to-day operation of society. This contribution of the cultural sphere would seem to be of particular importance when we are trying to trace the origins of qualitatively new directions of social reflection such as environmentalism.

We find both aspects of society covered in the theory of *hegemony* as first developed by Antonio Gramsci. Classical Marxist political theory had led people to believe that "ruling" classes could only survive and impose their rule by a combination of brazen force and transparently self-serving systems of thought and law, the result of which should have been constant and articulate class warfare. While serving as a functionary of the Italian Communist Party Gramsci noticed, however, that long-term domination was actually achieved by the much more subtle means of cultural hegemony. Workers were constantly being steered into nonrevolutionary modes of thought by everything from newspapers to cheap adventure novels, and their whole daily existence, too, was largely absorbed by "mainstream" activities that effectively reinforced the current organization of society. The bourgeois-capitalist industrial system was thus maintaining itself through hegemonic domination, a condition later described by Raymond Williams as

> a whole body of practices and expectations, over the whole of living: our senses and assignments of energy, our shaping perceptions of ourselves and our world. It is a lived system of meanings and values—constitutive and constituting—which as they are experienced as practices appear as reciprocally confirming. It thus constitutes a sense of reality for most people in the society . . . beyond which it is very difficult for [them] to move. It is, that is to say, in the strongest sense a "culture," but a culture which has also to be seen as the lived dominance and subordination of particular classes. (Williams 1977, 110)

The strength of this kind of domination lies precisely in its invisibility, or rather, in its subtle omnipresence. By "saturating" the most disparate and seemingly unimportant areas of life, it does not so much force adherence to itself as remove all indication that there are alternatives. In the terms of psychology, its values are internalized: again, not because they have been beaten into us, but because they have always been there as a structuring system of meaning, a "cultural order" (Williams 1977, 110, 93).

Fortunately there is a flip side to this. Because hegemonic domination relies for its effect on the absence of any challenge to its offered system of meaning, it is continually and exquisitely vulnerable to cultural innovation. New phenomena—ideas, fashions, subculture trends, genres—can and do arise constantly, creating a space in which alternative systems of meaning can be constructed. Rarely will these enter in a fully articulate form, though; instead, the perceptions, practices, and values that vary from the dominant mode must often germinate at an inarticulate level. Gramsci recognized this as a key stage of political change when he wrote in 1920 that a proletarian revolution "presupposes the formation of a new set of standards, a new psychology, new ways of feeling, thinking and living" (Gramsci 1985c, 41). Raymond Williams used a similar term, a "structure of feeling," to describe how a gradual process of explication can have a formative effect on a new group identity. In this phenomenon we get a sense of the cultural realm as a crystallization point for later, organized social movements, or as what Thomas Nipperdey called the "area of the proto-political" (*Bereich des Vorpolitischen*) (Nipperdey 1984, 180).

If we reflect for a moment on the overall historical development of bourgeois industrial civilization, it is clear that it has functioned through the propagation of fundamental, sine qua non dominant values. These include linchpins of the economic and legal framework such as private property and maximization of profit, homologous norms such as individualism and consumption, and the uniquely effective bundling of personal, technological, and social elements that is conveyed in the term *progress.* Not coincidentally these are the same values that have reinforced and legitimated environmentally detrimental behavior, in what we might call a "hegemony of environmental exploitation." True environmentalism, when it arises in this context, is an oppositional phenomenon: it will always be marked by an attack on these values, a deconstruction of their absoluteness, and their subsumption within a different framework. But that attack has to start somewhere. "Environmental values" and "environmental consciousness" do not simply present themselves as full-fledged and articulate convictions. New beliefs start small, generally, and even in an age full of ecological insight people need time in which to absorb and begin to apply it in ordinary life. Paradoxically, in order to even start this process they have to overcome their ordinary life to some extent: they have to be able to entertain a different point of view, for starters, and must then withstand manifold hegemonic pressures to conform and consume while working out possible new behaviors. Given these difficulties even today, there can be little doubt that environmentalism in its earlier, emergent phase was existentially dependent on strategies that could nullify or at least quiet the internalized imperatives and transparent "common sense" of the industrial profit system.

These results allow us to gain a whole new perspective on the aesthetic environmentalism of the Heimatschutz. The area of aesthetics has long been seen as a "realm apart" from bourgeois capitalist society. Especially with the Romantics it was here that the hegemonic values of rational utility could be most decidedly negated. Even if this area can often be "specialized" and turned into a marginal legitimation of the center, artistic and literary realms remain a potent source for emergent "structures of feeling," giving them both form and forum (on specialization see Eagleton 1990, 88; also Williams 1977, 114, 133). My hypothesis in these terms is that it was precisely the aestheticism of the Heimatschutz movement that allowed it to develop a new perception of the environment as a collectively maintained "homeland." This perception went beyond a merely descriptive function and simultaneously invested the newfound object with an alternative and sometimes virtually oppositional set of environmental and social values. If there was much here that was still "bourgeois," still tied to dominant values, then that is understandable, given the social base of the movement and the solid entrenchment of cultural norms. But the new direction taken in the vision of Heimat clearly constitutes that kind of "counterculture" innovation whose "importance . . . it would be wrong to overlook" and that, "while clearly affected by hegemonic limits and pressures, [is] at least in part [a] significant break beyond them" (Williams 1977, 114). What hegemony theory lets us see clearly is that a challenge issued in nominally aesthetic terms can well represent the leading edge of a much larger and more articulate political challenge. In the case of the Heimatschutz we can observe how a new aesthetic sensitivity toward the landscape played this pivotal role, opening up an increasingly sophisticated and increasingly significant debate on Germans' treatment of the environment.

CHAPTER 1

Aesthetic Perception and Environmental Reform in Modern Europe

Where do the sources of modern environmentally destructive behavior lie? It would be easy to say that simple material needs are to blame, but this would not take into account the fact that what different ages see as "necessary" is culturally variable. One could also plausibly argue that ecological ignorance and advanced technology are the culprits; nevertheless they, like physical need, are probably only necessary, not sufficient, components of destruction. In and of themselves they remain insufficient to explain the sustained growth and striking dynamism of environmental intervention over the last several centuries, something that can only have been maintained by a high level of social approbation. Approval and reinforcement came from many different sources within Western culture; as one member of the Heimatschutz movement speculated, environmental neglect could ultimately be traced to Christian doctrines emphasizing the absolute primacy of an immaterial soul (Krische 1918, 76). This particular debate about the environmental impact of Christianity rages on to the present (cf. Attfield 1983). Critics of the religious tradition certainly have a point when they say that biblical injunctions such as "Go forth and multiply" are recipes for environmental disaster. Nevertheless they were just recipes, and not the cake itself. If we are to understand the quantitatively and qualitatively different character of environmental damage in the modern age, we must focus on the new forces and discourses that arose in this period and were able to exploit the manipulative potential of rhetoric that had long been slumbering.

Modernity began to hit its stride in the sixteenth and seventeenth centuries. At the same time that an expansionist capitalist economy was beginning to open up vast new areas of the world to commerce and exploitation, thinkers at home were bringing about the "Scientific Revolution." It was perhaps the most influential intellectual epoch since that of the ancient Greeks, and certainly one of the defining moments of modern Western culture. The decisive development of philosophers and scientists such as Francis Bacon (1561–1626), Thomas Hobbes (1588–1679), René Descartes (1596–1650), and Robert Boyle (1627–91) was their origination of an intensely mechanistic, dualistic view of the world. In keeping with the dualism of Christianity, the mechanists attributed real value only to "souls" or mental substance. As such, this was nothing new. What was new was their re- and devaluation of matter. Now the natural world was expressly conceived of as an agglomeration of dead, inert particles, circulating and colliding according to simple laws of mechanics. Nature was nothing more than a large set of very small building blocks, sometimes arranged in one way, sometimes in another, but fundamentally disconnected from any larger or ethically significant entity. As Carolyn Merchant (1980) has documented, these thinkers of the Scientific Revolution discarded the medieval image of Nature as an all-encompassing organism, likening it instead to a machine. Nature could now be broken down into separate parts, each of which could individually be studied, tinkered with, and then put back together in an improved version. This was the precise point at which the theoretical model of a mechanism—probably a necessary feature of any science—was overextended and used to justify a program of technocratic control of nature (Worster 1985, 379). Bacon's scientistic utopia *New Atlantis* (1624) is the seminal example of this, describing an ideal state made possible by manipulating animal and plant breeds and the weather. Descartes pursued a similar program of making men "masters and possessors of nature" (cited from Merchant 1980, 188).

The mechanist thinkers proceeded from the fundamental assumption that nature in its present state was threatening, dangerous, or simply incomplete. "Improvement" and "progress" were the central values behind research into natural systems. It is not surprising to find that this "progress" was to be measured in the same terms of economic output that were beginning to dominate all of society. A given technique "worked" if it could increase profit at the end of the year. Already in the seventeenth century, then, a scientific-industrial complex had begun to arise, "a smoothly functioning relationship between the pure science of nature and the profitable domination of nature" (Schäfer 1985, 206).

If this sounds quite normal and unremarkable to us today, there is a reason: it is because we still fundamentally share the attitude toward nature that Bacon expressed at the beginning of the seventeenth century. The difference between this approach to the environment and that of earlier ages is marked. Rather than being enveloped in or part of an organism, humans in our modern "engineering culture," as Paul Shepard calls it, stand apart from, above, against the ground they walk on. Every sight is of a potential object, instrument, tool; every glance an incipient experiment in which properties are involuntarily catalogued, usefulness tested, and profitability of an improved version entertained. As we are just now beginning to realize, though, this "vision of 'improved' nature is . . . an ecological disaster because it identifies order with a limited kind of regularity [and] . . . sanctions the destruction and removal of any elements which do not seem to fit the pattern" (Shepard 1967, 235).

Mechanistic philosophical doctrines may have freed us to look upon the world as a set of building blocks, but it can be questioned whether so many people would have begun playing with the blocks if they had not been able to call them their own. Private property, however, has always been the central legal doctrine of liberal capitalist societies. The effects that it has had on the environment are worth considering. With good reason, when we think of land, it is of something that comes in pieces. Land, for us, is always someone's property. As with any other property, that means that land can be bought, sold, plowed, developed, mined, gardened, built on, blown up, burned over, neglected, or cherished by the owner as that person sees fit. The definition of private property under liberal capitalism is radical in this respect. In a fundamental way it cuts the world up into discontiguous segments and then erects walls around each one—not visible walls or fences (these are optional), but cultural walls that block and change our perception all the same. These consist in telling every one of us that we have no business in knowing or caring what goes on behind the property line; eventually, by dint of cultural conditioning, we come to believe that the use that is made of the environment is mostly someone else's right, and certainly not our responsibility.

This particular separation of the private and public spheres is a key part of the entire capitalist order. Property lines define that space in which the entrepreneur can amass private capital, indeed, it is where the legal owners are allowed to manipulate their social and natural environment almost unhindered. Free market economic theory sanctions this separation and this manipulation by arguing that private capital is the only important goal of an economic order. It is a tribute to the power and effectiveness of this theory that we are nowadays forced to call the nonprivate things, the things it ignores, "external"—as if workers' well-being, general education, social harmony, and the condition of the environment were somehow inessential or unimportant. But if capitalism is hard on the public sphere, it invades the private sphere as well with its unsilenceable demands for cash and profit. In a capitalist culture, no individual is completely free to ignore these demands, and this in turn means that no individually owned piece of the environment can escape the possibility of being reduced to a mere exchange value. Along with the belief that nature is essentially a meaningless jumble of building blocks, private property is thus one of the most securely anchored and one of the most environmentally harmful tenets of modern, "Western" civilization.

The words *belief* and *tenet* are perhaps ill-chosen in an age of optional religions and faddish philosophies, for our convictions with regard to technology and property are arrived at in a singularly untentative and unreflected manner. They are not optional, nor are they settled upon after a period of calm and reasoned reflection. Instead, we grow up with them: they are infused from all sides in a permanent process of acculturation, socialization, and education. For hundreds of years now they have colored our very perception, forcing us to see nature *as* instrument or *as* property instead of in some other fashion. The way in which these lenses through which we view the world have systematically been produced and reproduced can accurately be labeled *hegemonic:* within our modern culture, it has saturated the realm of "commonsense" ways to think about nature.

As a world we are just now beginning to realize that a good deal of this common sense must be rejected, and that we cannot, for example, arbitrarily transform one piece of the world without affecting many others in the bargain. This is approximately the same lesson as was contained in the traditional organic model, of course, but nowadays it is expressed in terms more appropriate to our exact functional understanding. Enlightened self-interest tells us that we must take the teachings of scientific ecology to heart and learn to identify, understand, and act according to our material dependence on the world. But if this kind of scientific self-interest is the wave of the future, it is evident that it was unable to contribute much in the past. Dates vary according to the commentator, but it seems well-established that the science of ecology began to arise as a self-conscious discipline only in the last decade of the nineteenth century (Worster 1985, 198; McIntosh 1985, 23). As a phenomenon of *mass* consciousness, ecological science probably did not exist before the Club of Rome report (1971). With a pessimistic look at how little has been accomplished since then, one could argue that popular acceptance of scientific ecology is still not here. At any rate, the theoretical apparatus and physical techniques necessary to the formation of ecology as a true science appear on the scene rather late in history, and then only haltingly. For most areas and situations ecological science can virtually be ruled out as an effective brake on environmental destruction before 1900.

This result can be used to construct a very rough historical typology of environmentalism. In the medieval era, whatever environmental restraint was practiced had a primarily ethical-religious character; in recent years, meanwhile, we have begun to protect the environment based on the pragmatic insights of scientific ecology. In the large historical gap between these, however, the relentless forces of a capitalist modernity raged almost unchecked. Nature had been atomized and stripped of its ethical significance by mechanistic philosophy; in addition it was now exploited more than ever by the calculating individuals produced and unleashed by society's liberal economic practice. In this situation it was left to the aesthetic realm to provide a sense of wholeness or interrelatedness in the world. Because I see this as the central tradition taken up by the Heimatschutz, I shall try to develop the idea in some detail.

The connection between natural aesthetics and the specifically modern philosophical position has not gone unnoticed. Joachim Ritter's 1962 essay "Landscape. On the Function of Aesthetics in Modern Society" documents the "simultaneity of scientific objectification and aesthetic representation in the relationship to nature" of modern man (Ritter 1962, 154). Nature's reduction to the status of an "object" had apparently reached something of a crisis stage by the eighteenth century. In response to the distance and isolation imposed on the natural world by science, Europeans developed a new and holistically oriented appreciation of nature as landscape. Ritter saw this aesthetic turn to be important because it "holds open the connection of humans to the surrounding/nurturing nature and gives it expression and visibility" in an age that otherwise condemns nature to the mechanistic status of an "object world" (161). Ritter himself, it must be said, was more interested in the human significance of the phenomenon: to him it represented one more step in the metaphysical and psychological development of the "richness of human existence" (163). Nevertheless it is possible to reverse his emphasis, and see potential benefits to the natural world. If aesthetics can bring humans closer to their environment or make them feel a part of it, this may well lead to a new ethical respect for nature.

Some would argue, of course, that the idea at the heart of this aesthetic environmentalism is philosophically flawed. Indeed, at first glance we do appear to have committed the "naturalistic fallacy" of confusing the mere perception of connections—an "is"—with the ethical import of those connections—an "ought." On the other hand, though, it seems to be true that the aesthetic perception of a true, fundamental unity implicitly carries with it the obligation of preservation. It is generally not considered right, for instance, to kill a living animal in the absence of any reason to do so. In this case an *is* goes a long way toward determining an *ought.* If aesthetic contemplation helps us perceive and define such a living, functioning unity, it can indeed provide the ethical basis for maintenance and protection.

Traditionally, of course, the aesthetic quality of nature was often linked with its ethical standing. For the Greeks, natural beauty was simply the sensible expression of innate goodness, a cooperation with the God-given order of the universe. In a similar vein, many Christians saw God's signature in the "Book of Nature" and revered the physical features of the Earth for their Maker's sake. It was characteristic of this traditional view that even the arduous and destructive aspects of nature were accepted, without much ado, as part of the greater plan of God. But the developments of the sixteenth and seventeenth centuries, with their increasing incentives and possibilities for manipulating nature, destroyed much of this traditional reverence.

Technological society's manipulative ambitions thus created a new urgency for an old theological problem. People now seriously wanted to know why God had fallen short of creating a different and, in their eyes, better Nature. Why are there unfruitful deserts? was one popular query; why do the rains fall into the oceans sometimes, rather than onto our fields? was another. Both questions reveal that men and women had absorbed the modern teaching that nature could be radically re-formed to human ends. More and more people found that the world around them could easily be more useful, more comfortable, and, of course, more profitable. The traditional philosopher's task of constructing a theodicy—a "justification of the ways of God" (Hegel)—thus became more and more difficult. To the modern sensibility this legitimation of the existing state of things might sound like rank apologism. To get away from such speculation has generally been seen as a healthy thing, a sign that people were no longer willing to accept repressive political regimes, for example, as a God-given necessity. But as much sense as it makes to argue for freedom and change in our human cultural realm, these may not always be the appropriate principles for our relations with the natural world. For if we accept that the world has a specific and ordered structure, what we need in order to understand it is an "apologetic" frame of mind, one that will allow us to look for possible ecological necessities along with our own desires for expedience and comfort. Maybe then we would not be as quick to declare that an area has no "worth," or that it needs to be "improved." As we learn more about our environment we can perhaps better appreciate the caution and humility recommended by Leibniz, since in nature, as perhaps in the great scheme of life, "that which we have been able to see hitherto is not a large enough piece for recognition of the beauty and the order of the whole" (Leibniz 1985, 207; see also Leibniz 1989).

It is peculiarly appropriate that Leibniz should come up in this context, for his ideas about man's role in the world make him an important transitional figure. A basic premise of his philosophizing was a fundamental acceptance of the Creator's world as it was. It was this sympathy with an older and perhaps "unmodern" orientation, though, that ultimately provided Leibniz with his sensitivity to the intensely self-centered concerns of his more "modern" contemporaries, whose "perpetual anthropomorphisms" with respect to God only mirrored their narrowly focused anthropocentricism with respect to the environment (1985, 196). Thus while Leibniz subscribed to the idea of "a certain constant and unbounded progress in the whole universe," and clearly saw that "the world . . . will become more and more cultivated," he nevertheless aimed at a kind of culture that would be gentler and more environmentally responsible than that of a Francis Bacon (1989, 154). Significantly, some of his more environmentally relevant convictions were directly related to his position on aesthetic matters. Diversity stood as a central principle uniting the two realms: diverse surroundings engage and please us, he wrote, while an exaggerated "uniformity brings forth disgust" (ibid., 153). Leibniz's philosophical system of monads applied this emphasis to the very structure of the world, a structure that was more sublime the greater the multiplicity of states and perceptions that it allowed. What Leibniz pointed toward was a vision of a tolerant, diverse, cultivated world in which humans know that they are not the measure of all things. It is a world where humans have come to realize that they participate in and have effects on organic unities of all different scales, and that inherently valuable life abounds even where it is not immediately obvious, for "there is nothing sterile and uncultivated in nature, even if many things seem that way to us" (1989, 171). With this balanced, arcadian spirit, Leibniz can be taken as the starting point for an aesthetically based critique of hegemonic utility.

Standing at the threshold of the eighteenth century, Leibniz got to know firsthand an era in which almost all of the major and minor trends of bourgeois civilization began to develop a powerful industrial synergy. A strong middle class had developed in the leading cities of France and England. The urban life-form in almost all its present-day aspects had established itself: bustling commerce, free artistic and political exchanges, an increasingly developed division of labor, and a corresponding differentiation in prestige all made their mark on the social scene; meanwhile, air and water pollution and a lack of contact with the rural countryside for many people increasingly characterized the eighteenth-century experience of the environment. This capitalist-urban lifestyle must be seen as the proximate cause for a new, countervailing turn to nature.

As the subsequent discussion will make clear, in many respects the eighteenth century can be regarded as the origin of modern conceptions of nature conservation. This is perhaps not surprising for a century that praised nothing as highly as Nature. But it is important to point out that in much of the eighteenth century, discussion of this topic shows itself to be quite unrelated to a specifically ecological consciousness. "Nature" was often instrumentalized as a metaphor for political life, for example. Because of the vagueness of the word, different social groups were able to integrate it into their vastly differing ideologies. On the one hand, representatives of the feudal order—nobles and clerics—could refer to "nature" as the God-given hierarchical structure of feudal society. On the other hand, bourgeois capitalists could point to the philosophical construct of "nature" as the state in which every individual had the right to pursue his own egotistical interests. Neither use of the word carried any particularly well-developed environmental meaning; in fact, with every reference to nature the two sides were essentially fighting over the right to be the preeminent exploiter of natural resources.

But if the mainstream of eighteenth-century societies remained firmly in the grip of what David Orr (1992) calls the "modern paradigm" of environmental exploitation, it must be noted that there were significant developments in other areas, notably in the aesthetic and philosophical realms. In his novels and treatises Jean-Jacques Rousseau caused people all over Europe to cast a critical eye on the stuffy conventions and even stuffier housing of the city, and he inspired in them a new counterenthusiasm for the simple life of a gentleman farmer. In Germany, Salomon Geßner's idylls and J. W. Goethe's *Sorrows of Young Werther* transmitted and reinforced this basic impulse. "Back to Nature" became the watchword of the day, and the number of urban dwellers seeking relief can be judged from the fact that tourism in the Swiss Alps now took on the status of a major industry. All of these things are evidence of a qualitatively new appreciation for the importance of Nature and are probably related in the long run to a kind of environmentalist attitude. But a close examination and correspondingly thorough appreciation of the natural world—as well as mankind's place in it—is largely lacking in many of these manifestations. "Nature" had quickly become a kind of style or convention possessing little true critical opposition to the dominant value of utility. Nevertheless there are three significant developments to be mentioned here: the physicotheologists, aesthetic theory, and the revolution in garden design that is associated with the landscape garden.

Physicotheology was very closely related to Leibniz's project of a natural theodicy and promoted the view that the world was a physical manifestation of God's goodness. What distinguished physicotheologists such as John Ray (*Wisdom of God Manifested in the Works of Creation,* 1691) and William Derham (*Physico-Theology,* 1713), however, was their more specific, even scientific, attempt to understand, gauge, and use God's gifts in Nature. Desiring to prove that Nature was ordered in the best possible manner, physicotheologists tried to outline the wisdom evident in natural phenomena, examining and reasoning about things ranging from the shape of mountains to the water cycle to plants' medicinal value. Gernot Böhme rightly says that this consciously directed attempt to trace interrelationships within the biosphere marks physicotheology as "by all rights . . . the historical predecessor of today's ecological thought" (Böhme 1989, 80–81).

Nevertheless it seems necessary to temper this recognition, as Carolyn Merchant and Donald Worster do, with a careful examination of the way in which such ecological knowledge was to be used by humankind. Merchant notes that physicotheology was intensively "managerial" in outlook; Worster similarly asserts that Derham and Ray never got beyond the dominant, exploitative view of nature. "Their . . . ecological model accurately mirrored the popular bourgeois mood. Its fundamental assumption was that the 'economy' of nature is designed by Providence to maximize production and efficiency" (Merchant 1980, 248; Worster 1985, 52). Such tendencies come clearly to the fore in the intellectually related economic school of the Physiocrats. These thinkers centered around François Quesnay explicitly stressed that ecological or agricultural knowledge should be used in order to maximize profits. The physicotheologists thus define an "unblushingly utilitarian" (Worster) and anthropocentric standpoint against which we can perhaps better judge the innovations in natural aesthetics that follow.

In the course of the eighteenth century a theory and practice of natural aesthetics arose in reaction to the excesses of "Reason" in both its rationalist-philosophical and capitalist-utilitarian forms. Philosophically this process keyed upon opening up and refining an area of human perception that had been dogmatically ignored since Descartes. The rationalist philosophers had tended to define humans—in the dualist tradition—exclusively as "thinking beings" (*res cogitans*) with no appreciable connection to the world of bodies, that is, their environment. As early as 1735, however, Alexander Baumgarten took up the study of aesthetics as a "lower faculty of cognition" via the senses (Franke 1972, 40). Leibniz had prepared the way for this with his observation of "*petites perceptions*" (subconscious sensation) and "confused ideas," two areas of human perception and knowledge that could serve to link minds with the world of bodies. Baumgarten's closer focus on sensations as a kind of knowledge narrowed the material–mental gap even further.

A centrally important feature of his philosophy was the way in which sensation was called upon to supplement the role of rational thinking. "Reason," Baumgarten said, "is more exact than the senses, but also poorer" (cited here from Gregor 1983, 365). By this he meant that science, through its dependence on conceptualizing, inherently abstracts and analyzes away from the concrete richness of the world. Sensation was able to come to the aid of this kind of rationality in two ways. For one thing, it could help reinforce and flesh out our individual rational conceptions. But perhaps more importantly, sensation also offered the unique faculty of association, binding together a mass of characteristics into a real-world object as an individual, a unity. In fact, only sensation could provide "extensive clarity," the recognition that certain things belonged together in forming a whole (cf. Gregor 1983, 367).

With this revaluation of sensation Baumgarten paved the way for an environmentalist understanding of aesthetics.[1] This is to be seen, first, in the finding that aesthetic perception is that faculty capable of perceiving a natural system in its property as an interconnected whole. But there is another important component of his theory, too. If we look at the strengths and weaknesses that Baumgarten assigns to aesthetic and rational realms respectively as a historical-cultural assessment (and not so much as a timeless philosophical statement), we begin to see that the aesthetic pursuit of environmentalism was uniquely adequate to the specific constellation of forces in pre–twentieth-century modern Europe. For much of the modern era, science has been synonymous with analysis, single-factor causation, and technological manipulation. This is precisely the kind of thinking that Baumgarten labeled "reason": one that is unable to piece things together, to see different factors or organisms as part of a whole. It is perhaps no accident that in this specifically mechanistic phase of science the study of aesthetic perception suddenly gains ground. It might be going too far to call this an instance of the "cunning of Nature" or to attribute it to some greater plan, but nevertheless it seems that *aesthetic* perception kept alive a sense of the natural world's connectedness until its unity could be theorized and proved at the *scientific* level of ecology.

Baumgarten himself was largely prevented from discussing the topic of natural beauty because the conventions of his discipline forced him to concentrate on rules for the production of artworks. But even in his discussion of artistic beauty we can find the contours of an aesthetic that would give new impetus to the appreciation of nature. Unity among a multiplicity of elements—a "*consensus phaenomenon*"—forms the core of Baumgarten's definition (cf. Franke 1972, 79). Here we find the forerunner of the coherent mood—an "atmosphere" or a *Stimmung*—that was to play a major role in defining the qualities of the landscape garden. But it is worthwhile noting that diversity—the "maximum fullness of notes consistent with order"—was the most important correlate of unity (Gregor 1983, 368).

An important application of these theories came with the landscape garden, a new type of garden that arose in response to the specifically modern conditions of the eighteenth century. The first landscape gardens came into being in England around 1720, under the direction of such garden architects as William Kent. After a time lag of some fifty years the fashion struck deep roots in Germany, however, and found some of its most avid practitioners and theorists there. Fürst Franz von Dessau's park in Wörlitz became one of the most visited and copied gardens of the modern era; Friedrich Ludwig von Sckell and Fürst Hermann von Pückler-Muskau designed numerous landscape gardens all over central Germany. What characterized the landscape garden can perhaps best be appreciated by looking at the dominant, "French" style of gardening against which it consciously reacted. The French style was that of a strict, geometrically shaped pattern of shrubbery surrounding a palace or chateau. Trees were pruned into spheroids, shrubs clipped to form columns, paths and plantings were laid out with mathematical precision. Gernot Böhme has aptly described the French garden as one fundamentally hostile to the idea of an independently living and developing nature: "In the French garden nature itself counted for nothing. Plants were simply material, building material . . . In this sort of situation the self-activity of nature can only appear as a disturbance or deviation" (Böhme 1989, 85). By way of contrast, the landscape garden was understood in the main as a way of displaying the beauties of an unforced, unarranged, unmanipulated, *natural* Nature.

This paradigm of nature led to specific aesthetic criteria such as irregularity, contrast, and diversity in the experience of the park. Such rules were conscious reactions against the excessively strict, geometric Baroque and French traditions. These aesthetic features constitute something much more than a mere stylistic innovation, however. They reveal a fundamental shift in the way Europeans could see nature. In contrast to the highly interventionist French style, the landscape garden encouraged a plant's self-development into a locally harmonious, unifying atmosphere, or *Stimmung. Stimmung* is doubly important as an aesthetic concept because it not only denotes a unity in the object, but also indicates that the surroundings in some sense connect with and exercise an influence on the human observer. This quality was highlighted by Germany's foremost garden theorist, Christian Cajus Lorenz Hirschfeld. His massive five-volume *Theory of Garden Art* (1779–85) emphasized that the goal of garden art was to bring out the objective "character" of a landscape and allow it to work its effects on the viewer/participant. This emphasis on cooperation rather than domination makes the landscape garden not only a "predecessor to the Heimatschutz movement," but also an essential foundation for today's ecological thinking (Gradmann 1910, 4; Böhme 1989).

Along with the cooperative theory of *Stimmung*, though, there is a second set of formal aspects of the landscape garden that gives it immense significance in a history of environmental reform. Here we are talking about the way in which the garden consciously refers to and tries to include the surrounding "working" landscape: "The garden is no longer a sealed-off unit," namely; "instead it reaches far out into the land" (Däumel 1961, 19). This process began in the English parks, where landscape gardeners revolutionized the feel of their gardens by eliminating the walls that had traditionally enclosed and protected the nobility's preserves. Of course, they often made up for this by building sunken ditches (known as *ha-has*) instead, but the aesthetic trend was established. It was reinforced by a series of other features. These included lookout points, paths that sometimes extended beyond the "official" borders of the park, and also the largely native vegetation that was chosen. The overall effect is to present the surrounding countryside to the viewer in a new light. Paul Shepard (1967, 88) has rightly noted that "the removal of garden walls was an extraordinary turn in the ecology of human consciousness." The landscape garden strives for conversion: it not only creates a mood or an outlook of ecological cooperation, it seeks to impress that outlook on the wider scope of our daily experience.

In this context it is important to point out the range of different interpretations of nature that was present even within the genre of the landscape garden. Some architects and theorists, following a Rousseauian-Romantic course, preferred to emphasize the powers and attractions of a nature gone completely wild. In architecture their taste ran to ruins, rude huts, and bridges constructed of wood that was bizarrely twisted and branched. This "romantic-sentimental" style, as it is known in garden history, was opposed by a more sober "classical" style.[2] By and large it is this classical style that had a greater influence on active reform movements. Rejecting the notion of an untouched nature—just as they rejected the possibility of a Rousseauian "noble savage"—the adherents of classical principles sought to reform modern life so as to bring out the best features of the irretreviably lost state of nature. Friedrich Schiller defined the goal thusly: "Our civilization should lead us on the path of Reason and Freedom back to Nature" (Schiller 1969, 5). Johann Wolfgang Goethe pursued this same objective of a refined or ennobled nature in his own gardening efforts in Weimar. In the park he helped construct along the Ilm, vegetation and architecture were brought into a restrained harmony, with neat, whitewashed buildings complementing a scenic, but by no means wild, landscape: the meadows were regularly cut for their crop of hay, and the paths were laid out with a conscious concern for the recreational needs of the small city. This kind of pragmatism was an extremely important characteristic of the classical attitude. Beauty and utility were not considered irreconcilable; instead, the highest aesthetic achievement would be one that arose out of and consisted in the patterns of everyday life. As we have seen, landscape gardeners tried to apply this conviction by uniting what had previously been the two separate realms of "artistic" garden and "utilitarian" countryside.[3] It is in this far-reaching attempt to "aestheticize" all of daily life that the classical landscape garden prefigures many subsequent environmental reform programs.

No description of the landscape garden would be complete without an indication of its political meaning. The French garden's ruthless shaping of plants had definite overtones of despotic absolutism. A character in Theodor von Hippel's novel *Lives on an Upward Path* (*Lebensläufe nach aufsteigender Linie,* 1778) draws this connection between the vegetable and political realms after hearing about an official who has trained a fig tree in the French style. To his female companion he confesses his consternation: "I don't know, my lady! if this official, who puts thumbscrews on a fig tree and twists it around, mightn't he also be the type to do the same to the peasants?—The only difference is that that fig tree doesn't breathe!" (cited here from Gerndt 1981, 115). This vignette provides a good example of the powerful associations linking progressive social thought with a kind of environmental concern (cf. Nash 1989). In this connection it is interesting to note that the landscape garden anticipates later movements in that it was, on occasion, an integral part of a truly comprehensive social and philosophical reform project.[4] This is particularly true of the first and perhaps best-known of the landscape gardens in Germany, Prince Franz von Dessau's park in Wörlitz.

The Wörlitz garden was arguably just one manifestation of this prince's genuine commitment to a reformed and enlightened brand of absolutism. Prince Franz gathered an entire circle of progressive bourgeois officials and artists around him and devoted significant efforts to reform pedagogy in the Dessau Philanthropinum, an elite school that put into practice Johann Bernhard Basedow's pathbreaking educational methods. Dessau assumed a leading position in the spread of popular education, too, under the leadership of Carl Gottfried Neuendorf. A kind of miniature Olympic Games for the entire youth of the land reinforced the pedagogues' emphasis on all-around development (Ross 1987, 143f.). Concrete environmental measures, bodily health, and individual moral betterment as the basis for overall social progress were all linked with the garden in a kind of green-tinged reform project that we will find continually repeated in the nineteenth century.

It is important to see that the Wörlitz garden not only tells us something about the new environmental or social ideas that took hold in Germany in the late eighteenth century, but that it also demonstrates the close link between such conceptions and the medium of aesthetics. If we look more closely at Baumgarten's project of aesthetics we see that the proposed existence within an individual of a unique faculty of cognition had a second and very important social corollary, namely in its implications for forming a realm of discourse. Significantly, this new area was defined against the existing features of a dominant rationality. In essence Baumgarten said that rational deduction and calculation may well command the main forum of social debate—but that aesthetic perception and the discussion of aesthetic concerns may constitute a separate and equally valid approach to the world. This was a momentous step in the construction of an "aesthetic realm" within modern bourgeois society.

The aesthetic realm and the inherent question of how it relates to the wider run of social life (where the procedures and ideals of rationalism presumably obtain) have been a frequent source of contention ever since. In one way or another every major cultural or artistic movement has been concerned to develop a position on the issue. Very few commentators of the twentieth century have been willing to accept the rigid divide between aesthetic and rational faculties that Baumgarten set up. But this rejection of Baumgarten is less an indictment of his theory than it is a comment on the character of our modern civilization. Baumgarten was an optimist in the best eighteenth-century tradition. He saw no conflict between a rational assessment of the world and that world's corresponding aesthetic impression. A useful or sensible thing would automatically be beautiful, and an attractive object, in Baumgarten's eyes, would turn out to be one that was eminently suited to its position in the world.

But the real world—*our* real world—is not the harmonious blend of faculties as envisioned or assumed by the philosopher; in particular we can observe that all too frequently the economic or legal "rationality" of bourgeois capitalism flatly contradicts aesthetic values. In this situation the theory of a separate aesthetic realm has had a double-edged effect. On the one hand it has prevented people from taking seriously the discrepancies between what their aesthetic sense tells them and what is considered "rational" by the obtaining rules of decision making. One of the clearest effects of the separation between the realms has thus been to severely limit the power of aesthetic critiques. Nevertheless it is possible to see in this cloud a silver lining. For the very creation of a separate realm of discourse is a potentially liberating development. Namely, the aesthetic realm is one in which "rational" rules are not immediately and unthinkingly applicable; indeed, in an atmosphere that emphasizes the separateness of the two realms it becomes possible to entertain entirely different conceptions of value and behavior.[5] Of course this is only a possibility, and given the overwhelming influence of cultural norms we should not be surprised if only a small part of aesthetic discussion and production turns out to have a critical or alternative thrust to it. But Baumgarten's theory is important precisely because it pointed out that the aesthetic forum exists; it thus paved the way for further important developments in the eighteenth century.

Baumgarten may have identified the new realm of aesthetics, but it was others who began to see its full significance in a societal context. As Terry Eagleton points out, the aesthetic realm became an important feature of conceptions of bourgeois social order. Rousseau, for example, implied that the aesthetic exercised a key social function in his famous words of praise for law "which is not graven on tablets of marble or brass, but on the hearts of citizens." "Pieties, sentiments and affections," Eagleton notes, were seen here as "the ultimate binding force of the bourgeois social order" (Eagleton 1990, 20).

Despite Rousseau's prophetic early words, it was left to Germany's Friedrich Schiller to provide the first coherent theoretical statement of this "bourgeois-aesthetic" position. His *Letters on the Aesthetic Education of Mankind,* written in 1795, were aimed at establishing aesthetic sensibilities as an all-important dimension of human existence. The ramifications explicitly included the political realm. Schiller's letters to Prince Augustenberg of Denmark candidly argue for a program of gradual reform through aesthetic education as the only real alternative to the Terror of the French Revolution. Schiller theorized that the Terror was not just a quirk of French history, but rather seemed to anticipate the logical course of politics in any modern society. He saw the source of the problem in a disharmonious human character, one where the "mere nature" of the human body alternated with the cold and even "perverse" "unnature" that all too often resulted from modern education and the systematic division of labor (Fifth Letter).[6] Even the most "civilized" state and finely reasoned constitution could not overcome these inner character flaws of its subjects and officials. The trick would be to heal these flaws from within, without a revolutionary change of regime: "The living clock-mechanism of the state must be overhauled as it runs; it is a question of changing each spinning cogwheel during its very rotation" (Third Letter).

This is the point at which aesthetic education took on its political importance. Schiller proceeded from the assumption that, in Germany and the greater part of Europe, peaceful political change would have to take place under the aegis of an enlightened nobility. Here he saw the greatest danger not in raw egoism, but in the "barbaric" application of rules and the extension of an iron rationality into all aspects of life (cf. Thirteenth Letter). This clearly subsumed capitalist economic imperatives as well: already in the second letter Schiller had observed that "*Utility* is the greatest idol of the times." Given this feared excess of abstract rationality, Schiller theorized that the key to further human improvement lay in a process of aesthetic education—the "education of the faculty of sensation" (Eighth Letter). In the following letters Schiller goes on to explicate the connections between sensate-aesthetic refinement and moral betterment, but the essential point remains unaltered: an aesthetic sensibility is seen as the Archimedean point from which to reform the entire workings of society, and this project draws its legitimacy precisely from the fact that the aesthetic realm is not beholden to the rationality and the order that obtain in the rest of modern societal life.

As noted above, Schiller was primarily worried about the excessive political rationality exercised by absolutist princes and their bureaucratic apparatuses. But his argument can easily be applied to the more general confrontation with an all-encompassing bourgeois economic rationality. In such a situation the aesthetic realm is important because it holds out the promise of an "immunity" to "everything that is positive and which human conventions have produced." Although we need not agree with Schiller that the aesthetic realm gives us an *absolute* immunity to the obtaining social order, it is important to see that in a condition of hegemonic domination it opens up a relatively free space: "The political lawgiver can close off this area, but he cannot rule therein" (Ninth Letter).

From this point many commentators have gone on to critique the logic and convincingness of the paradigm that Schiller set up. Terry Eagleton in particular has recently read the *Letters* as a misguided attempt to realize freedom in a nonpolitical and ultimately "utopian," nonreal realm; all the emphasis on aesthetics is seen to fit into a program of restraining desire (which stands also for the political drive of the lower classes) and of continuing the real-life dominance of conservative bourgeois or absolutist reason. Taking a decidedly and even one-sidedly materialist position, Eagleton continually emphasizes that the role of aesthetics here is that of a subtle subversion of healthy natural drives, "the law introjected and habituated," or as a tactic in which "reason plays along with the senses, inscrib[ing] them formally from the inside as a kind of fifth columnist in the enemy camp" (Eagleton 1990, 114, 105).

But this reading presents an unfairly dark version of how the cultural and aesthetic realm fits into society. Schiller's hopes for the aesthetic are remarkably similar to later hopes for a counterhegemonic potential; in much the same philosophical register as Schiller, Antonio Gramsci was interested in the political efficacy of a culturally instilled "rational spontaneity" (Gramsci 1985a, 22; similarities admitted by Eagleton 1990, 106). In Gramsci's thinking it becomes clear that the cultural process is not only one of domination, but that it also presents opportunities for innovation and resistance: in other words, for the articulation of the kind of "desires" that Eagleton only sees as being squelched. The text of the *Letters* unfortunately obscures this issue in those passages where it sounds like raw Nature is to be "overcome" or, in a historical sense, left behind. A close reading reveals, however, that Schiller consistently wanted to achieve a synthesis with ethical behavior, as, for example, when "desire widens its scope and raises itself to love" (Twenty-Seventh Letter). Schiller, unlike Eagleton, accepts the notion that a true society is always something more than individual material fulfillment and indeed may consist in the restriction of certain desires. What culture—or "the aesthetic"—allows is precisely the negotiation of a balance between the two, and it does this in particular by providing a forum in which incipient desires can be articulated against the dictates of the dominant brand of social rationality.

In reading Schiller we have to be aware that instrumental rationality, as a kind of technical, economic, or even political functionality, should be considered under the same category as raw egoism, or "mere nature." When Schiller writes of "savages," these can very well be men who have highly complicated machines at their disposal. The definition hinges on the source of motivation. A savage, in Schiller's vocabulary, is one who is ruled by his animal passions (hunger, lust, greed, etc.) irrespective of the means employed. Schiller, along with Hobbes and Kant, believed that the vast majority of civil society was driven by such basic instincts. Moreover, we have to consider that much of the "order" in the bourgeois "state of nature"—the entire common-sense view of what is rational behavior—is likewise only a brokered power relation between antagonistic sets of self-serving individuals. This is a society that is far from rational in the moral sense of Reason (*Vernunft*). When Schiller writes of refining humans through aesthetic education, though, he is contemplating precisely this transition: from a kind of brute technical rationality of expedience, on the one hand, to a society in which the fulfillment of needs would coincide with deeply moral behavior.

It may seem to us today that Schiller aimed very high with his scheme: too high, perhaps, for the jaded citizens of the late twentieth century. But, as we shall see, the idea of aesthetic education found many adherents in the nineteenth century. The large amount of interest can be explained by the way his project spoke to many of the concerns and priorities of bourgeois citizens. Its educational emphasis dovetailed with the great faith that this class put in the idea of *Bildung;* moreover it addressed the growing proletarian presence. The "lower and more numerous classes" had in fact been one of the chief objects of concern for Schiller (Fifth Letter). By the middle decades of the nineteenth century the associated problems of pauperism and a presumed moral decline figured large in the consciousness of many middle-class Germans.

Schiller's *Letters* proposed a solution to this incipient class problem. In a sketch that can be understood as either historical or psychological, Schiller described the progression from raw nature to full-blown Reason as a series of small, concrete steps. These steps take humankind gradually away from the mere fulfillment of sensual desire and give us a more distanced insight into the form and formability of the world around us. A key point is reached with the aestheticization of the household, which Schiller thought would help transform even the most surly proletarian:

> Just as form will come closer to him in an external sense, in his living quarters, in his household appointments, his clothing, so, too, will it finally begin to take hold of the man himself—at first transforming merely the external man, but in the end also the inner one. (Twenty-Seventh Letter)

The revelation of this passage is the simple conviction that the environment makes the person (or more precisely, that the process of a person's aesthetic refinement of his/her environment is simultaneously the process by which the inner self becomes more moral). The power of such a view from an organizational perspective is that it allows one to start small, with the most common features of daily life. Rather than a direct assault on the power structures of society, its method is, as Schiller wrote, a "revolution in sensibility" (Twenty-Seventh Letter). It is at precisely this level of a "structure of feeling" that more recent theorists see new or alternative paradigms of value being able to generate, perhaps beginning with the way people live, eat, and clothe themselves.

We should learn to apply this understanding of cultural politics to the Heimatschutz, which often styled itself quite consciously in the Schillerian tradition as an "aesthetic education of the people" (Spahr 1913, 2). Ernst Rudorff often cited Schiller, and his fellow Heimatschützer recognized, in turn, the "Schillerian spirit" in his writings (Hagen 1914–15, 237). Although their vocabulary was often updated to sound more physiological, Heimatschützer retained Schiller's central conviction that the environment's aesthetic effect would make itself felt in the character and state of mind of an area's inhabitants. As a Schiller-quoting teacher and member wrote in 1911, the eye was perhaps the greatest determinant of one's emotional life; through it the features of one's world would be "photographed into" the soul itself:

> Thus it is not irrelevant, whether one is to become a human being under dark chimneys with their even blacker clouds of smoke or in free and bright nature . . . How very different becomes the spirit of a person who grows up under the influence of bright, cheerful colors, whose emotional world has been brightened and softened by congenial local surroundings and joyful impressions of nature. This pleasant world can give him much more than school or church can, for it predisposes him to freedom, to innocence, to cheerfulness. (Mugler 1912, 77–78)

Of course much was added to Schiller's theory over the years. When the Silesian Association for Heimatschutz challenged a fellow German to think of the "close relationships between the external world and his own life," a whole new level of concreteness was implied (Schlesischer Bund 1911). After a century of advances in biological science people began to see that the environment had not only morally or educationally formative effects, but also fairly specific material and health advantages.

In the Dessau reform project we have already seen an early example of the way in which such concrete interests could be combined with the aesthetic demands of a landscape ideal. Parks such as the one at Wörlitz thus serve to illustrate the positive heritage of environmental concern that the Enlightenment has to offer us, one that extends into the nineteenth century with the innovative curricula developed at a few schools in central Germany.[7] But in acknowledging these pathbreaking attempts to link aesthetic, ecological, and social issues we must also see that the projects remained limited in their practical effects. Relying on a strategy of personal exposure and a kind of individual environmental *Bildung,* they made little headway toward institutionalizing and promulgating their ideas, and they did not succeed in challenging wider segments of society with their respective visions. In this respect the early nineteenth-century movement for "land beautification" or "rural embellishment" (*Landesverschönerung*) represents an important new phase of aesthetic environmentalism. As Gerd Däumel has argued, it became the "true heir" to the classical landscape garden. For here "the garden ideal was projected out onto the entire landscape," and reformers went to work with the ambitious ideal of "ordering and beautifying humans' entire environment in the style of a *Gesamtkunstwerk*" (Däumel 1961, 30, 41).

As the name implies, *Landesverschönerung* was a movement that derived a great amount of its legitimacy and direction from an aesthetic sensibility. Its goal was, after all, to make the country "beautiful," a goal that entailed many specific measures to plant trees, maintain hedges, and otherwise dress up locales. The implicit aims of the landscape garden were thus made explicit by the land beautifiers, who crusaded under the cry "Germany, all of Germany one great garden!" (Schuderoff 1825, 64).

Nor was it merely a question of dimensions that distinguished the new movement. There was an important qualitative difference to be seen as well in its tendency to absorb and cooperate with a range of technical and hygienic disciplines. A number of these had typically been included in the idea of "land cultivation" (*Landeskultur*). With its roots in eighteenth-century administrative reform and cameralist science, this latter movement had never evidenced much environmental concern and in fact tended to push for large-scale civil engineering projects in order to increase agricultural production. Munich's Freiherr von Stengel, for example, had seen his draining of the marshes along the Danube as a great civilizational deed, a "transformation [of] empty bottomless steppes into fields pregnant with harvests." Nor did von Stengel shy away from claiming aesthetic worth for his drastic landscape transformation, celebrating the fact that "the gloomy abode of lone water- and marsh-birds" had been "made into a garden" (cited in Däumel 1961, 35). Here again we have the key word "garden," but it is easy to see that it is now dangerously close to providing legitimation for merely expedient interventions into the natural world.

In all fairness it must be remembered that von Stengel's statements are taken from the era before 1800, a time in which human impact on Germany's land was still only a small portion of what it would be later. In addition we should realize that the program of land cultivation, as well as its successor, the *Landesverschönerung* movement, was informed by a relatively liberal political and social orientation. Von Stengel had wanted to contribute to the "general welfare of the country" with his public works projects, and his fellow reformer Joseph von Hazzi had argued in his *Catechism of the Bavarian Laws on Land Cultivation* (1804) for "free property rights," "free culture," and "freedom to do business." Such Enlightenment-based sentiments were by no means self-evident in a country that was still only just emerging from feudal structures. It may therefore have been uniquely advantageous for the technical reformers that they could now ally themselves with a garden-based aesthetic vision to produce the new program of *Landesverschönerung.*

There were actually two centers to this movement of the 1820s, each of which was only loosely connected to the other. One branch of the movement was based in Berlin, where the landscape gardener Peter Joseph Lenné (1789–1866) inspired the Association for the Promotion of Gardening in the Prussian States and promoted planning ideas within the Prussian bureaucracy (cf. Buchwald and Engelhardt 1968, 100). The second group was based in Munich and consisted largely of civil servants within the Bavarian state administration. These included not only the aforementioned von Hazzi, but also the architect Gustav Vorherr, who proved to be the cause's most tireless, albeit somewhat single-minded, publicist and proponent.[8] In his words *Landesverschönerung* would be a far-reaching program to improve the quality of life for ordinary citizens, one that would encompass everything from city planning to agricultural extension work to architecture. Its aim was moreover explicitly egalitarian: "Everyone, even the most common burgher and farmer should breathe in lovely fields, as well as in beautiful living quarters, villages and cities!" Vorherr added further emphasis to this sentiment by calling for state expenditures to help Bavaria's "little people" in their long-neglected villages and minor towns; by contrast, he said, "the residences of the great ones . . . should come last" (cited here from Däumel 1961, 53).

The bottom-up approach described here demonstrates some of the popular-egalitarian political tilt of land beautification. It is interesting to note that Vorherr put the fetish of technological progress in the same category as wasteful royal ostentation. Both tended to draw funds away from the real needs of average people, with the result being an incongruous and unjust kind of nonsimultaneity along class lines:

> Boldly we seek roads through the air and neglect to construct passable paths for our villages; we dig harbors and new streambeds, but lay no pipes for the tapping of fresh springs; we erect palaces, towers that climb to heaven, troubled little with the living quarters of the millions, which are after all, together with the nearest surroundings, that which makes the first lasting impression on a person, and that which continually has an effect on him. (cited here from Däumel 1961, 54)

The closing phrase here is very characteristic of nineteenth-century reform movements in the wake of Schiller. Again and again one finds the unshakable conviction that the "nearest surroundings" make the individual. And although at first this was understood to include mainly the built environment, it soon became a habit to extend consideration to the wider green surroundings. The important thing to remember is that this conviction owed much of its strength to an aesthetic sensibility.

We see the particular importance of aesthetics precisely in those measures that went beyond quantifiable utilitarian benefits to provide for less well articulated and more diffuse needs of the environment. Nowhere was this aesthetic bonus more apparent than in measures affecting agriculture, a sector of the German economy in which the ideas of rationalization and profit-maximizing had very quickly taken hold. Many of the *Landesverschönerer* had experience with farmers and fully sympathized with their economic aims. It did not appear to conflict with this, however, to assert that aesthetic values were also worth considering. As the authors Voit (1822, 37) and Barth (1831, 82) pointed out, nothing was more beautiful than a "cultivated land" full of "luxuriant fields." Here beauty and utility went hand in hand. Indeed, Gerd Däumel (1961, 87) has noted that the joining of the two standards is perhaps a central philosophical characteristic of the *Landesverschönerung* movement: "Again and again one finds the tendency to combine the useful with the pleasant, to assume a kind of identity of useful and beautiful."

There are problems with this axiom, however. Saying that the useful is also the beautiful is superfluous, in a way. Nothing is actually gained by the second term in the comparison, because in modern societies economic utility—based in part on technological and scientific knowledge—has always been the unchallenged paradigm of decision making. At best it would amount to an aesthetic legitimation of what people would do anyway. Things get more interesting if one proposes the relation in the reverse order: if one says, namely, that the beautiful is also the useful. For there have always been uncertainties with the idea of usefulness: costs and benefits are not always clear even in the short term, and the utilitarian calculus can become hopelessly complex if one tries honestly to assess long-term and large-area effects of an action. But if beauty is now accepted as the shorthand version of true utility, we can trust in our aesthetic sense to guide us to do the right thing. The application of this axiom to a landscape could have an immense ecological impact. We would not have to provide detailed proof, for example, that mixed forests are more economical than pine or fir monocultures, nor would we be faced with having to document that they are better water regulators or preservers of a more diverse biota. Instead we could rely on aesthetic evidence and begin to make changes right away.

In this respect it is significant that the *Landesverschönerung* movement promoted a number of ecologically significant measures that could only partially be supported by direct utilitarian argument, including planting the edges of streams and ponds with willows or alders and employing hedgerows in fields. Such measures were inevitably given two-pronged justifications, as when riverbank plantings were said to be "well suited to holding the soil," and that, in addition to a possible crop of wood, the result would "often [be] a picturesque scene, and a gentle point of rest for the eye." This kind of willingness to see factors beyond the most immediate effect on crops and income gave the *Landesverschönerer* a good basis for appreciating ecological connections. As the same man stated in 1824, "Not only individual products of the farmer should be taken into account and promoted, but rather the whole household (*Haushaltung*) of a landscape should be examined and respected" (Voit, cited here from Däumel 1961, 89–90).

It was with precisely this aesthetic-holistic conviction that the *Landesverschönerer* defended and promoted hedgerows. The Prussian circle was particularly productive here: starting with Albrecht Thaer and going through Peter Joseph Lenné, Prussian landscape reform was synonymous with an "economic embellishment of fields" (Buchwald and Engelhardt 1968, 100). The depth of the insights behind this program are illustrated in an 1845 brochure by the agricultural expert Alexander von Lengerke titled "On Living Field Enclosures and Their Introduction." A whole range of various effects was noted: moisture retention, protection against cold winds, and fruit and wood production were seen as some of the everyday utilitarian benefits; in a gesture to the Prussian army, the defensive significance of hedgerows was pointed out; and finally there was the aesthetic value of the diverse vegetation. Significantly, Lengerke looked forward to the day when all German farmers would recognize the way in which aesthetic appearances signaled the degree to which things were functionally optimized. In this sense he predicted that in the future farmers would come to respect "the power of the beautiful" (cited here from Barthelmeß 1988, 213).

We have come to see *Landesverschönerung* as the practical culmination of the ideals first expressed in the classical landscape garden. The main points of that common ideal are centered around the identity of beauty and utility, or, in other words, the creation of a working landscape that is beautiful and garden-like in its entirety. But the nineteenth-century society in which the land beautification idea arose inevitably infused the project with its own priorities. When forced to choose between beauty and utility, even the most outspoken supporters of *Landesverschönerung* were prone to choose the latter. Progress—not the least kind of it being social progress—seemed to them to depend upon increasing production first, and beautifying or preserving second. There were areas in which the two factors coincided, as some of the arguments in favor of hedgerows tried to show. But in a culture that respected only utility, it was perhaps counterproductive to assert that utility and beauty are entirely the same thing. It is indicative that aesthetic concerns were often banished from the great expanse of productive acreage and allowed free rein only in the cosmetic decoration of small-town streets, or the building of a playground. The generally small, local "beautification societies" (*Verschönerungsvereine*) of the latter half of the nineteenth century managed to preserve only this much-reduced version of *Verschönerung*. Their work was the logical form of expression for social-ecological ideas under capitalism: a postage-stamp park, perhaps, or a bench paid for with the scraped-together donations of a few civic-minded citizens. Everywhere else the landscape was increasingly shaped by the needs of "rationalized" agriculture and a growing industrial sector. Following the teachings of Justus Liebig, fields were pumped full of chemical fertilizers, and harvests were carried out by ever more capital-intensive machinery, which in turn demanded farther-reaching rationalizations of the production process. Egged on by the conventional wisdom of agricultural extension services, farmers strove to eliminate all hedges, nooks, wet spots, and edge habitat, often combining legal and environmental streamlining in so-called separation or field rationalization procedures. If anything, then, it was the lack of reliance on aesthetic criteria or the unwillingness to apply them in a broad landscape vision that doomed the *Landesverschönerung* to ecologically ineffectual action.

So it was that the total environmental reform that was supposed to emanate from the classical landscape vision never came about. Instead, under the combined pressures of industrialization, urbanization, and capitalization, Germans of the nineteenth century began to exploit the world around them at an unprecedented rate. Environmental degradation worsened, but public perception thereof was singularly muted. Rolf-Peter Sieferle has observed that the relative lack of interest in the problem was related to material interests, the sub-critical level of degradation, and a complex set of coping behaviors:

> The advantages of big industry were so convincing that environmental degradation was accepted. This remained, after all, always limited: individual trees died, a few fields became poisoned and unproductive; one could no longer bathe in rivers, at least in the immediate vicinity of industrial centers; drinking water, too, became unpotable. But not all were equally affected by these problems. Those who could afford to found other locations not too far away that were still intact. The noise and smell of the factories did not come into the villa-studded suburbs. The workers, on the other hand, were exposed to environmental poisons not only at their place of work, but also in their living quarters, which were usually in the immediate vicinity of the factories. Nevertheless they considered themselves compensated in the form of a job and a decent wage. The security of their livelihood was in every instance more important than a healthy environment. (Sieferle 1984, 145; see also Mieck 1981)

Thus, although strategies varied according to class and related factors such as education and income, it seems that for most of the nineteenth century the greater portion of the populace did not engage in any pro-environmental action. This fact leads one to suspect that people were essentially unaware of environmental degradation as a criticizable phenomenon. Sieferle neglects this facet of the problem in his analysis; for him, all classes seem to have been transparently aware of what a "healthy environment" was, and they consciously developed strategies to cope with its absence. But it seems to me that this was not the case at all. In his reception of Kant, Heinrich von Kleist despaired about a world in which everyone could be wearing green glasses without knowing it; in the same way it could be said that for much of the nineteenth century, people looked through smoke-colored glasses. Their environment was simply an outgrowth of the dominant way of life, and few were able to see that the smoke, water pollution, and land abuse of the time were indeed something both admitting and needful of change.

Protests against the ruination of aesthetic, environmental, and historic treasures were not unheard of, but they remained largely on the level of local, ad-hoc reactions to events. In Germany some of the best-publicized incidents focused on the scenic Rhine valley. The year 1827 saw a violent protest against stone-quarrying that threatened the castle and cliff in the Siebengebirge known as the *Drachenfels.* Although this picturesque spot was eventually saved by the Prussian king's decision to seize the property by eminent domain, much of the surrounding area continued to be quarried. In 1886 a local group called the Association for Saving the Siebengebirge was formed and managed to call attention once again to the problem. But arguments focusing on aesthetic, recreational, or civic value were greeted with contempt by the leaders of the stone industry, who stood firmly by their property rights:

> We share the enthusiasm for this charming little piece of German soil, but even more highly do we estimate the idea of *property* and the right to *do with this as we see fit,* one that may not be infringed upon without penalty. We will not have anyone interfering with our property rights, which in our law-abiding state stand under solid and secure protection. To such efforts our answer reads: Humbug! (emphasis in original; cited here from Barthelmeß 1988, 133)

The owners knew that they had a powerful legal and political argument. Ever mindful of the source of its tax money, no government of the nineteenth century wanted to be seen as anticapitalist or as having too free a hand with property seizures. Royal intervention in this 1880s dispute was thus limited to approving a money-raising lottery for the purchase of the land. We should also recognize that the publicity and positive results achieved in these cases were unusual. They are the exceptions that prove the rule of nineteenth-century environmental neglect.

As the quarry-owners' answer suggests, one basic reason why so little was done has to be sought in the solidly entrenched economic order of private property and capitalism. These structures were reinforced in turn by their association with technological and social progress. With the positive values of society thusly co-opted, it comes as no surprise that at mid-century even environmentally aware and skeptical citizens were typically unable to conceive of a real alternative to this system. Their opposition took on a strangely veiled or internally broken form, with a tendency toward escapism, resignation, and nostalgia. Karl Leberecht Immermann's novel *The Epigones* (1836) is a case in point. Although the book portrays the advance of nineteenth-century capitalism from the critical perspective of a disappointed romantic, it envisions no answer to the pollution and landscape destruction it so clearly abhors. The only practical demonstration of an environmental awareness is the decision of a sensitive factory owner to dismantle his inherited business—a desperate individual act devoid of wider societal impact. In his eyes, resignation and the attempt to preserve a personal refuge are the only strategies available:

> With the speed of a storm the present is heading toward the condition of a dry mechanism. We cannot hinder its career, but cannot be blamed if we fence off a little green place for ourselves and our families, and secure this island for as long as possible against the impact of the industrial waves that go crashing by. (Cited here from Sieferle 1984, 117)

A cultural history of German environmentalism would have to give great significance to two of the phrases used here: "a little green place" and the related image of an "island." In their ambivalent mixture of resignation and implicit aesthetic protest they epitomize the latter half of the nineteenth century. Even a cursory examination of this theme reveals that the "green island" theory, as we may call it, was shared in many aspects of cultural life.

Perhaps its most important proponents were the critical bourgeois-realist authors such as Immermann and Wilhelm Raabe. Raabe especially pursued a strategy in which memory and the environment could both provide "green islands" of refuge within the increasingly regimented life of an urbanized and industrialized Germany. These remembered environments are continually played out against the society of the present—not in an unbearable tone that says that everything was better previously, but rather in a more nuanced and reflective way that suffuses the reality of the present with a powerful sense of loss. This strategy figures prominently in *The Vogelsang Documents* (*Die Akten des Vogelsangs,* 1895), but is carried out perhaps most strongly in *Pfister's Mill* (*Pfisters Mühle,* 1884).

A short passage early in the book gives us access to the thoughts of a middle-aged man who is about to sell the once popular but now badly polluted riverfront property on which he grew up:

> We have already taken up housekeeping on the shaded side of our street in the great city of Berlin, and there I have longed all too frequently for the sunlight of the Heimat of my youth . . . And I have taken a look at the ground plan and the rest of the design of the great factory which the owners-to-be will erect in this location, and I know how little light and warmth the brick walls and high smokestacks will leave here, too, in just one more year. (Raabe 1985, 20)

The resignation in these lines can be explained in part by the overwhelming power of Germany's technological-industrial system, one that first polluted the area (in defiance of individual protests) and then found it to be a useful location for production. Nevertheless the book implies that this takeover was abetted by an inner weakness or collaboration on the part of those affected. The narrator's best friend becomes an industrial chemist, tied to the very same production processes his neighbors are fighting; and while the narrator's father is driven into his grave by the transformation, the narrator himself learns to disregard environmental problems and "get on with" his life in the city. Even though he is not happy about the changes, he too will be profiting from the industrial transformation of what was once a pleasant country setting. Given this kind of profound moral ambiguity it is not surprising that Raabe offered his environmental vision as a nonbinding individual preference, a memory or a place that the modern world could ignore or treasure as it pleased (cf. Craig 1978, 79). But diffidence and self-effacement were unable to challenge directly the wider cultural norms of the nineteenth century, and any isolated "green island" was doomed to slip under the rising tide of grey. What was needed was a universalizing ethic such as that connected with the idea of Heimat: one that demanded its changes everywhere, and most radically in those places that enjoyed the least prestige and endured the greatest utilitarian pressures. It is perhaps only logical that the most important precursor of the Heimatschutz aesthetic technique can be found in the middle of Germany's most industrial cities.

If we think about environmentalism as an achievement in everyday perception, and not just as an object-directed concentration on nature, it becomes clear that we must give new recognition to efforts to improve the built environment. Programs such as reformed worker housing, new sewage systems, and city planning were admittedly oriented toward explicitly urban situations, but they nevertheless constitute an important demonstration of how the aesthetic realm could be an essential element in the organization and concrete political enactment of a kind of environmental reform. As we shall see, the movement for housing reform cannot be understood without this aesthetic component.

The nineteenth century was marked by a steady increase in the number of workers seeking housing in cities or near factories. Invariably the quarters they received were unhygienic, overcrowded, and ill-equipped. Friedrich Engels's *Report on the Condition of the Working Classes in Manchester* is perhaps the best-known contemporary analysis of this situation; nevertheless one didn't have to be a committed Socialist to realize that something was gravely wrong with this aspect of the laissez-faire economy. *The Bitter Cry of Outcast London,* a pamphlet published by Arthur Mearns in 1883, was the bourgeois counterpart to Engels's work, describing the horrific housing conditions in such detail and with such moral indignation that it ignited a nationwide debate in Britain on the issue. By the last third of the nineteenth century almost all European countries had experienced similar revelations, with the result that even defenders of capitalism and members of the upper classes were beginning to attribute great significance to the problem.

In Germany attempts to better housing conditions for the working class date back at least to Victor Aimé Huber and the Berlin Charitable Building Society, founded in 1847. Over the next twenty years attention to the problem tended to wax and wane in relation to the overall business climate, but by the early 1870s an ever-widening segment of the educated bourgeoisie had become convinced of the need for reform. The Association for Social Politics (*Verein für Sozialpolitik*), a kind of liberal think tank, provided strong academic support with a series of reports in the 1880s, and the turn of the century saw a multiplicity of projects launched by organizations such as the Association for Housing Reform and the German Garden City Society (on these and the role of the *VfS* see Bullock and Read 1985, 67f.). These private associations came in addition to middle-class efforts to enlarge the role of the state and municipalities in social welfare (for comparable efforts in England see Gould 1988, 96). Regardless of the delivery system, though, the attempt to create better housing for workers and related efforts to guide sanitation and institute city planning were central components of evolutionary bourgeois social reform.

It is too easy to see this process as nothing more than a crude political ploy in which the upper classes tried to maintain their social power by meliorating conditions and thereby reducing the relative appeal of revolutionary Socialism. Although it captures a certain basic truth of the situation, this explanatory model of rigid class opposition fails to appreciate how cultural, scientific, technological, and grassroots political processes were continually helping to reshape the social order from below. If the bourgeoisie was able to dominate in the nineteenth century, this was partly because it was reasonably successful in incorporating new demands—even Socialist demands—into its particular version of the social order. A history of this time period will thus pay close attention not only to the way in which the dominant classes were able to prevent change and to impose their definition of morality, but also to the way in which they adapted to change and developed forward-looking ideas. Housing reform was one such adaptation, and it is worth noting that aesthetic argumentation played a significant role in gaining acceptance for this progressive idea in a climate of extreme individualism.

Like Schiller, the bourgeois reformers of the nineteenth century were convinced that everyday surroundings had a powerful effect on people; although they worried a lot about the negative possibilities inherent in this relation, they nevertheless retained the optimistic conviction that moral, social, and political betterment would be the result of a more beautiful setting for daily life. Their argumentation was thus, not surprisingly, filled with aesthetic analogies. Present housing was not simply "inadequate": it was filthy, vermin-ridden, soot-covered, loud, dank, stuffy, and smelly. These drawbacks were contrasted against a vision of what rooms in new, reformed housing might be like: namely, full of sunshine, light, airy, dry, warm, and clean. In looking at this schematic argument it is important to acknowledge the simultaneity of several registers of meaning. "Clean" and "filthy," for example, are not merely descriptive of an aesthetic impression; they were and are loaded with moral connotations as well. Many reformers hoped that the clean and more spacious apartment would allow for a better family life and eliminate the sexually corrupting atmosphere that was rumored to dominate in overcrowded worker housing.

Bourgeois reform in the nineteenth century is inseparable from its argumentative eclecticism. Indeed, I would contend that its mixing of aesthetic, moral, and functional reasoning gave it its specific dynamism: an impassioned debate in these more "cultural" terms helped create pressure for more concrete considerations and also helped publicize these results. In the area of housing we can see that aesthetically minded reformers soon developed a symbiosis with the rapidly expanding medical profession. It provided them with a set of powerful scientific arguments: miasmic contagion, bad air, and bacteria were concrete dangers that lurked in substandard buildings. Through the Association for Public Health (*Verein für öffentliche Gesundheitspflege,* founded in 1873) and a network of memberships in other organizations, doctors helped push through large-scale sanitation and planning projects designed to secure the health of the community as a whole.

Of course, it is often questioned whether this *Hygienebewegung,* or public health movement, was motivated by anything more than the enlightened self-interest of the propertied classes. Leaders such as England's Edwin Chadwick (1800–1890) and Germany's Rudolf Virchow (1821–1902), for example, displayed an unabashedly "bourgeois-utilitarian stance" in pointing out that the healthier people who would be produced by better housing would simultaneously be better and more efficient workers (Berndt 1987, 142f.).[9] Before we condemn this kind of economic instrumentalism, though, we should remember that public health was in many respects a progressive and oppositional demand, one that had to be pushed through against the organized resistance of capitalists and their moral supporters, who preached—then as now—about "individual responsibility" and the necessity of pulling oneself up by one's bootstraps. In this situation it made sense to speak the economic-utilitarian language of those from whom concessions had to be wrung. As a result a good portion of the advances that were made in the nineteenth century can be attributed to the skill with which reformers "conformed" to this dominant paradigm in order to widen it.

Nevertheless it should be noted that the aesthetic component played an important and sometimes determining role in public health politics. This was particularly evident in the early stages of a given discipline, when either no relevant scientific-functional theory existed, or when the theory was not yet fully institutionalized. Early gutter and sewage systems are a good illustration of how aesthetics "rounded out" the functional level. When these systems were first being installed no one scientific theory existed from which adequate specifications could be derived. The most widely accepted explanation proposed that disease was engendered by mysterious gaseous miasmas; nevertheless the overall configuration of the sewage system built to overcome this danger, including the distance that effluents were carried away from the city, was almost always determined by the boundaries of what was considered aesthetically acceptable. As Susanne Hauser (1992) has observed, sewage systems and other infrastructure projects rose and fell on their ability to tap into values such as "purity," "order," and "beauty." Aesthetics thus provided much of the greater cultural predisposition to proceed with those technical programs that we today often interpret in purely functional terms. Nowhere was the role of aesthetics more central than in the case of housing reform, where, as Brian Ladd has argued, the catchphrase "*Licht und Luft!*" ("light and air!") was instrumental in furthering "a myriad of plans and projects" (Ladd 1990, 45).[10]

"Light and air!" was probably such an effective phrase because it provided a tangible vision of reform, a feeling to which civic boosters and bourgeois taxpayers could relate. With this unique ability to mobilize support, aesthetic discussion was a central component of many of Germany's advances in basic social welfare. Only later would this influence decline, when public health programs had acquired a professionalized bureaucracy and a more widely accepted, scientific raison d'être. But even then the planners and specialists were often forced—by political necessity—to speak in the more common language of the culture, and the most successful of them did this with consummate skill.

The movement for housing reform and urban planning thus provides us with an important point of reference in studying the Heimatschutz movement. Without denying the great importance of functional and medical argumentation in the establishment of these programs, one can say that aesthetic arguments were enormously influential in preparing the way, building support, and reinforcing the idea at an affective level. We can speculate that aesthetic argumentation was also closely tied to the socially liberal direction of these programs. Unlike the realm of politics, where measures are typically carefully weighed in terms of which groups they will benefit, the eighteenth-century tradition of aesthetics looked at things from the standpoint of a universal subject. When applied to the real world this could have an egalitarian result. If a clean, dry apartment would produce the universally positive and spiritually reforming experience of "light and air," then there was an implicit ethical demand that these conditions be extended to all citizens. The establishment of an aesthetic of "light and air" thus gave the reformers a powerful language to say what previously could not be said in bourgeois-individualistic Europe: that the working poor deserved decent housing, and that everybody deserved a cleaner and less polluted city environment.

The Heimatschutz can be seen as an extension of this model of housing reform to the landscape. Both movements had their social bases in the educated bourgeoisie; both pursued a concrete economic reform of capitalism; both showed great sensitivity to the role of a person's physical environment in the Schillerian tradition; both also retained varying degrees of Schiller's philanthropic paternalism toward the lower classes. The historical progression in which they stood is clear: from an initial theoretical concern with inner *Bildung,* aesthetic reform became both more concrete and more comprehensive over the course of the nineteenth century. By the end of the nineteenth century the scope of investigation had widened to include not only people's apartments, but also their neighborhoods, the city, and the surrounding countryside; meanwhile, each element of this entire ensemble of relations was subjected to an increasingly rigorous material analysis. As a student of medicine, Schiller himself might well have approved of this development.

The trend toward more science was not uniform, however. Housing reform was able to produce significant quantities of scientific and medical evidence to support its measures; but as we shall see in greater detail in chapter 3, little such functional theory was available to justify landscape reform. On the other hand the Heimatschutz was able to fall back on a very well-established body of aesthetics. The German cultural tradition was rich with the praise of Nature, and among the educated bourgeoisie a strategy of aesthetic discourse could be activated in a political campaign for landscape protection. For its material it drew on an eclectic mix of conceptions: Goethe, Romanticism, and newer sources were often tossed together with little regard for their nuances of meaning. In fact we should probably understand their application to new problems as the formation of a new meaning. But the organizing principle was the important thing: for a not inconsiderable portion of the German bourgeoisie, an aesthetic mode of discussion proved to be the key to awakening environmental perception, and further helped to convince middle-class citizens of the necessity of sweeping reforms in their capitalist-dominated society.

The precedent for this aesthetic environmentalism is to be found in England, for in almost every step of their development John Ruskin (1819–1900) and William Morris (1834–96) anticipated the mutations, branchings, and slow radicalization of nineteenth-century bourgeois aesthetic interest. Their starting point was an appreciation of everyday craftmanship and vernacular artistic ability in the Middle Ages; it was this that first led them to reflect on the societal conditions for artistic production. The further they pursued this problematic the more they came to see that artistic creation was the outgrowth of an aesthetically satisfying physical existence. What had started out as a concern for individual works of art thus developed into an "all-encompassing concept of design" that "linked the natural and domestic surroundings of the workmen with the practice of his craft" (Spear 1984, 138). The path whereby this artistic concern could be transformed into environmental activism is traced out in Morris's simple injunction: "Those who are to make beautiful things must live in a beautiful place" (Morris 1914b, 170).

This ambitiously holistic demand for beauty led inevitably to a critique of nineteenth-century English society and of capitalism as an economic system. Both Ruskin and Morris undertook to preserve many different facets of England's rich heritage after recognizing that "the present course of civilization will destroy the beauty of life" (Morris 1914a, 53).

But already in their first efforts to save unappreciated buildings they came, as E. P. Thompson rightly notes, "directly into conflict with the property sanctions of capitalist society." For Morris especially, a respect for old buildings thus "quickened and deepened his insight into the destructive philistinism of capitalist society" (Thompson 1977, 231, 233). Both men displayed an equally acute sense for the endangered beauty of the countryside. Their writings are full of deep-felt rage at the industrial system of greed and exploitation that had reduced the once richly appointed and lovely home of man to a barren, joyless hovel. In an 1880 speech Morris lamented the despoliation of "possessions which should be common to all of us, of the green grass, and the leaves, and the waters, of the very light and air of heaven"—all of which "the Century of Commerce has been too busy to pay any heed to." Never one to mince words, he castigated his cultured bourgeois listeners for their hypocrisy in the face of such environmental degradation: "I say that their love of art is a mere pretence: how can you care about the image of a landscape when you show by your deeds that you don't care for the landscape itself?" (Morris 1914a, 70, 71).

Here it is clear that Morris was able to build on aesthetic sensibilities and cultural traditions in arguing for environmental-social reform. In his literary writings this strategy was often applied positively, evoking an ideal time or place. *News from Nowhere* (1890) has aptly been labeled a "beautiful description of the English landscape under an ecological society" (O'Sullivan 1990, 8). His speeches, on the other hand, often make use of a negative strategy. In a calculated attempt to shock the educated out of their complacency, the speeches withhold the socially desirable status of having good aesthetic sense from anyone who refuses to recognize environmental demands. As Morris put it bluntly in one such speech: "Anyone who cuts down a tree wantonly or carelessly . . . need make no pretence of caring about art" (1914a, 73). Conversely, a person with true aesthetic sensibility would be obliged to engage in a top-to-bottom "cleaning of England." By the 1880s the "Century of Commerce" had already managed to ruin large areas of the country; in Morris's estimation there was as a result "work enough for the next century to repair the blunders of recklessness" (1914b, 170, 173). But for anyone concerned about the overall "aesthetic" quality of life, the hard truth of the matter was clear: "If we accept art we must atone for what we have done and pay the cost of it. We must turn this land from the grimy back-yard of a workshop into a garden" (Morris 1914a, 61).

As we have seen, aesthetic impulses were central in guiding both Ruskin and Morris to their environmental insights. It is important to see that these were not superficial, but truly radical and critical of the society in which they were developed. In particular the aesthetic-holistic viewpoint gave rise to an increasingly articulate denunciation of capitalist-utilitarian economic thinking. Ruskin's axiom in *Unto this Last* (1862)—"THERE IS NO WEALTH BUT LIFE"—was calculated precisely to expose the biases of nineteenth-century free-market economics, which began and ended with the accumulation of capital. By contrast, Ruskin found that "the final outcome and consummation of all wealth is in the producing of as many as possible full-breathed, bright-eyed, and happy-hearted human creatures." To the extent that accumulation took away from this goal—for instance, by driving a system that forced workers to accept the existential minimum wage—it was not only not wealth, but in fact a positive unvalue, a social crime, or, in Ruskin's neologism, "illth" (Ruskin 1985, 211).

As the twentieth century draws to its environmentally dubious close we are beginning to realize that this simple concept of illth was in fact a "brilliant economic insight" (Spear 1984, 145). Ruskin was pointing a warning finger at the dishonest bookkeeping of capitalism, a system that blindly added up the individual entrepreneur's costs and profits without any regard for the overall *external* impact of his activity. One obvious externality was of course the worker, something that the affirmative science of economics studiously ignored:

> [In] the science of Political Economy, the plus quantities, or—if I may be allowed to coin an awkward plural—the pluses, make a very positive and venerable appearance in the world, so that every one is eager to learn the science which produces results so magnificent; whereas the minuses have, on the other hand, a tendency to retire into back streets, and other places of shade,—or even to get themselves wholly and finally put out of sight in graves: which renders the algebra of this science peculiar, and difficultly legible; a large number of its negative signs being written by the account-keeper in a kind of red ink, which starvation thins, and makes strangely pale, or even quite invisible. (Ruskin 1985, 213)

Ruskin wrote here in terms worthy of a Marxist. But he actually went beyond Marx in explicitly applying these insights to contemporary ecological problems. It was a false and misguided political economy, he wrote, that could "imagine precious and beneficient things, such as air, light, and cleanliness to be valueless"; in fact it was one of the prime commandments of a true economy "that land shall not be wantonly allowed to run to waste, that streams shall not be poisoned by the persons through whose properties they pass, nor air be rendered unwholesome beyond given limits" (Ruskin 1985, 209; and as cited in Spear 1984, 145).

Such clear and well-founded pronouncements more than justify seeing the aesthetic reformer John Ruskin as a "forerunner of the contemporary ecological movement" (Danahay 1991, 62; also Lloyd 1989; Fuller 1985; Spear 1984). What today's ecologists often seem to forget, though, is that the main historical carrier of their environmental concern was a finely tuned and cultivated aesthetic sensibility. The following passage remains one of the most powerful and lyrical evocations of an environmental ethic in the English language:

> No scene is continually and untiringly loved, but one rich by joyful human labour; smooth in field; fair in garden; full in orchard; trim, sweet, and frequent in homestead; ringing with voices of vivid existence. No air is sweet that is silent; it is only sweet when full of low currents of under sound—triplets of birds, and murmur and chirp of insects, and deep-toned words of men, and wayward trebles of childhood. As the art of life is learned, it will be found that all lovely things are also necessary;—the wild flower by the wayside, as well as the tended corn; and the wild birds and creatures of the forest, as well as the tended cattle; because man doth not live by bread only, but also by the desert manna; by every wondrous word and unknowable work of God. (Ruskin 1985, 226)

Three things about this passage are worth remarking. As the last lines indicate, Ruskin possessed a strong respect for diversity: not merely the utilitarian has value, but also the wild, even the seemingly hostile and barren areas of the Earth. The reference to "unknowable works of God" in fact brings us full circle, back to Leibniz and his environmental theodicy. The second thing to note is the fact that, while wilderness is given its due, it is also said that ordinary, productive human activity should be a realm of beauty and variegated life. The message of environmentalism is thus not aimed solely at exotic areas of wilderness, but rather is also directed straight at the home, the field, and by extension even the city and the factory.

The third and final thing that should be noted about the above passage is not so much its message but the way in which that message is brought across. From the standpoint of ethics alone, Ruskin could merely have stated that one ought to keep up one's house and that birds are a desirable accompaniment. Instead he constructed a powerful image, a panoramic country scene that simultaneously possesses an incredible depth of acoustic sensations. More than any discursive exposition, it is this kind of vivid scene in its aesthetic immediacy that lay at the heart of Ruskin's environmental conviction.

The technique is given an important philosophical justification. If "all lovely things are necessary," as Ruskin states, then an attribution of worth that takes place in the aesthetic sphere is as weighty as any scientific or rational explanation. Such a statement might be hard to defend in the abstract, but in the historical context it made a lot of sense. Ruskin could not afford to wait for science to prove him right (cf. Ruskin 1906). Instead he took the aesthetic tools at hand and tried to use them to construct a socially viable index of the land's condition. It is no accident that the Heimatschützer repeatedly cited this very passage by Ruskin.

William Morris went beyond Ruskin in the comprehensiveness and utter logical consequence of his mature vision. While sharing the older man's rage at capitalism's trail of destroyed buildings, devastated landscapes, and impoverished workers, Morris was more politically astute and went on to embrace Socialism as the only possible societal alternative to laissez-faire capitalism. But even here it is clear that the aesthetic and artistic component in his thinking was absolutely crucial. As E. P. Thompson (1977) demonstrated, Morris's life must be seen as an organic progression from the purely "romantic" interests that characterized his youth to the larger social issues of his later years. Morris did not leave aesthetic interests behind as he moved into day-to-day political battles; instead he built on them, tirelessly pursuing a vision of a just society in which environmental beauty could at last come to full fruition. Every step of the way this aesthetic sensibility was an all-important source of critique and hope; romanticism in this sense was eminently and radically political, for it "defined a quality of living which it was the whole purpose of political change to make possible" (Williams 1983, 157; in full agreement, Thompson 1977, 647). Morris has nevertheless been regularly misunderstood by utilitarians on both the right and the left. Neither free-marketeers nor the socialist advocates of mere redistribution could work up much enthusiasm for what seemed to them to be an unnecessarily "utopian" or "romantic" vision of life—not realizing that in their common emphasis on material and technological gain lay the seeds of environmental disaster.

Morris saw through the smoke-filled world of the economists more than a century ago, all the way to a different world in which land, air, and water would no longer be shamelessly exploited, but rather would always be protected for their inherent, life-giving, and life-sustaining beauty. Rather than being an "enemy of nature" and a creator of "feverish ragged squalor," man could then aspire to a new relation to the land,

> a life in which year by year the land of his fathers shall grow dearer and fairer to a man as he gets to know it better and better, although his times be cast in a place where nature wears her everyday clothes, no queen but a thrifty housewife: so that when he goes into other

> lands richer of startling beauty and wild romance, he will fare not as a man driven by dullness that nothing can brighten, by weariness that no idleness can soothe, but as a pilgrim who has left his home a while that he may come back stored with new pictures and tales of the life of other men. (Morris 1969, 49–50; see also writings from 1894 in MacKail 1899, 305f.)

The vision here is of a profound integration of material and emotional ties to the land. It is striking that Morris seizes on the word *home* in order to describe this more environmentally friendly configuration of the future.[11] A fundamental feature of this society is its ability to do without fruitless romanticism: instead of travelling to far-off lands in search of compensation, people of the future will be able to find fulfillment and security in their own immediate and unspoiled surroundings. Germans would later seek to convey all of this in their related concept of Heimat, that "steadfast home" to which even the lowliest worker was entitled, and of which he would, in Morris's words, "never weary" (1969, 49–50). But the parallels do not cease there. If we look at the impact of Morris and Ruskin on a wider segment of British society we soon discover many of the same societal elements using very similar philosophies and organizational strategies to combat environmental degradation.

One of my overall concerns in this first chapter is to provide a historical and argumentative context for the Heimatschutz movement. I hope that once this context is taken into account the aesthetic approach of the Heimatschutz will no longer be dismissed as a bizarre phenomenon of German "irrationalism," but instead can be understood as a typical, historically necessary form of early environmentalism. The case of England confirms this latter viewpoint, for here we find aesthetic concern at the very center of the powerfully critical environmental visions of John Ruskin and William Morris. These two men provide a particularly compelling model for understanding the Heimatschutz because they were not content to merely write about their insights. Instead they allowed their aesthetic convictions to take them into a whole range of practical activities designed to help maintain a livable environment. Their united effect on the development of an English environmental movement would be difficult to overestimate. Along with a few other key figures and disciples such as Octavia Hill, Robert Hunter, George John Shaw Lefevre, Hardwicke Rawnsley, and James Bryce, Ruskin and Morris participated to some extent in every major English organization in the latter half of the nineteenth century that was concerned with preserving buildings, providing green space in cities, conserving landscapes, and combatting pollution (cf. Richards 1978, 82).

A characteristic feature of the English situation was the proliferation of a large number of groups focused along the lines of different issues. Interest in the landscape was thus spread out over a series of different organizations: the Commons Preservation Society, founded 1865, which fought to maintain unbuilt common areas against privatization and development; the National Footpath Preservation Society, founded in the early 1880s to assert public rights of way in the countryside (later amalgamated with the CPS); the Selborne League, founded in 1885, interested in zoological and botanical preservation as well as forest preserves; the Scapa Society for Prevention of Disfigurement in Town and Country, founded in 1893 (the name deriving from the acronym for its original title, the National Society for Checking the Abuses of Public Advertising); meanwhile in the cities there were the various Open Spaces groups, the Kyrle Society, founded in 1876 to promote philanthropic beautification schemes, and the Metropolitan Public Gardens Association, founded in 1882 (see thorough account in Ranlett 1983). Ad hoc committees were also formed in a large number of cases, one of the most famous being Ruskin's successful campaign against a railroad extension in the Lake District. Complementing the landscape organizations above was also the Society for the Protection of Ancient Buildings, founded in 1877 under the inspiration of Morris, which also prepared the way for one of the lasting contributions of this early phase of British conservation, the National Trust for Places of Natural Beauty and Historic Interest, an endeavor that got off the ground in 1895.

The large overlaps that we see in this spectrum of reform associations were reflected in the organizations' relatively homogeneous layer of bourgeois membership.[12] The Victorian environmental movement has been aptly described as an "intertwining structure" in which officers, members, and office space were often shared two, three, or even four ways between various groups. As an example, by 1912 the Scapa Society shared the same offices at 25 Victoria Street, London, with the Commons Preservation Society and the National Trust; it was also being led by Lawrence Chubb, Secretary of the CPS (cf. Ranlett 1983). With these close connections organizations could form that were new in little more than name or specific objective. When the Scapa Society was founded in 1893, for example, it was largely made up of "the officers and members of older societies, such as the Selborne, the Kyrle, the Commons Preservation, the Metropolitan Public Gardens Association, and the Society for the Protection of Ancient Buildings." Aesthetic sensibilities were clearly one of the strongest motive forces behind such efforts, which, when united, could well label themselves a "great national party for the defence of out-of-door amenities" (Evans 1926, 50, 76).

The Scapa Society can serve as an example of how even a superficial-sounding concern to "save the scenery" from garish advertisements was often tied in with fundamental deliberations about what was truly good for the nation and its people (Evans 1926, 216). As with the slightly later Heimatschutz movement, its opposition to excessive advertising was grounded in the conviction that preserving "elements of interest and beauty"—whether the "picturesque simplicity of rural and river scenery" or the "dignity and propriety of aspect in towns"—was an issue of "national importance." Following Ruskin, the Scapa Society pointed out that the capitalist "passion for profitable enterprise" and its inherent disregard for the environment was based on "false economic notions." At fault was an incomplete cost/benefit analysis:

> If a thing can be sold over the counter or in the auction room, or figure in the price lists of the mercantile world, it is wealth, but if it is the simple gift of nature and has involved no effort for its production, and is free to all, it has no place in the calculations of the purely business philosopher. The unbought and unpurchased blessings do not enter into statements of the national income. (Evans 1926, 341, 138)

The appeal to a sense of the "nation" here should be understood as an attempt to build up an awareness of the public or external realm in explicit contrast to the private wealth emphasized in capitalist economics. This "national" understanding would be the basis for what Ruskin called "the real science of political economy," which would at last categorize public goods "such as air, light, and cleanliness" as "precious and beneficent things" (Ruskin 1985, 209).

Although the size of the membership was, as one member candidly admitted, "small," the Scapa Society was able to exercise a relatively strong influence through social ties to government decision makers and through its journal, *A Beautiful World.* Its insistence that the landscape represented a "national asset of incalculable value" laid the groundwork for a 1907 law, the "Advertisements Regulation Act." This act empowered local authorities to restrict certain egregious uses of advertising, namely, any structure over twelve feet high, and any advertising that might "affect injuriously the amenities of a public park or pleasure promenade or to disfigure the natural beauty of a landscape." Although the second provision's broad wording could conceivably have included nearly any rural landscape, that same inherent uncertainty of its terms meant that it could only be enforced in cases that involved scenery of exceptional beauty. As the Society's Secretary later put it, "on the daily round" of life in the ordinary landscape "nothing could be done under the Act" (Evans 1926, 323, 217, 83ff.).

In this fatal weakness, as well as in its more positive aspects, the English law of 1907 was strikingly similar to legislation that was enacted in Germany in 1902 and extended, under Heimatschutz pressure, in 1907. In both countries environmental concern was chiefly expressed by bourgeois associations with strong aesthetic and philanthropic motivations (cf. Jefferies 1995, 59).[13] In addition we can say that the German Heimatschutz movement was almost as strongly influenced by Ruskin and Morris as was the comparable English environmental movement; one writer accorded them the role of "the true herolds of the Heimatschutz" (Gradmann 1910, 4). Translations of Ruskin and Morris were "more popular than ever" in Germany around 1900; at least four men published new works adapting Morris's ideas (Muthesius 1974, 151). All four of these men were early supporters and members of the *Bund Heimatschutz;* for them and their fellow reformers, England constituted nothing less than the "Heimat of the Heimatschutz" (Fuchs 1909b, 5). Against much earlier research, then, it seems that "one must be careful not to view Heimatschutz as a German 'peculiarity'" (Jefferies 1995, 60). Both countries were clearly dealing with the same sets of problems brought on by industrialized capitalism, and in very similar ways.

As compelling as these similarities are, they should not blind us to some interesting differences, chief among which was the relative speed of Germany's response. England's industrialization had preceded Germany's by at least fifty years, and permanent, nationwide groups to protect land- and cityscapes were forming there thirty to forty years before comparable groups in Germany. Nevertheless, it seems accurate to say that Germany's environmental movement, centered on the Heimatschutz, was already beginning to pull ahead of England by about 1907. Two factors seem to explain why in England "there arose no movement devoted to the maintenance of nature and the protection of old buildings or groups of buildings that was comparable to the German Heimatschutz" (Muthesius 1974, 132). One is that the English movement suffered from basic structural drawbacks, repeatedly being hampered by its unwieldy multiplicity of bodies and differing agendas. As John Ranlett has observed, what the English failed to institutionalize was precisely "the acknowledgement that all the various voluntary societies . . . were working on associated aspects of the same problem" (Ranlett 1983, 216; similar assessment reached by Muthesius 1974, 132). In the late 1920s Richardson Evans, the mastermind of the English Scapa Society, was in fact still looking forward to the day when a strong and truly popular "network of voluntary associations" might "cover the country."

Evans went on to suggest the second factor behind Germany's relative success by perceptively noting that a broad, positive conception of environmentalism would be the key to any future increase in effectiveness. His own group, Scapa, had unfortunately "confined its interest to the negative function of repressing advertising disfigurement"; only with the aid of hindsight did Evans see that this strategy was "less efficient even for that particular purpose than one that had regard for the whole field of kindred objects, including those of constructive benevolence" (Evans 1926, 191). It is no coincidence that the program pursued by the Heimatschutz possessed these very qualities. Thus, if the German environmental movement grew faster and, in some ways, achieved greater public presence than its English counterpart, this can ultimately be attributed to the comprehensiveness of the ideal of Heimat, a vision of a possible home whose social, economic, and even ecological dimensions were all integrated, and that thus allowed Germans from many different walks of life to begin working toward its realization.

Summary

In Western European societies the history of modernity is the history of a dominant culture of capitalism, individualism, and utility. This particular hegemony of values has been a decisive enabler of the pollution and plundering of the globe. Conversely, an ecologically responsible form of society will be dependent on the appearance of a new scheme of values: not values in the abstract, but values that inform, structure, and grow out of a concrete culture, a whole way of life. In this first chapter we have seen how aesthetic perception and aesthetic discourse historically served as incubators and carriers of environmental values. Bourgeois society often ghettoizes the aesthetic realm as the site of irrelevant impermissibilities, but this special role also allows the pre-articulation of new structures of feeling and the development of opposition to dominant values. The attribution of beauty to ordinary, working landscapes in particular was a momentous step in the fight against the destructive practices of modernity because this new way of seeing "occupied" those areas, granted them a social value—however tenuous—and raised the question—however fleetingly—about whether or not the dictates of short-term profit making were appropriate.

No one would have appreciated the significance of this development better than William Morris himself, who more than once remarked on the "subtle slavery" of capitalist industrial society; in this respect it compared unfavorably to despotic regimes of the past, its mastery being "harder to overthrow, because supported by such a dense mass of commonplace well-being and comfort" (Morris 1914a, 65). In this situation a committed Socialist and revolutionary such as Morris saw the overweening importance of education and cultural change. Anticipating Gramsci, Morris saw clearly that, in the face of a hostile dominant culture, the process of constructing a counterhegemonic environmental culture was absolutely crucial. His commitment came in spite of the fact that this was no easy or rapid task: "I cannot help but saying that there is nothing for it but for us to slowly build up some kind of public opinion which shall allow us to have our own way, and that, of course, will be a very long job indeed" (1914a, 17).

Over the course of his life Morris accordingly devoted countless thousands of hours to consciousness-raising through causes that many orthodox Marxists still dismiss as "cosmetic" or "utopian" or hopelessly "bourgeois" (e.g., Gould 1988, 16).[14] Contrary to this kind of dogmatic purism, Morris saw organizations such as the Society for the Protection of Ancient Buildings, the Kyrle Society, and the Commons Preservation Society not as "palliatives of an unendurable state of things," but as powerfully formative "tokens of what we desire" (Morris 1947, 259). In his last public speech before his death, at a meeting of the Scapa Society in 1896, he took up this issue of cultural change. Always a realist, Morris soberly observed that environmental notions were a long way from being ingrained in the populace:

> We have to remember that the enormous majority of the people of the country do not care one straw about natural beauty. They have, I allow, a certain pleasure in wandering about in the fields and enjoying the fresh air, but, as for looking at nature in detail with anything like observation, it is a fact that the greater part of the people of this country are entirely without eyes! Unless you can use your eyes, and unless the use of your eyes makes you suffer, nothing will be done. (Cf. *A Beautiful World* Dec. 1896, 16)

Morris demonstrates here how "romanticism" and aesthetic sensibility were perhaps central to any "realistic" campaign for improvement. It was the eyes that were the key; true change would only be possible in a population that had learned to see, to distinguish the aesthetically and environmentally good from its opposite. In brief this describes the historical and cultural role of the Heimatschutz.

CHAPTER 2

The Heimatschutz in Wilhelmine Society

As we saw in the concluding sections of chapter 1, aestheticism formed a tremendously important avenue for the popular articulation of environmental concern in England, a country whose advanced state of industrial development led, in logical fashion, to some of the first widespread critiques of that development. In Germany a very similar set of material and cultural factors soon helped give rise to the Heimatschutz. Beginning in particular with the boom brought on by the end of the Franco-Prussian war, Germany's well-organized and increasingly sophisticated capitalist-industrial system built over large chunks of land near the cities and intensified its exploitation of natural resources in even the remotest corners of the country. The basic economic figures tell a tale of relentless expansion: from 1870 to 1914 the overall population increased from approximately forty million to sixty-eight million; major cities absorbed almost all of the growth and grew in physical size by factors ranging from two to twenty. In heavy industry over the same time frame anthracite production went up ten times, the production of iron went up twenty times, that of steel twenty-five times; meanwhile, scientific research and advanced electricity and rail networks allowed the new German chemical and electrical industries to boom, with the result that they captured over fifty percent of the world market for the respective classes of products. Rationalization and intensification characterized the agricultural sector as well: over the latter half of the nineteenth century forests in Germany were increasingly converted to high-return pine and fir monocultures; agribusiness also took hold with a vengeance in the fields, where the rate of application of most chemical fertilizers more than quadrupled, and where fallow-lying acreage declined by more than two thirds during the Second Empire (Nipperdey 1990, 9, 193, 228, 235f.; Ladd 1990, 213f.; Pflug 1969, 249f.). All of these activities combined to transform the face of the land in Germany, whose air and water were likewise burdened with heretofore unknown quantities of pollutants (Rommelspacher 1987; Dominick 1992, 78).

In a nutshell we can say that Germany, in the last decades of the nineteenth century, reflected the same conditions of modern, capitalist "progress" that English society had achieved slightly earlier. The landscape destruction and environmental degradation entailed in this development have to be seen as the fundamental material cause of the kind of environmentalist concern that surfaced in the Heimatschutz. Nevertheless it is important to remind ourselves that problems are always mediated through structures of public or social perception. The mere fact of pollution or landscape transformation is just that: a fact, a quantum of data awaiting interpretation as to its wider practical significance and/or ethical importance. Without this kind of framework the fact itself may not even be absorbed into a person's consciousness, instead being discarded as irrelevant. The cultural contribution of environmentalist groups like the Heimatschutz is to provide this framework of interest and interpretation, to lift previously ignored facts into the light of consciousness and to inspire social action.

None of this work can be done overnight. Humans are creatures of habit, and societies function precisely through the creation of stable, hegemonic systems of interpretation that are often directly and materially opposed to new truths and new priorities. If the Heimatschutz was able to achieve a decisive breakthrough for environmentalist thinking in the first decade of the twentieth century, it is equally true that this breakthrough built on a number of discourses, practices, and motives that were operative within German society in the latter half of the nineteenth century. We have already discussed one of these in the previous chapter: the critical-resignative strategy of the "green island," a construct that held out the hope that a bourgeois individual or family might preserve its own oasis of calm and beauty against the liquidating forces of capitalism. This narrowly focused defensive perspective was buttressed by a series of wider concerns in Germany around 1900, though, which helped draw attention to the environment as an important factor of public life and of the future.

Parts of this awareness had their roots in older writings such as those by Wilhelm Heinrich Riehl. Riehl was an astute old curmudgeon of a writer who "stands like an erratic block in the progress-infatuated era" of the nineteenth century (Sieferle 1984, 149). His 1852 demand for a "right to wilderness" contained a sharp critique of the domesticating and monotonizing tendencies of capitalism, and praised the politically conservative effect that would emanate from all of the peasants, loggers, and fishermen who daily engage in elemental battle with nature. Rolf-Peter Sieferle rightly emphasizes that this betokened concern "less for nature than for an idealized, estate-based condition of society" (Sieferle 1984, 149). Riehl's many invocations of the peasant estate (*Bauernstand*) as an all-important bulwark against change—a "power of inertia" within society—indeed distinguish his writings from later environmental conceptions (Riehl 1861, 50). When his writings experienced something of a revival around the turn of the century it was obvious that the increasingly urban and industrial population of Germany had managed to separate Riehl's perceptive environmental critique from its original sociopolitical intention, which was directed straight against the very idea of a metropolitan civilization.

Riehl's political interpretation of the environment dovetailed to some extent with the powerful and widespread ideas of "decadence" and "degeneration." The 1880s and 1890s produced a flood of writings by men such as Otto Ammon and Max Nordau who feared that modern, urban society was upsetting the natural, biological foundations of human life. These *Großstadtkritiker* diagnosed a new epidemic of "nervousness" as well as a decline in reproductive capacity or life span; all of these phenomena seemed to demonstrate that the urban resident's lack of contact with nature and fresh air could lead to a long-term decline of the human species. Possibly more compelling, in nationalists' eyes, was the specter of a measurable decline in the German armed services. The medical statistics gathered during entrance screening comprised one of the few sources of health data on a broad cross-section of the German population; all the more important, then, was the apparent fact that smaller cities and rural areas produced healthier recruits (Lees 1975; Hamann and Hermand 1977; Bergmann 1970). Although they were much debated, such medical data helped reinforce the widespread but vague idea that well-ventilated green surroundings constituted a key component of public health.

These functional convictions were strongly supported by arguments that drew attention to the aesthetic or cultural effects of the wider built environment. Historic preservationists played a key role in developing this kind of awareness from its rudimentary state in the mid-1800s. By the year 1900, specialists in architectural preservation and art-historical matters were able to begin putting on their own yearly conference, called the *Tag für Denkmalpflege.* The last decade of the century was particularly significant for its changing emphasis in preservation. Under the influence of John Ruskin and William Morris and their anti-purist "Anti-Scrape" philosophy, architectural historians and city planners began to lay greater worth on preserving a building's entire historical record. Instead of "scraping" away all the changes and restoring a building back to some theoretically pure style, now all of the additions made by various ages would also be preserved. In Germany the wider preservation philosophy coincided with a new interest in previously underappreciated architectural epochs. Masterpieces of the German Baroque period came in for a more positive revaluation through the work of Cornelius Gurlitt and others; meanwhile, the simple and straightforward bourgeois structures of the late eighteenth century were held up by a number of architects and critics as a healthy antidote to mass-produced pastiches in the so-called historicist style. The net effect of all these related developments was to attribute value to a far wider variety of architectural structures and situations than had previously been the case.

A central feature of this revolution in architectural appreciation was not so much its choice of object as it was a wider angle of view. The last decade of the nineteenth century was the time of the discovery of an urban fabric (Breitling 1981, 52; Ladd 1990, 122f.). Camillo Sitte's seminal 1889 book *City Planning according to Artistic Principles* (*Der Städtebau nach seinen künstlerischen Grundsätzen*) inspired a whole generation of planners and architecture buffs to reject the "meager and unimaginative character of modern city plans" (Collins and Collins 1986, 221; see further Ladd 1990; Wurzer 1992; Fehl 1995). Instead they were urged to consider the value of aesthetic variety and intimacy in building neighborhood character. This was perhaps a form of "antimodernism," but it was nevertheless a reflective and reform-minded one that wanted merely to retain the affective diversity of older city settings. "Naturalness," texture, and historic context were now at the center of a new kind of aesthetic interest:

> Old cities and old streets have a special charm for anyone who is open to artistic sensations. Even when they are not "beautiful" they attract one's mind and please with an unforced disorder that did not come about as the uniform result of the artist's deliberations, but rather through the natural accretion and transformation of structures along a crooked lane over centuries, a gradual growth into the status of an urban street. (Weiß 1894, 225)

A sense of the ubiquity of history worked hand in hand here with an aesthetic perspective that demanded to see the whole street as a meaningful unit. One result of this new way of seeing was that limited, object-oriented preservation began to give way to one in which whole "fields of care" were developed. In the particular parlance of German preservation this process was tracked as a continual expansion of the scope of the term "monument," a status that contemporaries began to accord to entire city centers and ensembles of bourgeois structures (Koshar 1993, 17).[1] The kind of wide-angle, large-scale frame of reference that we see here reinforced trends in the building and planning of new construction, which for the first time devoted attention to the impact of an entire "streetscape" (*Straßengefüge*).

The last decades of the nineteenth century were thus an important time in which people collectively opened their eyes to the built environment and thus effectively anticipated a phase of actively protecting it as their own Heimat. A growing interest marked Germans' relationship to the natural environment, too—based here primarily on the enormous upsurge in outdoor recreational activities. Fifty years of industrialization had succeeded in establishing a nearly self-contained urban-technological lifestyle for many of Germany's citizens. For compensation and relaxation they turned to nature. Although the options varied by class and pocketbook, millions of Germans took to the hills, fields, and rivers in a bewildering array of leisure-time or vacation activities. Outdoor sports were on the rise. Tourism to places of natural beauty climbed precipitously with better infrastructure and transportation. A modestly well-off bourgeois family might spend a short vacation in a commercially run inn or *Sommerfrische,* and Sunday excursions became a common practice among even the poorer workers.[2] Hiking and mountaineering were organized through countless local or regional associations, some of which had tens of thousands of registered members. Adults were not alone in the passion for hiking, for there were also the newly founded bands of youth in the *Wandervogel* movement. Hundreds of thousands of people engaged in "life-reform" (*Lebensreform*) activities as well: these were vegetarians, nudists, temperance advocates, clothing- and lifestyle-reformers. Still others engaged themselves in pragmatic efforts to integrate nature into the urban setting: the *Deutsche Gartenstadtgesellschaft* was founded in 1902 to promote a greener style of building, and a similar constellation of planners and social reformers pressed for more and better parks in German cities. All of these activities reveal that the turn of the century was a time in which interest in nature was exploding, something that goes a long way toward explaining the successful creation of the Heimatschutz.

Still, the general atmospheric interest in nature as a "compensatory space" for recreation was far from making environmentalism inevitable. Even today only a fraction of those hikers, bicyclists, walkers, swimmers, anglers, and hunters who enjoy their exposure to nature go on to the logically necessary step of defending that realm through political and personal engagement. At the turn of the century this same kind of human inertia must be imagined in a historical situation where the level of environmental awareness was incomparably lower. And even in those cases where an awareness was able to arise, the existing recreational and other associations provided little in the way of a forum for environmental action (cf. Fuchs 1909a). A concerned member of a large hiking club, the *Sauer-*

ländischer Gebirgs-Verein, noted in 1912 that the leadership "had its hands full" with trail-maintenance duties; "the necessary step of deliverance" could only consist in going outside the club and "founding a Heimatschutzverein for the Sauerland" (Ist keiner da? 1912). By 1912, of course, the Heimatschutz already existed and was able to provide this person with an organized perspective. Before 1904 it was a rare German who could not only perceive the importance of environmental issues, but also envision an organization that would stand up for them in public.

Ernst Rudorff did just that. Born in 1840, the Berlin native spent many of his summers on a family estate in rural Lower Saxony. His privileged bourgeois upbringing included personal exposure to some of Germany's most famous late Romantics: the brothers Grimm, the Arnims, and Ludwig Tieck (Schoenichen 1954, 129). With a traditional German humanist schooling and through further acquaintances with leading musicians—Clara Schumann and Johannes Brahms, for example—Rudorff's life from early childhood through his adult career as a music professor was permeated by a powerful aesthetic sensibility (biography in Knaut 1991). It is not unreasonable to believe that this "romantic" influence was a key to Rudorff's environmental perception. Much like his contemporaries John Ruskin and William Morris, Rudorff had an aesthetic sense that guided him in a far-reaching critique of a falsely rational and unnecessarily uniformized modernity. The area around his family's property in Lower Saxony gave him ample material for his diagnosis, for as early as 1878 he published a broadside against "landscape ruination" due to field rationalizations in Berlin's *Post* (Hartung 1991, 49; Knaut 1991, 24). Two years later Rudorff published what is generally accepted to be the seminal statement of Heimatschutz ideals in the *Preußische Jahrbücher.* His article "On the Relationship of Modern Life to Nature" is notable for the aesthetic mode in which it surveyed the countryside of rationalized progress:

> They are doing their best to make the pleasant and variegated countryside into a schematic plan—one that is as bare, clean-shorn, and regularly parcelled-out as possible. Any tip of woods that juts out gets shaved away out of a love for the idea of a nice straight line; any meadow that stretches into a forest is planted up; even inside the forests they no longer tolerate any clearings or glades into which deer could run. Streams which have the bad habit of winding their way along a curved bed are now forced to flow straight ahead in ditches; . . . all the hedgerows and free-standing trees are sacrificed to the axe, then, too. (Rudorff 1880, 262)

For Rudorff the aesthetic contours of the land clearly illustrated the dominant societal paradigm that controlled modernity's relationship with nature. Late nineteenth-century Germany was under the sway of a thoroughgoing hegemony, "a more and more exclusive dominance of the realistic view of life," by which Rudorff meant the one-sided pursuit of short-term individual advantage (264).

Against the bleak "realism" of this system of profit and exploitation Rudorff set an "ideal co-ownership of God's Earth" (275). "Idealism," in Rudorff's sense of the word, was defined first and foremost by reference to everything that nineteenth-century capitalist rationality was not. In particular this meant a different ethical basis. Capitalist "realism" had its ethical basis in the individual and consequently valorized an absolute sense of individual ownership of the environment. An ideal sense of ownership, on the other hand, promised a less egoistic use of the earth. Here the ethical basis would not be restricted to one privileged individual, but could well include the animals of the area and the land itself. Such a wider ethic would be expressed in a diverse landscape: instead of a landowner's optimally rationalized and monocultured production area, it would be one where streams could assert the "bad habit" of taking up space with supposedly unproductive twists and turns, and one whose forests provided deer with the clearings they needed for sustenance.

Along with the ecological aspect of Rudorff's idealism it is important to see that his idea of "co-ownership" carried some very concrete social implications. Taken to its logical extreme it would have entailed a more communitarian sense of property. Rudorff, however, shied away from any such revolutionary implications; all he wanted to do was to provide a kind of "compensation for the necessary chasm between property-owners and non-property owners" (275). Passages such as this one have moved critics to label Rudorff an arch-reactionary. It is one of the difficulties of dealing with this figure that the accusations are well founded. Rudorff came from a background of privilege and saw the class system as simply "necessary"; moreover it often appears that he was actually incapable of understanding the repression and odium to which those below him were exposed. In an 1897 article he went so far as to complain that the railroad had made it more difficult to get stall boys and callously dismissed those men and women who had gone to the city as "profit- and fun-seeking folk" (Rudorff 1926, 43). In the main he therefore despised even the most basic, labor-saving forms of technological progress and also feared "the unnatural, indeed, monstrous and dangerous growth of the great cities" (Rudorff 1880, 270). In all of these aspects it is clear that Rudorff opposed modern societal developments on a more or less uninformed and self-serving basis, and as such he deserves the appellation "reactionary."

But Ernst Rudorff's particular brand of conservatism also had a socially conscious aspect. Inspired by Riehl, he opposed the Manchesterian idea of simply letting entrepreneurial initiative determine the shape of society and, by extension, the land. Riehl had repeatedly emphasized that environmental decisions have social and psychological consequences. In particular he had championed an "agricultural scheme full of variation" (*vielgestaltiger Landesanbau*) as the best guarantee of truly "individual social formations" (Riehl 1861, 54). Divorced from Riehl's original political intention, it is an insight that still has value today. Environmental psychologists are just beginning to understand the "identity-confirming" function of a diverse and aesthetically sensual environment, one that has far-reaching implications for the smooth running of a modern society (Schimank 1983, 46, 79). Whatever his other faults, Ernst Rudorff intuitively shared this insight. Already in 1880 he saw that enormous psychological pressures stemmed from the same modern rationality that was monotonizing and economizing the landscape to death. Here he could even feel his way into the shoes of a worker—for under the absolute property laws of modern capitalism, everyone had to assume the position of a worker in the landscape:

> What an unbearable feeling: having to look upon the earth as a conglomeration of individual bits of property, having to consider the world blocked off with the exception of the public highway and those few spots that the mercy of others has been so good as to allot to me for my vacation! (275)

Rudorff did not supply a general solution to this central problem of modern society, although, as mentioned above, his concept of "ideal co-ownership" carried with it radical implications—some of which would be developed by the later Heimatschutz movement. Instead he approached it within the familiar terms of bourgeois politics, accepting the "necessity" of class differences and seeking to make workers' lives more tolerable. Given the importance he attached to an experience of nature he therefore advised his fellow *burgher* to

> give the municipal facilities a more generous form; instead of mass rent barracks build more and more individual houses; and even if it is just in the form of a garden, put a piece of nature in the immediate vicinity of everyone, so that the poison of the atmosphere does not by itself force him to seek his fortune far away. (276)

In order to make this program work Rudorff even recommended that workers be invited to use railroads for their garden visits to the outskirts of town. The image that one often receives in the secondary literature of Rudorff as a dyed-in-the-wool reactionary or an "eccentric and extreme figure" stands in remarkable contrast to the kind of reform perspective opened up in this passage (Applegate 1990, 104). These are not the words of a cramped and mean-spirited throwback to feudal times. In fact they could well have come from the pens of the progressive bourgeois reformers in the Garden City movement—people who would arrive on the scene more than twenty years later. In this respect we should also recall the passage in which the well-traveled Rudorff praised the English landscape:

> Here we have the factual evidence that a highly-developed civilization does not necessarily have to lead to a maltreatment of nature . . . In this country of factories and rationalized agriculture there lives along with the practical mind a feeling for the charm of the landscape that is so deep and so generally cultivated that the country does not look like some warehouse for commercial products, but rather like a garden. (Rudorff 1880, 270)

From today's perspective we might ask whether Rudorff was critical enough of the English landscape, which by 1880 was certainly degraded enough in some parts to deserve an emphatic critique. But the fact that Rudorff could see the outlines of an ideal in England, the most industrialized country of the world, gives an important indication that a program of reform, and not absolute reaction, was at the heart of his agenda.

In this sense it is allowable to say that Rudorff's essay of 1880 constitutes a "milestone in the history of Heimat and nature protection" (Schoenichen 1954, 153). Whether one agrees with all of Rudorff's fears and predictions or not, it is safe to say that the essay was remarkable for its ability to give political, economic, and social contour to the problem of land use. Even more remarkable is the fact that it did not stop there. It proceeded to suggest the possibility of reform and in fact became quite specific about what measures could be undertaken right away. This is clearest where Rudorff came to speak of managing the rural landscape. First on his list was the practice of stream regulation, "a barbarous act not only from the perspective of aesthetics," but also one whose "too-rapid water drainage . . . brings with it the most decisive real disadvantages" (271). Here Rudorff suggested that the state authorities responsible for overseeing agricultural rationalization put a stop to the practice. The same agencies would be in a position to issue further guidelines protecting "all the individual transition zones between forest and meadow and field" and also to "dispense with the system of absolute straightness and perpendicularity in laying out roads." Finally Rudorff called for a no-net-loss rule with regard to hedgerows, one that would require the planting of new hedges when old ones were torn up; indeed, he suggested that meadows and gardens of all sorts be hedged. Even individual trees and bushes ought to be protected as much as possible, he advised:

> not only in order to promote the picturesqueness of the landscape, but at the same time in order to take steps to maintain the bird population, whose nesting sites are on the verge of being almost completely eliminated by the modern practice of field rationalization. (272)

Taken together, these suggestions having to do with stream regulation, hedgerow retention, and edge habitat constitute the outlines of a concrete reform program. It is worth noting that the aesthetic language of landscape beauty has already here begun to absorb functional environmental concerns such as water retention and the provision of bird habitat. It is also important to see that the essay consciously targeted government technical offices and training programs as key determiners of land use.

As we have seen, Rudorff's 1880 article contained much of what was later to become the domain of the Heimatschutz movement. But his ideas remained just that, and thus Rudorff appears to have been increasingly concerned with the pragmatic question of how to organize support for his program. Several attempts to graft the Heimatschutz agenda onto existing organizations failed. Rudorff's 1888 petition to the Central Association of German History and Antiquity Associations (*Gesamtverein der deutschen Geschichts- und Alterthumsvereine*) to include the "protection of landscape nature" was rejected because the preservationists "feared being accused of exceeding their competence" (Knaut 1991, 37). The year 1892 found Rudorff trying his luck with the General German Association (*Allgemeiner Deutscher Verein*). Here he proposed the creation of a special committee devoted to the "Protection of Germany's Landscape Nature and the Historical Monuments" (Moewes 1926, 40f.). The proposal was accepted, but the committee was nevertheless abolished soon after its founding. Nevertheless this attempt was significant, since it gave the project some new, concrete dimensions. Legislation requiring an aesthetic impact statement for construction projects was to be pursued, as was a program of architect training. Local tourism and beautification societies were also targeted as important local organs (Knaut 1991, 32).

In 1897 Rudorff changed his tactics and boldly proposed the establishment of a totally new organization that could nevertheless draw together existing tendencies. Rudorff referred specifically to Freiburg's

Association for the Preservation of Folk Dress (*Verein zur Erhaltung der Volkstrachten*), planning and preservation advocates who stood behind the ideas of Camillo Sitte, vernacular architectural interests, and those who were committed to the idea of American-style national parks or conservation areas. Governmental structures were welcomed as well, with the Province of Saxony's Commission on the Research and Protection of Monuments (*Kommission zur Erforschung und zum Schutz der Denkmäler*) singled out for praise. Tying together all of these scattered initiatives with a canon of literary and artistic references, Rudorff's article in the respected journal *Die Grenzboten* at last achieved a greater resonance for his ideas, something that may well have been due to the increasingly ubiquitous scars of Germany's industrialization; nevertheless it may also have had something to do with the striking title that Rudorff now used for his reform project. The word *Heimatschutz* had previously been used only to designate a force for the military defense of the homeland; now, though, it was supposed to galvanize citizens into a program of environmental defense (Rudorff 1926).

In the past many commentators have seen this martial vocabulary along with Rudorff's other appeals to Germanness as a sign of his reprehensible nationalism. But these critics do not give enough weight to the fact that Rudorff was in fact creatively twisting the meaning of *Heimatschutz,* appealing to national sentiments only in order to redirect them. Rudorff made it clear that he considered the usual run of nationalist enthusiasm in the Second Empire a waste of time: if people really wanted to "promote love of the homeland and the Fatherland," he said, let them work on a truly practical program of saving architecture and landscapes. "That would be a way," Rudorff concluded, "more solid than fireworks and flower garlands and all those fine speeches with which people these days are always going overboard in celebrating patriotic holidays" (Rudorff 1926, 34).

These lines set up a cultural opposition with great significance for a historical assessment of the Heimatschutz. When one reads of a patriotism that consisted of an excess of fireworks and flower garlands and fine speeches, one cannot help thinking of Diederich Heßling, the epitome of blustering and speechifying nationalists in Wilhelmine Germany from Heinrich Mann's *Man of Straw* (*Der Untertan,* 1916). On a purely personal level it is perhaps worth remembering that Rudorff distanced himself from that kind of blind pseudopatriotism. This distance becomes meaningful on a broader historiographical level if we recall how Thomas Nipperdey used Diederich Heßling as the starting point for his essay challenging the popular notion of Wilhelmine Germany as a bootlicking "Untertanengesellschaft" (Nipperdey 1986). As he suggested, such clichés of Prussian nationalism do not adequately reflect the many complexities of the Wilhelmine bourgeoisie, a class of individuals who were in many cases critical of official politics and often deeply committed to social reforms. Along with a better understanding of the Heimatschutz aesthetic method, it is this differentiated view of the German middle classes that provides an important foundation for revaluating the social and political position of the thoroughly bourgeois Heimatschutz movement. As this will be done in detail further below, it suffices here to note that there is no reason to reject the Heimatschutz out of hand due to its objective class position.

This note of caution applies even when we see that the Heimatschutz appealed to the ideas of "Germanness" and the *Volk.* Such rhetoric was used already in 1897 to describe the future Heimatschutz organization. It was to be, Rudorff wrote,

> a gathering-together of all like-minded people whose objective it is to maintain German folkdom (*deutsches Volkstum*) unweakened and unspoiled, and—what is inseparable from this—to protect the German Heimat, with its monuments and the beauty of its nature, from further disparagement. (Rudorff 1926, 24)

Rudorff's use of the term *Volkstum* sends up red flags to the modern reader, who is sensitized, and in some cases oversensitized, to the later use of such terms by the Nazis. It is my opinion that Rudorff and the great majority of the organized Heimatschützer who used this kind of language before World War I did not have a genetic-racial understanding of *Volk.* In their mouths the word *Volkstum* was highly cultural, connoting the continuity of traditions of art, craft, language, or religion.[3] This kind of nationalism was perhaps even more widespread than the explicitly patriotic sort mentioned above. Precisely because the symbolisms of nationhood were endemic in bourgeois Wilhelmine society it is important to distinguish the meaning and aims of each respective usage. As we shall see, in the main the program of Heimat represented an extremely sober and pragmatic application of German national feeling. The nation was not a trivial matter to most Heimatschützer, and if there were moments of overbrimming national rhetoric and raw anger mixed into their appeals, these only indicate the depth of ethical commitment that they brought to their unequal battle against an entrenched capitalist system.

The fact that this battle could be started at all was due, of course, to a groundswell of wider cultural developments. As we saw above, a whole palette of nature interest had come into existence by the years around 1900: along with painters' colonies, sunbathers, and hikers, there was also a growing scientific interest in preserving rare habitat. In much of this Germany participated in an overall international trend. The United States' national park idea inspired an 1898 proposal for similar legislation in Germany; meanwhile, out of the Englishman Ebenezer Howard's ideas there arose a German branch of the Garden City movement (Schoenichen 1954; Bollerey et al. 1990). This latter movement is significant because it illustrates the kind of organizational potential that was contained in a technically conversant and socially aware segment of the German middle class. Attention should also be paid to those organizations that devoted themselves to the care and protection of historic buildings, for the "preservationist public" amassed here showed particular understanding for the aesthetically driven idea of Heimat protection (Koshar 1993, 14).

If we take a closer look at efforts to preserve Germany's architectural heritage in the decade preceding the Heimatschutz we can see that they took on an increasingly well-organized character, even as its concerns widened to encompass larger aesthetic ensembles. A significant number of the people involved were state employees, often in the technical or artistic branches that were responsible for official preservation programs. These close contacts to state agencies occasionally led to formalized institutions such as the Committee for the Fostering of Heimat-Style Building in Saxony and Thuringia (*Ausschuß zur Pflege der heimatlichen Bauweise in Sachsen und Thüringen*), a semigovernmental body founded in 1903 (Jefferies 1995, 62). In Prussia as well, support for preservation was reflected in efforts at the state level. The so-called *Verunstaltungsgesetz,* or "disfigurement law," of 1902 recognized the importance of landscapes of "exceptional beauty"; it forbade the erection of advertisements that would impair such scenery. In these same years around 1900 the executive branch, moved by the calls of preservationists, was also working on a law regarding the protection of monuments and historical finds (see documents in PGStA MdgUMA Rep. 84a, 10746). Controversial provisions in the proposed law allowing for the seizure of private property demonstrate that a qualitatively new sense of importance was being attributed to aesthetically "environmental" factors by governmental offices around the turn of the century.

The growing interest in Rudorff's explicit program of Heimatschutz was also furthered by support from the much more vaguely defined "Heimat movement" (*Heimatbewegung*) of the time. Regional Heimat organizations were forming several years before the call went out in late 1903 to form a *Bund Heimatschutz;* the two biggest were the Heimat League Lower Saxony, founded in 1898 in Hannover, and the Bavarian Association for Folk Art and Folklore, founded in 1902. In addition there was a wide range of organizations that often traced their origins back to the 1860s or 1870s. Historical societies, beautification societies, local museums, genealogy fans, and scientific clubs often anticipated the Heimat stance by wanting to protect the "distinctiveness" of the region or town, to act as stewards of local tradition, and to contribute to a general sense of civic pride. Ringbeck (1991, 217) thus credits these organizations with "contributing decisively" to the successful uptake of Heimatschutz ideas.

The central difference between this earlier type of Heimat interest and that which blossomed at the turn of the century seems to have been the mode in which the activities were carried out. The new spirit of the Heimatbewegung has rightly been described as an activist one:

> If in the nineteenth century love of Heimat and the consciousness of roots manifested themselves both individually and collectively primarily in the form of historical research and the historical interest of the educated bourgeoisie, then this passive, antiquarian-descriptive basic stance now changed bit by bit into a conserving activism. (Hartung 1991, 72)

Accepting this description, we can say that Heimat enthusiasm had been building up for some time in the late nineteenth century. By around 1900 a kind of turning point had been reached.

This timing suggests that the new interest in Heimat was closely related to the commitments that had been made in the key decade of the 1890s, ten years in which Germany's decisive turn from an "agricultural state" to an "industrial state" was executed. By 1900, in the midst of a powerful economic expansion, it was clear that Germany was firmly on the road to becoming an urban, industrial society. Uniformizing and modernizing tendencies would from now on dominate the cultural scene, and the very certitude of this dominance energized the protectors and appreciators of traditional, historic, and regional cultural elements. The range of conceptions that emerged from this confrontation should not be minimized; the slogan of Heimat, in particular, was taken up by people with very different aims and emphases. But whether as Heimatbewegung, Heimatschutz, or a wide variety of "Heimat art" (*Heimatkunst*), the turn of the century was ripe for Heimat ideas that in general promised to attenuate the now powerful forces of industrial modernity.

Within this spectrum the Heimatschutz constituted a recognizably independent phenomenon. Its activist reform program dedicated to protecting and preserving the material basis of everyday aesthetic experience set it apart from the wider run of Heimat interest (similar distinction in Knaut 1993, 22f.).[4] If we are careful not to read too much into the association, though, it is easy to see that the Heimatschutz was able to draw on preexisting forms of the Heimatbewegung in its initial stage of organization.[5] The literary movement for Heimatkunst, for instance, had begun to flower around 1900; much of the activity centered around the publisher G. H. Meyer. Being deeply involved in the promotion of the new fad of Heimat literature, Meyer sought to encourage the public's interest wherever possible. In 1901 he republished Ernst Rudorff's 1897 article "Heimatschutz" as a brochure and introduced Rudorff to his fellow Brandenburger Robert Mielke. Mielke, an admirer of Ruskin and Morris and advocate for folk art, had been interested in forming a "German Cultural Association"; he now agreed to join forces with Rudorff and pursue the idea of an organization devoted to Heimatschutz (Mielke 1900; Mielke 1929).

The ensuing efforts to arouse support built heavily on networks of associates in Berlin and in Saxony and Thuringia. In Berlin the two men recruited Oskar Hoßfeld, a high official in the Prussian Ministry of Public Works and simultaneously the editor of Germany's central preservationist journal, *Die Denkmalpflege.* Meanwhile, journalists, publishers, and art critics were being introduced to the proposed organization through contacts based in central Germany. Here it was above all the publisher Ferdinand Avenarius who helped spread the word about Heimatschutz (Kratzsch 1969; Knaut 1993, 18). Some would later go so far as to label Avenarius "one of the first leaders of the Heimatschutz movement" ("Weg mit dem" 1912). His journal *Die Kunstwart,* a mainstream, conservatively progressive magazine of arts, culture, and lifestyle, possessed "extraordinary influence" in bourgeois reform circles of the time; as the left-liberal Theodor Heuss observed, it was a true "cultural power" whose tastes were reflected in furnishings and apartments of educated civil servants all over Germany (Viehöfer 1988, 112). Its effect was reinforced by a unique parallel associational structure, the "Dürer League" (*Dürerbund*) (cf. Schumann 1919). Together, *Kunstwart* and *Dürerbund* spoke to a body of readers who shared many of the Heimatschutz's reform ambitions.

The writers who were involved with Avenarius's *Kunstwart* became some of the earliest and most stalwart advocates of Heimatschutz. It is likely that Robert Mielke knew the man who was to become the first chairman of the *Bund Heimatschutz,* the architect Paul Schultze-Naumburg, through their common work as contributors to the journal. Schultze-Naumburg had gained a certain amount of fame in reform circles with his 1902 anti-corset book on the "reform dress"; on top of this he was, by 1903, the chairman of the Committee for the Fostering of Heimat-Style Building in Saxony and Thuringia.[6] A former painter of landscapes and a man with deep interests in arts and crafts, Schultze-Naumburg appears to have been the kind of all-around cultural reformer that the Heimatschutz needed. Sometime in the spring of 1903 the triumvirate of Mielke, Hoßfeld, and Rudorff decided to offer him the chairmanship, and by July of that year Schultze-Naumburg had accepted (on this and other particulars of the founding see the detailed exposition in Knaut 1993). A number of other leading figures announced their willingness to participate in September during the yearly conference of preservationist interests, the *Tag für Denkmalpflege.* These included L. F. Karl Schmidt (1853–1922), an official in Saxony's Finance Ministry who was simultaneously a leading member of the Committee for the Fostering of Heimat-Style Building in Saxony and Thuringia; Carl Johannes Fuchs (1865–1939), professor of national economy in Tübingen; and Karl Rehorst (1866–1919), an architectural preservationist from Halle who would later assume duties for the city of Cologne. By this time the Heimatschutz had also assured itself of the support of Hugo Conwentz, a natural-history museum director from Danzig and Germany's leading figure in scientific efforts to catalog and preserve exceptional resources.

The role of scientific organizations in preparing the way for the Heimatschutz is a largely unexplored chapter. According to Robert Mielke, the initial rapprochement with Hugo Conwentz was a difficult process, but this probably reflects nothing more than Conwentz's personal inflexibility and rigid emphasis on cataloguing and preservation (Mielke 1929). In Saxony it is evident that Heimatschutz and science got off to a much more cordial start. In the first months of 1904 there was a rush of activity in anticipation of the founding assembly. An extraordinary meeting of a leading general scientific association, the Isis Scientific Society, in Dresden, on 26 February 1904, shows that science groups as well as more culturally oriented ones were all part of the same associational network that gave birth to the Heimatschutz. The meeting was cosponsored by two other groups: one was the Dresden Goethe League, an association of citizens organized around artistic and free-speech issues; the second was the Association for Earth Science (*Verein für Erdkunde*) in Dresden. Invitations had further been extended to all of the Saxon state ministries, the agricultural board (*Landeskulturrat*), professors at the nearby forestry academy in Tharandt, and other scientific and hiking groups. After the keynote speech by Hugo Conwentz, the chairman of the *Isis,* plant geographer and botanist Oscar Drude, pledged Saxony's scientific community to the goals of conservation. Choosing his words carefully so as not to provoke dissent, he stated that "all those gathered here" were interested in a "protection of our heimatlich landscape from those harms that cannot be considered immediately necessary for the vibrant economic life of the present" (Sitzungsberichte der Naturwissenschaftlichen Gesellschaft Isis 1904, 13). One month later Drude encouraged his fellow members of *Isis* to attend the founding assembly of the *Bund Heimatschutz.* At least eight members followed his lead and joined the national organization individually within the next three years; meanwhile, many more became members of Saxony's own State Association for Heimatschutz. The Isis itself formed its own Heimatschutz committee and became active in local conservation measures.[7] As we shall see in chapter 3, over the course of the next fourteen years scientifically interested individuals were key contributors to the conservation program of the Heimatschutz.

The same city of Dresden in which the Isis reigned was the setting for the founding of the national *Bund Heimatschutz.* Dresden seems to have been chosen because of the outpouring of support from Schultze-Naumburg's Committee in 1903 and the otherwise high level of interest in the central German lands of Saxony and Thuringia; the Kingdom of Saxony was to remain one of the hotbeds of Heimatschutz activity through World War I and beyond. Ernst Rudorff had summarized the aims of the proposed *Bund Heimatschutz* in a short appeal that was mailed first to a selected group and then published more widely with the solicited endorsements of over two hundred prominent Germans ("Aufruf" 1904). The list not only reflected the kind of recruitment efforts that had been going on behind the scenes, but also provided an effective demonstration of the support for the ideas of Heimatschutz among some of Germany's leading cultural and artistic circles. The signers included architects, planners, and critics such as Theodor Fischer, Karl Henrici, Karl Hinckeldeyn, Ludwig Hoffmann, Alfred Lichtwark, Hermann Muthesius, and Paul Wallot; academic and art-historical experts such as Cornelius Gurlitt, Karl Lamprecht, Friedrich Ratzel, numerous museum directors, and all of the Prussian Provincial Conservators; writers such as Wilhelm Bölsche, Friedrich Lienhard, Heinrich Seidel, and Johannes Trojan; and, finally, an impressive array of artists that included the young Worpsweder painters Fritz Mackensen and Heinrich Vogeler as well as established artists like Hans Thoma, Fritz von Uhde, and Graf von Kalckreuth. Many of the signers subsequently attended the meeting in Dresden on 30 March 1904, at which the *Bund Heimatschutz* was officially constituted.

Before going into the history of the *Bund Heimatschutz* the reader should probably be reminded that the idea of Heimatschutz and the associated ideal of Heimat are not necessarily one hundred percent congruent with or reducible to this one organization, particularly because many of the regional Heimatschutz organizations often asserted their independence from national-level policies. Moreover we have to consider the wider effect of Heimatschutz influence on entirely different institutions such as state agencies and its transmission to the public at large. Nevertheless the positions taken by the *Bund Heimatschutz* were of decisive importance for the general public understanding of Heimat and Heimatschutz, and thus an examination of this organization is called for. Its early history, which is the segment under consideration here, can be broken down into two rough phases: the first from 1904 to 1907, and the second from 1908 through the end of World War I.

The welter of reform and preservation organizations had been a bonus for the Heimatschutz during its founding because it provided networks of contacts; on the other hand, though, the newcomer could now not afford to give the impression that it was taking members away from preexisting regional groups or particular areas of activity. A presentation held in Frankfurt an der Oder was indicative of the sensitivities involved, for there a Heimatschutz supporter had to assure his listeners that "the League does not intend to interfere in the membership or practical work of the associations that already exist for art, animal protection, local beautification, history and other things" (cf. *Brandenburgia* 1905–6, 9f.). With such factors in mind Rudorff and his fellow founding members had therefore conceived of the *Bund Heimatschutz* as a "centerpoint for all related efforts" (SBPK Nachlaß Mielke, Mappe Rudorff, 80). It was an eclectic mixture of interested individuals, old regional Heimat groups, new regional Heimatschutz groups, and issue-oriented organizations. Under the overall goal of "protecting the unique natural and historically developed character of the German Heimat," the work was at first divided into six different issue areas: monument preservation, vernacular architecture, aesthetically oriented landscape protection, scientifically oriented species and site protection, folk crafts, and folk customs including festivals and dress (cf. "Satzungen" 1904–5).

These six "workgroups" made up the organizational basis of the *Bund.* Like spokes to a hub, only the chair of each workgroup participated in the thirteen-member executive board. Four members at large provided the only other representation for independent regional or issue groups; the remaining three members were executive officers. The board was given great latitude to act according to its own judgment; the only controlling instance was an annual conference. But here the large organizations were easily outvoted: they were allowed four votes maximum, while any one dues-paying individual member was also given a vote. Without a clear avenue to power, influence, or support, it is perhaps no wonder that several of the largest Heimat organizations, for example those in Lower Saxony and Bavaria, chose to remain at arm's length from the central *Bund.* Even in Saxony, the birthplace of the *Bund Heimatschutz,* and where the regional association was a loyal member of the national *Bund,* there was a feeling that the national level was drawing members and money to itself that ought to go to the region (cf. *Mitteilungen des Bundes Heimatschutz* 1905–6, 152).

Despite the failings of its somewhat "Bonapartistic" structure, the first years of the *Bund Heimatschutz* were productive ones in terms of setting a high-profile public agenda. An unfortunately losing battle to prevent a dam project on the Rhine near the South Baden village of Laufenburg brought in waves of publicity and new members in 1905. The press was almost universally favorable, and the Heimatschutz, along with a number of members of the Prussian *Landtag,* began to press the government for a law that would finally protect the ambience and character of scenic locations. Already on the books was a 1902 law against "disfigurement" of a landscape, but it was narrowly worded to apply only to the impact of advertising billboards on exceptionally scenic areas. The legislative history behind it made it clear that the law was only intended to protect a few canonical points of tourist interest along the Rhine valley. Under the influence of the Heimatschutz's new demand for protection of scenic ensembles everywhere, though, the Prussian government drafted a new law that would allow for the review and possible rejection of construction or other projects on aesthetic grounds.

Prussian ministerial documents demonstrate that the legislative process leading up to what would be known as the "Disfigurement Law" (*Verunstaltungsgesetz*) had begun already in 1902, thus before the Heimatschutz can claim credit for originating the idea (for this and following see PGStA MdgUMA Rep 84a 10746). The case that kicked off the procedure was a petition of 25 March 1902, submitted by a Magdeburg architect named Moebius. It described the increasing ugliness of Magdeburg's main street, the Breitenweg, and traced this development to a lack of legal handholds on construction activity. As a solution the petition suggested police powers and/or laws to help protect existing buildings recognized to have lasting value or "significance for the character of a townscape or landscape." The petition was signed by a range of aesthetically and historically interested associations, including the Magdeburg Architects and Engineers Association, the Association for the Preservation of Monuments in the Province of Saxony, an Association for History and Ancient History, an art society, and an association of garden architects. The petition circulated in the Prussian houses of parliament in the spring of 1903 accompanied by supporting petitions from the *Tag der Denkmalpflege* and from a long list of local historical and architects' organizations.

On the side of the government little appears to have happened in response to the petitions until September 1903. The Ministry of Justice had been putting off the matter, so the Ministry of Public Works, together with assistance from the Ministry of Culture, took on the task. Here it was Oskar Hoßfeld, a founding member of the Heimatschutz, who was centrally involved. By mid-April of 1904 a first draft was ready, and on 29 April 1904 the first round of review began. A fundamental concern was the measure's "drastic impact on the affected property owner"; representatives from the other ministries also desired clearer specification of the areas to be protected, the standard of ugliness, and the procedures for decision, enforcement, and appeal. On 9 December 1904 the Interior Ministry repeated the call for very clear, uniform guidelines. It noted that the Rhenish Center Party had been looking for ways in which to increase local "self-administration" (*Selbstverwaltung*) and would use this issue to attack regulation procedures imposed by Prussia on Rhenish communities. As did most German states, Prussia assigned a wide range of tasks to "police" authorities, including technical inspection duties connected with construction and commerce. Wanting to reduce possible frictions, the Interior Ministry suggested that standards for intervention be made "*fully independent* of the initiative of police agencies." Part of the year 1905 was spent accommodating these concerns; a panel of experts was mandated in order to give objectivity to police decisions, and appeals procedures were given community presence. But the government also took this opportunity to drop a fairly progressive requirement that would have required property owners to install higher-quality living quarters when they renovated, and it also eliminated the applicability of the law to nonurban areas. To the chagrin of the Heimatschutz movement, which would later petition to restore the provision, this particular change was considered "insignificant," since a case involving construction "outside unified towns . . . would have to belong to the rarest of instances." As late as January of 1906 the Justice Ministry had not tired of repeating its complaint that the measure amounted to a "painful intervention into private property," and eventually forced the addition of a clause stating that added costs for aesthetic measures could not be too high. The Finance Ministry showed its unfamiliarity with developments by returning to the argument that such things were in principle "not police tasks in the actual sense."

When the government submitted its law to parliament for review, members of the upper house forced it to include not only "streets and squares" under the law's protection, but also "exceptionally scenic countryside" (*landschaftlich hervorragende Gegenden*) as well (Hahn 1905–6). The Heimatschutz pointed out that this still did not go far enough: this specification of *landschaftlich hervorragend* was the same as that in the 1902 law and would therefore limit the bill's effectiveness to only a tiny fraction of the landscape. A petition was made up demanding the words "scenically beautiful" (*landschaftlich schön*) instead, a lower threshold of protection that would apply to a much greater area. The *Bund Heimatschutz* was backed up on this by accompanying petitions from the *Tag der Denkmalpflege,* the Rhenish Association for the Care of Monuments and Heimatschutz, and the leading garden architects' association, the *Deutsche Gesellschaft für Gartenkunst* in Hannover (Glogau 1907).

On the other hand, governmental and parliamentary offices were ringing with howls of protest from property owners' associations. The Provincial League of Silesian House- and Landowners' Associations, for example, warned that the projected law "threatened private landowners with very alarming limitations on property"; moreover, it added, police powers were not supposed to be used to pursue "merely aesthetic considerations," but rather only the least-common-denominator factors of "peace, security, and order." What the property owners really feared was that they would no longer be allowed to build as profitably as they had planned; as an absolute minimum, then, they demanded compensation (PGStA Rep 84a 10747). Industrial and commercial circles chimed in, asking that the bill be rejected outright or at least suitably gutted (see petitions in HdA 20, Leg.per. 227). By 1907 the public discussion of landscape, construction, development, city planning, and property rights was in full swing, and it was due in great part to the efforts of the *Bund Heimatschutz.*

Despite the excitement and prestige that went along with this clamor, though, more realistic Heimatschützer such as R. Hahn took on an almost desperate tone as they encouraged their fellow members to keep pushing for the best law they could get—for "what is not achieved now cannot be made up for later. Let us therefore not come to a stillstand halfway there, but rather let us make sure that this 'Heimatschutz' becomes a sufficient one and takes effect everywhere" (Hahn 1905–6, 187). Hahn's impassioned critique of the government's proposal conjured up once more the full and indivisible ideal of Heimat:

> Protection must not be limited merely to exceptionally beautiful areas and sites famous from artworks, but rather must encompass the entire land. Who wants to live in a place that consists only of factories, department stores, modern rent barracks and ostentatious villas, and that has kept none of the intimacy and beauty of the old German cities and villages? Who wants to spend his days in a region in which forests and bushes have been destroyed for miles around and where a person can only walk freely on the highway over which automobiles race? A short vacation trip into exceptionally beautiful areas cannot make up for vegetating in such an environment. Therefore this "Heimatschutz" must be a universal one. (Hahn 1905–6, 186)

The demands that Hahn laid out are the same ones that make the idea of Heimat a powerful environmentalist category even today. For they attacked not only the halfhearted measures of Prussian ministries, but also the entire modern fragmentation of the world: its functional division into areas of work versus play, lower class versus upper class, and utility versus beauty. Modernism teaches us that we cannot have both at any one time or place, and thereby it accelerates that continuous oscillation and geographic fragmentation from which modern capitalism lives. Taken to its logical conclusions, Heimat, by contrast, is a far-reaching call for integration, for not consenting to the divisions and concessions imposed on the social and natural worlds. The early phase of the Heimatschutz can be thought of as the period in which this demand in its specifically landscape-oriented form was presented to the German public for the first time, a period in which even radical change seemed possible—if only because very little had yet been instituted.

The division of Heimatschutz history into a second phase beginning in 1908 has its justification in two main events. One is the enactment in Prussia of the final version of the much-discussed "Disfigurement Law" on 15 July 1907. Along with the important administrative decrees that followed in January 1908, the opportunities and shortcomings of this law upon which the Heimatschutz had concentrated so much effort constituted a new and influential point of reference for the movement. The law as enacted largely conformed to the changes recommended by parliament. It was a piece of enabling legislation: police were allowed to intervene against construction that "grossly disfigured" (*gröblich verunstaltet*) town streets or squares, as well as exceptionally scenic countryside; in addition local administrative bodies, assisted by advisory councils, were given the right to formulate guidelines prophylactically protecting the character of clearly designated streets or buildings. Finally, the area's *Regierungspräsident* was to decide which stretches of landscape were *landschaftlich hervorragend,* and thus a priori protectable (cf. Jefferies 1995, 85f.). Unfortunately, the provisions went no further; in particular, the many *merely* scenic areas of Prussia remained unprotected, and the standard of "gross disfigurement" was in itself vague and open to challenge. But there could be no hope now of obtaining better wording, for after the flurry of excitement surrounding the 1907 law, legal and social inertia set in. Other German states did follow with similar laws, of course; many were based directly on the Prussian statute, but some, such as the law enacted in Saxony on 10 March 1909, contained more generous wording ("Zehnter Tag" 1909).[8] Nevertheless the paradoxical effect was everywhere more or less the same: these laws that were celebrated by the Heimatschutz (and many critics) as its greatest achievement ended up making it very difficult for the Heimatschutz to proceed on to further and possibly more fundamental changes.

The law's heavy focus on urban construction also succeeded in permanently shifting the attention of many Heimatschutz members away from its environmentally significant concern with landscape. Divergent tendencies had been present within the organization almost from the very beginning: Ernst Rudorff's more radical aesthetic reverence for the natural ensemble was not shared, for instance, by Paul Schultze-Naumburg, the architect and chairman of the *Bund Heimatschutz.* Even before the movement got off the ground Rudorff therefore criticized Schultze-Naumburg for having needlessly narrowed the idea of Heimatschutz in a speech in Erfurt:

> That he gave his presentation the title "On Heimatschutz" surprised me . . . In fact he spoke *not at all* or at least *hardly at all* on "Heimatschutz" in the wider sense that I have of it, but rather mainly only about architecture . . . The landscape was dealt with *much* too cursorily, hardly even touched upon, to have earned that thematic designation. (emphasis in original; letter of 23 October 1903 to Robert Mielke in SBPK Nachlaß Mielke, Mappe Rudorff)

These differences with respect to the extent and ambition of the Heimatschutz idea were only exacerbated in the years following the enactment of the 1907 law. Fritz Koch, the new general manager of the *Bund,* found it necessary to warn that "the Heimatschutz project must be preserved and carried forward in its entirety and must not be allowed to split up into a bunch of individual technical areas" (Koch 1909). Koch viewed the related trend toward a technocratic thinking with extreme misgiving. He found that "the scientific and technical part of Heimatschutz work" was "overrated" and that experts also had a tendency to reduce things to their own particular specialty; more helpful were a good understanding of the broad cultural significance of the Heimatschutz movement, personal initiative, and pragmatic political skills.

Such warnings were increasingly necessary as specializing forces within and without the Heimatschutz tugged at its holistic landscape vision from all sides. Paradoxically, as Koch pointed out, one reason for the centrifugal trend was governmental acceptance. The Prussian government essentially co-opted part of the Heimatschutz in 1906 by establishing an office for *Naturschutz,* or nature protection. As we shall see in greater detail in chapter 3, this State Office for the Care of Monuments of Nature under the directorship of Hugo Conwentz pursued a highly specific brand of scientific preservation that drew energy away from broader concerns;

the same was true of a private initiative to establish nature preserves. In a very similar manner there was a drain toward the increasingly powerful architectural wing. As we shall see below, the 1907 law opened up a new space: experts in local architecture and preservation were in demand as consultants, and it was only natural that many architecturally conversant Heimatschützer stepped into the new area, never to return. The growing architectural emphasis was underscored when the *Bund Heimatschutz* decided in 1909 to begin holding its large biennial general meetings in conjunction with the preservationists' *Tag der Denkmalpflege.* If in the following years the once-comprehensive understanding of Heimat often tended to get lost in the shuffle, this was due in large part to such institutionally reinforced technical specializations.

The era after 1907 also had positive ramifications for the Heimatschutz, though. The law legitimated the movement in the eyes of those who might earlier have simply dismissed it, and as Birgitta Ringbeck has pointed out, the accompanying decrees of 10 January 1908 and 19 January 1908 were especially effective in bringing government into the Heimatschutz fold, and vice versa. They gave considerable power to "willing experts" (that is, the local Heimatschützer!) who were expected to help government agencies with their aesthetic decisions (Ringbeck 1991). Works by Rudorff and Schultze-Naumburg were listed as recommended background reading. With the Heimatschutz being given such official recognition, the principle of governmental inclusion took hold in other areas as well: a 1912 Prussian guideline for road builders, for instance, suggested that the local Heimatschutz group be consulted as a matter of course before any final layout decisions were made (PGStA Rep 87B 3134, 15; "Heimatschutz und Wegebau 1915"). Once approved, Heimatschutz objectives were impressed upon the bureaucracy with typically Prussian thoroughness. *Oberpräsidenten* were directed by the 1908 decrees to support private groups such as the Heimatschutz "as a matter of personal interest." Further down the ladder as well, civil servants in the technical branch were to promote "Heimat-style building" not only in the office, but also by participating "in all efforts in this direction . . . outside official duties" (Ringbeck 1991, 221; text of decrees in *Heimatschutz* 1908, 1–6). In the following years Heimatschutz, at least in the specifically urban-aesthetic and architectural sense of the 1907 law, became an accepted feature both of public-affairs discussions and municipal and state government duties.

The other major reason for identifying 1908 as a turning point is the changes that were made in the structure of the *Bund Heimatschutz.* Andreas Knaut (1993) has painted a detailed picture of the personal and organizational tensions that had been building within the movement since its founding. A main point of contention was the role of the national *Bund Heimatschutz* executive board in relation to the regional organizations, many of which were dissatisfied with the system of passing on membership money to a central board over which they had little control and whose assistance in day-to-day matters was limited (cf. Hedemann-Heespen 1914, 30). To a certain extent this division was deepened by the outspoken centralist attitude and apparently inefficient work of Robert Mielke, the general manager of the *Bund.* Mielke was finally edged out of his position in May 1907 and replaced by Fritz Koch, a young lawyer in the Thuringian city of Meiningen (cf. *Mitteilungen des Bundes Heimatschutz* 1907, 61; Koch 1908). By the time of the annual meeting in Mannheim in September 1907, some movement was already apparent. Koch came out in favor of accommodating the individual regional associations, and the *Bund* proceeded to put its money where its mouth was: it approved a request from Saxony to hand over all of its single members in that area to the regional organization (Koch 1908).

At the annual meeting in Lübeck in 1908, a clear hierarchy was established in which regional Heimatschutz organizations were recognized as the structural basis of the movement. Membership moneys would largely remain with the regional level; only a small portion of the dues would be transmitted. Regional groups were also now more able to directly manage the central *Bund* through "representatives' meetings"; the annual plenary meetings, on the other hand, were reduced to once every two years.

In response to both publicity and the better administrative conditions for regional associations, new Heimatschutz associations sprang up all over Germany; even preexisting organizations showed impressive gains. In 1911 a leading Heimatschutz official estimated that there were approximately 15,000 individual members in this far-flung network (Otto 1983; *Mitteilungen des Bundes Heimatschutz* 1912, 4). After that not even the leaders of the movement seem to have had a solid idea of how many people were involved. My data show that the Heimatschutz continued to grow right up to the outbreak of World War I. Table 1, "Regional Heimatschutz," shows a total recorded membership of over 26,500 individual members in 1914 (assuming in cases where data is only available for 1912 or earlier that membership did not drop). Moreover, it must be emphasized that the table is by no means complete. Data is still lacking on organizations in the small states of Thuringia and Saxe-Anhalt as well as Hesse and parts of Westfalia and Schleswig-Holstein. It thus seems probable that the actual total membership could have been as much as 30,000 by the beginning of World War I.

The industrial culture of the late nineteenth century was aggressively ubiquitous. The same basic drive for capital, markets, and productive

TABLE 1. Regional Heimatschutz Organizations

Region, Organization	Number of Members in: 1908	1910	1912	1914
Baden, Württemberg				
1. Badische Heimat[1]	—	—	1,503	(1,503)
2. Bund für Heimatschutz in Württemberg und Hohenzollern[2]		2,600	3,500	~3,500
Bavaria				
1. Bayerischer Verein für Volkskunst und Volkskunde,[3] later Bayr. Landesverein für Hs.	(>2,000)	—	—	4,363
Brandenburg				
1. Verein Heimatschutz in Brandenburg—Landesgruppe des Bundes Heimatschutz[4]	—	550	580	632
Hesse[5]				
1. Hessischer Bund für Heimatschutz[6] (Darmstadt)		—	—	161
2. Verein für Naturdenkmal- und Heimatschutz in Kurhessen[7] (Kassel)		—	—	—
3. Frankfurter Verein f. Heimatschutz;[8] also Ortsgruppe Frankfurt		—	—	—
4. Ortsgruppe Marburg[9]		—	—	—
5. Ortsgruppe Schmalkalden[10]		(69)	(69)	(69)
Mecklenburg/Pomerania				
1. Heimatbund Mecklenburg[11]		1,615	—	1,936
2. Landesverein Pommern des Bundes Heimatschutz[12]		—	—	450
Lower Saxony[13]				
1. Heimatbund Niedersachsen (Hannover)		—	~800	1,015
2. Verein für niedersächsisches Volkstum (Bremen)		—	789	~800
3. Braunschweigischer Landesverein für Heimatschutz		—	850	~850
4. Verein für Heimatschutz und Heimatgeschichte Leer[14]		243	639	810
Rhineland/Westfalia				
1. Rheinischer Verein für Denkmalpflege und Heimatschutz[15] (Cologne)	1,155	1,459	1,767	1,680
2. Bund Heimatschutz im Bergischen Land[16] (Solingen)		—	330	800
3. Heimatschutz im Sauerland[17] (Dortmund)		—	—	—
4. Lippischer Bund für Heimatschutz und Heimatpflege[18] (Detmold)	164	200	—	~400
5. Minden-Ravensbergischer Hauptverein für Heimatschutz[19]		—	—	—

TABLE 1—*Continued*

Region, Organization	Number of Members in: 1908	1910	1912	1914
6. Verein für Heimatkunde und Heimatschutz in Siegerlande[20] (Siegen)		—	1,400	(1,400)
Saxony				
1. Landesverein Sächsischer Heimatschutz[21]	664	1,880	2,932	~3,650
Schleswig-Holstein area				
1. Schleswig-Holsteinischer Landesverein für Heimatschutz[22] (Kiel)		185	241	269
2. Ortsgruppe Heimatschutz in den Elbgemeinden[23]		100	110	113
3. Ortsgruppe Heimatschutz Husum[24]		40	55	35
4. Ortsgruppe Heimatschutz für den Kreis Herzogtum Lauenburg und Umgebung[25]		168	252	257
5. Ortsgruppe Heimatschutz Neumünster[26]		—	65	96
6. Verein Heimatschutz Geestlande[27]		—	—	—
7. Verein Heimatschutz im Hamburger Staatsgebiet[28]		—	380	~425
8. Verein für Heimatschutz Lübeck[29]	100	(100)	(100)	(100)
Silesia				
1. Schlesischer Bund für Heimatschutz[30] (Breslau)			379	493
Thuringia/Saxon-Anhalt[31]				
1. Ortsgruppe Jena[32]		—	—	(171)
2. Ortsgruppe Weimar[33]		—	—	—
3. Bund Heimatschutz Verein Kreis Gotha also Landesverein Gotha[34]		—	165	(165)
4. Bund Heimatschutz für das Eisenacher Land, also Ortsgruppe Eisenach des B.H.[35]		—	>60	139
5. Landesverein Reuß[36] (Gera)	149	(149)	211	(211)
6. Landesverein für den Regiergungsbezirk Erfurt		—	—	—
7. Landesverein Magdeburg-Merseburg[37]		—	—	—
8. Ortsgruppe Artern "Aratora"[38]			111	117

1. List of members as of October 1911 in *Badische Heimat* 1911, 74–96.

2. Founded 12 March 1909 as the Württembergischer Bund für Heimatschutz. Became the Bund für Heimatschutz in Württemberg und Hohenzollern in 1910. Membership cf. *Mitteilungen des Bundes für Heimatschutz in Württemberg und Hohenzollern* April 1911, 22; also Dec. 1911, 29.

3. Founded 1902. Membership total for end of 1913 in *Bayerischer Heimatschutz* 1914, 30.

4. Founded 1907. 1914 membership in Mielke (1914b, 60). One source indicates that Brandenburg had at least one local Heimatschutz organization well before its regional association: an "*Ortsgruppe für Heimatschutz*" in Küstrin is mentioned as existing already in 1905, and preparations were being made for an organization based in the Frankfurt/Oder and Niederlausitz area (cf. *Brandenburgia* 1905–6, 10f., also *Mitteilungen des Bundes Heimatschutz* 1905–6, 76f.).

5. The most complete summary of the rather fragmentary history of Heimatschutz groups in Hesse is Wolff (1984).

6. First evidence of activity in October 1913; a later *Aufruf* from May 1914 contains the names of 162 supporters (cf. Hessisches StA Darmstadt G 15 Friedberg, B51; G 15 Erbach, B12)

(continued)

TABLE 1—*Continued*

7. First founded in 1911 as the "Verein für Naturdenkmalschutz in Kurhessen und Waldeck"; by 1912 it had split, yielding a "Waldeckischer Bund für Heimatschutz" in Arolsen.

8. Reported active in 1910 with regard to preserving one of Frankfurt's Main river bridges.

9. Represented 1913 by Architect BDA Spahr.

10. Founded 14 April 1905. Membership was given as 67 in 1905, but no other indications of activity have come to my attention (Mielke 1906b, 4).

11. Founded 15 Jan. 1906. Membership list for 1909 included at end of *Mecklenburg. Zeitschrift des Heimatbundes Mecklenburg*1909.

12. Membership in *Pommersche Heimat (Beilage zum Pommerschen Genossenschaftsblatt)* June 1915, 22.

13. The most thorough investigation of Heimat and Heimatschutz groups in Lower Saxony remains Hartung (1991).

14. Membership total from *Verein für Heimatschutz und Heimatsgeschichte Leer-Ostfriesland. Dritter Jahresbericht* and *Fünfter Jahresbericht.*

15. Founded 20 Oct. 1906. Yearly membership totals in *Mitteilungen des Rheinischen Vereins für Denkmalpflege und Heimatschutz;* list of members as of 1915 in *Mitteilungen*... 1916, 389–427.

16. Founded 17 Sept. 1911. Membership figures taken from "Bergischer Heimatschutz," *Solinger Zeitung* 18 Sept. 1911; also a 1912 petition to the Regierungspräsident Düsseldorf, Stadtarchiv Solingen Ve 005-001. The association appears to have been almost entirely focused on the legal battle against polluters of the Wupper.

17. Founded by February 1914 according to notice in *Siegerland* 1914, 132.

18. Early membership in Lippscher Bund (1909) and (1910); membership for 1915 conservatively inferred from dues.

19. Founding reported in an article reprinted from the *Generalanzeiger,* "Verein für Heimatschutz und Denkmalpflege," *Wanderers Freund* 12, 9 (1907), 68f. Represented 1913 by Provinzialkonservator Ludorff, Münster.

20. Membership list in *Siegerland* 1912, 122–32 and 227–30.

21. Membership totals in *Mitteilungen des Landesvereins Sächsischer Heimatschutz.* List of members in 1911 from Stadtarchiv Dresden, Akte Heimatschutz/R. Weise.

22. Founded 15 October 1908. Membership from Hedemann-Heespen (1914).

23. Membership from Hedemann-Heespen (1914).

24. Membership from Hedemann-Heespen (1914).

25. Membership total and list in *Jahresbericht (3) für 1913—Ortsgruppe Heimatschutz für den Kreis Herzogtum Lauenburg und Umgebung,* 34–38.

26. Membership from Hedemann-Heespen (1914).

27. Identified in *Jahresbericht (3) für 1913—Ortsgruppe Heimatschutz für den Kreis Herzogtum Lauenburg und Umgebung,* 39.

28. Founded 29 December 1906. Lit of members as of 1912 in StA Hamburg A 507/13, Kapsel 1. 1917 membership was 425.

29. Represented 1913 by Oberlehrer/Dipl.-Ing. Mahn. Vorsitz 1914: Oberstabsarzt Dr. Hansen.

30. Founded 12 June 1910. Membership total for 1913 in *Mitteilungen des Schlesischen Bundes für Heimatschutz* June 1913.

31. The most complete listing of Heimatschutz groups in this heterogeneous area is in Koch (1912c).

32. Founded simultaneously with *Bund Heimatschutz* on 30 March 1904. Membership figure for 1913 according to Knaut (1993, 185). The role of the publisher Eugen Diederichs in this group is mentioned by Viehöfer (1988, 114f.).

33. Founded 19 May 1906. Further documents exist in the Stadtarchiv Weimar but were inaccessible at the time of writing.

34. Founded 3 May 1911. Cf. *Heimatschutzfragen* 1913 (2): 11.

35. The Ortsgruppe Eisenach was founded 19 October 1911. Membership information from *Heimatschutzfragen* 1914 (1):2; 1913 membership list from documents in Stadtarchiv Eisenach.

36. Founded 19 April 1907. Founding and initial membership of 76 reported in *Mitteilungen des Bundes Heimatschutz* 1907, 50; see further *Fürstlich-Reuß Geraer Zeitung* 9 June 1909; *Heimatschutzfragen* 1911 (1):2.

37. Apparently founded in the latter half of 1910. Cf. short notice in *Heimatschutz* 1910:64.

38. Membership total and list in *Zeitschrift des Vereins für Heimatkunde und Heimatschutz von Artern und Umgegend Aratora* 1913–14, 233–35. Affiliated with Landesverein Magdeburg-Merseburg.

capacity reshaped landscapes and societies all over Western Europe, and thus it is not surprising to find that people from many different countries developed Heimatschutz-style solutions to the problem. In their "Call for the Founding of a Bund Heimatschutz" the organizers around Ernst Rudorff explicitly acknowledged similar movements that existed already in England (apparently meaning the National Trust) and in France (*the Societé pour la protection des paysages*). Following upon the tremendous wave of activity unleashed in Germany still other countries followed suit. In Switzerland the *Schweizerische Vereinigung für Heimatschutz* was founded on 1 July 1905 and grew to nearly 6,000 members by 1912 (Boerlin 1912; Boerlin 1931; Le Dinh 1992); and in Austria a number of separate regional associations sprang up: by approximately 1909 or 1910 there were organizations in Tirolia, Salzburg, Steyr, and the Steiermark (Brückler 1989). Already in 1911 the Salzburg Heimatschützer hosted a common Austrian–German conference on Heimatschutz and preservation (*Stenographischer Bericht* 1911). By 1912 the various regional organizations had agreed to join forces in a *Verband österreichischer Heimatschutzvereine,* electing Karl Giannoni of Vienna to be its first general manager ("Verband österreichischer" 1912). Heimatschutz-style efforts to preserve historic buildings and neighborhoods were also occasionally mentioned for Sweden and for the Netherlands, where a *Bond Heemschut* was formed in 1911 (Koch 1912a, 74–5; *Heimatschutz-Chonik* 1918 [5/6], 15).

The international dimension of Heimatschutz activities led to an increasingly greater degree of recognition and cooperation between national organizations. Already in 1909 the *Societé pour la protection des paysages* invited delegations to attend an international Heimatschutz meeting in Paris. Three years later the Germans responded by hosting the Second International Heimatschutz Congress in Stuttgart. As Carl Johannes Fuchs noted in his opening comments, Heimatschutz was international in two different senses of the word. It was first of all international in that the problem being faced was common to all nations:

> The really central problem of the Heimatschutz is simply one and the same in all modern civilized states: it is the struggle against a capitalism that mercilessly destroys what has grown up over time along with its beauty. It is this which ultimately lies at the bottom of all Heimatschutz issues. (Fuchs 1912a, 4)

The enemy in each different country was "different only by degree; in its essence, though, the same" (Fuchs 1912a). Heimatschutz associations thus had good reason to get together and compare notes and strategies.

This pragmatic sense of internationalism dominated, but it implied a second kind of relationship "between [different] nations." For all of the Heimatschutz associations sought to maintain some of the uniqueness and geographically bound character of their regions. Dr. Münchgesang, a leading official in Prussia's Ministry of Public Works, addressed this side of the equation in his remarks to the International Conference. He made the simple observation that domestic Heimatschutz and the respect it entailed for one's own culture inspired an understanding for the desire of others to maintain their traditions (cf. *Heimatschutz* 1912, 63). Fuchs challenged his listeners to reconcile the apparent "contradiction" that this second perspective seemed to entail: "Can one go about something that is as national as Heimatschutz internationally?" he asked. Two years before Germany and Austria were to launch a world war, in front of an audience that included government representatives from all corners of Europe (plus one from Japan), it seemed possible. Fuchs saw no contradictions here, only difference:

> For only that person who loves and appreciates his own Heimat—not in crude, arrogant chauvinism, but with a refined perspective and with a knowledge of its cultural significance—only he will also respect the Heimat and uniqueness of others. (Fuchs 1912a)

Fritz Koch also emphasized the cooperative aspects of international Heimatschutz. Less than a year before the start of World War I he was writing in the journal of the German peace movement about the essential connections between "Heimatschutz and International Agreement" (Koch 1914; on the movement itself see Chickering 1973). In the brief article he rejected the idea that Heimatschutz efforts represented a "national or even particularist standpoint" in political questions and went on to point out that Heimatschutz groups participated in "a special reciprocal relationship with international life."

Ever the concrete thinker, Koch emphasized that active international agreement was "of the greatest value for a whole series of Heimatschutz issues that require the cooperation of interested circles in several states." Border areas would be one of the most obvious arenas that would benefit from coordination, Koch wrote: international cooperation would be key to properly handling dam projects on boundary rivers, such as the one at Laufenburg, as well as cross-border nature-protection areas, such as the Hohes Venn between Germany and Belgium. In still other cases the ideals of Heimatschutz made international agreement absolutely necessary. A consensus between nations would be vital to preserving animal species, he noted; "large marine mammals such as whales and rare species of seal," as well as birds, were obvious examples. More Heimatschutz would be key to these animals' survival, but then the progress of Heimat thinking itself "would profit greatly" from comparing strategies in different countries. Heimatschutz thus mirrored relations as a whole, Koch closed: for "how much better things in the world would be if peoples would mutually get to know each other and their arrangements!"

At home in the German Heimatschutz movement something closer to the reverse of this process was going on, of course: in the post-1908 period of decentralization the individual regions concentrated on building up their own organizations. It was a painful time for the *Bund Heimatschutz* executive board. The amount of money at its disposal shrank drastically, and with it the ability to play an active role in day-to-day events; indeed, it seemed that it could now fulfill "merely lobbying and publicity duties" (Knaut 1991, 45). This may be a slight exaggeration, though: the *Bund* still constituted an organizational and informational center for the smaller regional organizations, and by 1910 Fritz Koch was better able to assert his presence, having been assured an annual donated salary of 5,000 marks. Various government agencies were also approached for direct monetary subsidies, but until the war years the amount of money received was relatively small, on the order of 1,500 marks in any given year (cf. Koch 1908; Knaut 1993, 106, 143).[9] The Prussian bureaucracy in particular certainly did not shower the Heimatschutz with money: it invariably gave far less than requested, if anything at all, and insisted rigorously on the formalities of its donation. The Agricultural Ministry gave 300 marks yearly starting in 1909 and probably spent at least that much again in checking to see whether all of its libraries had received copies of the Heimatschutz journal (PGStA HA I Rep. 87B 3133, 62a–g, 161–68). Other agencies and even other German states were asked if they were being solicited as well. Smaller states almost always restricted their funding to local organizations, though (StA Dresden, MdI 17519/100).

As the central *Bund* was in decline a new attempt to ensure a supraregional presence of the Heimatschutz was made by establishing a "Foundation for Heimatschutz" in June 1911. The donor of the 1,000 marks in seed money was the Thuringian native Carl Martin Wuttig, a lawyer in Berlin-Charlottenburg for the Raiffeisen agricultural cooperative movement and a regular attendee for that body at Heimatschutz conferences. Wuttig wrote, however, that the money had been put at his disposal by friends, so the actual source of the financing is unclear. Officials charged with overseeing the foundation at first balked at the low level of capital, but they were eventually won over by Wuttig's status as a reserve officer and "thoroughly trustworthy" person (letters of April and May 1911, TStA Meiningen MdI 2021). Fritz Koch was named as the director of the foundation, which was incorporated in Sachsen-Coburg precisely so that Koch could remain in his home city of Meiningen while carrying out his duties. Two main reasons for the new foundation were mentioned. For one, Wuttig and others were concerned about tensions between some of the regional associations and the national-level *Bund Heimatschutz* and wanted to create a "further solidly established structure" for the Heimatschutz "in the case that the presently existing organization of the Heimatschutz for whatever reason can no longer serve the project in a productive fashion" (letter of 8 May 1911). For another it appears that Wuttig thought that Fritz Koch, "the universally admired general manager" of the *Bund Heimatschutz,* was not being fully utilized (letter of 27 May 1911). While this in part no doubt referred to the executive board's reduced powers, it can also be guessed that Koch was unhappy with the *Bund Heimatschutz*'s gradual slide toward emphasizing questions of architecture and away from the practical, engaged side of landscape reform. His earlier articles favored this dimension of the Heimatschutz, and in his later work for the foundation the same concentration is evident.

It is unclear whether the Foundation for Heimatschutz received the additional funds for its endowment that Wuttig believed he would be able to raise. Wilmowski stepped in with funds to keep the operation afloat, and Koch did mention a "special donation" having been made in 1912 (Knaut 1993, 167; Koch 1912a, 69f.). In an early report, however, Koch had to admit to fiscal overseers of the foundation that the lack of money did "not yet allow any fancy Heimatschutz work" (letter of 23 Dec. 1911, StA Coburg Min D 18, 116–17). In spite of these difficulties the Foundation for Heimatschutz gave Koch the freedom to pursue projects that were nominally unconnected to the *Bund Heimatschutz* and its oversight. Already in the first six months he managed to establish a collection of slides and photographs relating to Heimatschutz cases, including enlargements suitable for public exhibitions; along the way he produced and edited a quarterly Heimatschutz journal for Thuringia, the *Heimatschutzfragen;* he also started a collection of plans and drawings relating to regional architecture for Thuringia and neighboring parts of Franconia. One further activity was the publication of several series of so-called Heimatschutz postcards, picture postcards of scenic or interesting features of an area whose express goal was to "spread the Heimatschutz idea as far as possible" (letter of 23 December 1911, StA Coburg Min D 18, 116–17).

Koch was also very successful in encouraging a wider distribution for the materials he put together. A suggestion in early 1912 that the Ducal Saxon State Ministry purchase 300 subscriptions to his *Heimatschutzfragen* and distribute them was taken up almost immediately (StA Coburg Min D 18, 126–28). Records in Gotha confirm that copies of the

Heimatschutzfragen were sent out by the Staatsministerium there to a number of municipalities with the injunction to "be active in promoting this important program"; the Heimatschutz literature was accordingly distributed to city libraries, municipal building departments, foresters, and other officials (StadtA Gotha, Akte Heimatschutz, 17). Within a short while the Foundation for Heimatschutz also became something of an environmental-law clearinghouse for many of the smaller cities and regions all over Germany: officials were asked to submit copies of any regulations they came up with, and in return Koch transmitted relevant edicts and guidelines being formulated in other parts of the country (StadtA Gotha, Akte Heimatschutz, 10). Regional Heimatschutz journals fulfilled a similar function for their more restricted area. The accessible publication of such guidelines was an important step in creating an environmental public: it gave non-experts and interested parties the chance to familiarize themselves with the often only internal rules that governed the condition of their environment and thus prepared them to change those rules. With dozens of such small but well-placed initiatives, Fritz Koch showed how an organization with only moderate means could still have a great impact on a local or regional program of Heimatschutz.

Nevertheless, by 1914 the originally well-focused *Bund Heimatschutz* had grown and subdivided into an unwieldy welter of different-sized regional and local groups with respectively different needs. With the simultaneous trend toward technical fragmentation in mind, Robert Mielke, now a leader of the Brandenburg Heimatschutz group, gave a decidedly gloomy assessment: "The *Bund Heimatschutz* has failed" (Mielke 1914). In spite of the booming growth of the movement, the league had been severely crippled by the fact that large organizations had been "standing on the sidelines for years." And relations between some regional organizations and the central *Bund* had become on occasion "quite unpleasant" (Mielke 1914b). It was an open secret that the State Association in Saxony, for example, resented the national *Bund* and was continually trying to "trip up" Fritz Koch (cf. letter of Otto Ubbelohde, UB Tübingen MD 875, 378). These continuing strains took their toll: within a few months Paul Schultze-Naumburg stepped down as chairman after ten full years; Fritz Koch also resigned from his position as general manager. Going into the war years a new team of leaders took over who were simply, for lack of a better description, less broadly idealistic than the men who had started the movement. Freiherr Tilo von Wilmoski, the new chairman, was picked—among other things—for his connections to wealthy elites; the new general manager was Werner Lindner, a man with an engineering background who merely accelerated the trend in the *Bund Heimatschutz* toward a dialectic of specialism in which architectural matters and calls for nature preserves between them squeezed the lifeblood out of the kind of comprehensive landscape vision that had given the early years their philosophical firepower.[10]

This flux in *Bund Heimatschutz* policies may serve to remind us that a purely institutional history often only goes so far in explaining the motives and meanings tied up with its work; at the same time it should not provide an excuse to conduct an abstract history of ideas. What the idea of Heimat meant in a wider cultural sense can only be understood fully if we look at the specific social and political positioning of the Heimatschutz within the framework of Wilhelmine society. The first question is a simple one: who participated in the Heimatschutz movement?

Until very recently, researchers repeated only the most general characterizations pertaining to the social base of the movement, if indeed the question interested them at all. Many critics have assumed that the program of Heimatschutz appealed to the same segments of the population interested in turn-of-the-century life-reform efforts; however the identification of these with "parts of the *Bildungsbürgertum*" remains rather vague (Krabbe 1969; Andersen 1987, 143). A more specific examination of a group from the Heimatbewegung led Jörn Christiansen to the same layer of "the educated middle class," but more specifically to the "decisive role" of Germany's poorly paid elementary school teachers, the non–university-educated *Volksschullehrer* (Christiansen 1980, 15).[11] As we shall see, though, school teachers did not play a dominant role in the Heimatschutz. Other researchers have assumed a congruence with the social location of the founding members, a better-educated group of "men from the sciences, administration, and the arts" (Linse 1986, 20). Only very recently has there been a more thorough examination of regional groupings in Lower Saxony (Hartung 1991). Since this covers only a small part of Germany, though, the vast majority of the Heimatschutz movement has remained uncharted territory.

I have tried to throw some more light on the subject by analyzing the membership lists of as many Heimatschutz organizations as possible. I have done this in two different ways. In table 2, labeled "Occupational Overview," the data on membership are arranged according to rough economic sectors. This arrangement makes it possible to compare my results to those obtained by Werner Hartung in his study of the Heimat movement in Lower Saxony. Together the figures cover a total base of 10,656 individuals in thirteen different Heimatschutz organizations. Civil service and the cultural and educational sector together comprise over 42.5 percent of the total membership, which means that the movement did indeed derive a decisive amount of its support from the *Bildungsbürgertum.* The main occupations here were those of a civil servant involved in general administration or justice, or those of a teacher or pastor. If one adds in the members who were medical doctors, other health workers, artists, and journalists, this segment of relatively well-educated citizens would almost comprise the absolute majority with 49.8 percent. If this is somewhat lower than previously imagined, it is due to the considerable number of members who worked in the commercial and manufacturing sectors. Hartung's research had indicated only weak participation here; if we look at a wider cross section, though, we find that just over thirty percent of a typical Heimatschutz organization came from the area of business and commerce. A few main types of occupation account for the lion's share of the members in this category. In the industrial sector it was either a person in a construction-related job (architect or developer/builder) or a *Fabrikant,* a manufacturer, most often without any more precise information as to the branch. In the commercial sector the most common occupations were "businessman" (*Kaufmann*) or a more prestigious titled commercial position such as "Commercial Councillor" (*Kommerzienrat*). Women accounted for a relatively small proportion of the Heimatschutz. The figure given includes only those women who did not list an occupation such as *Lehrerin.* Since employed women were the exception, though, the given percentage (4.5 percent) is fairly accurate. Many of the women appear to have had a husband or relative who was a member. The agricultural sector was underrepresented at 5.2 percent.

All in all it is clear that the Heimatschutz was a middle-class, white-collar phenomenon. I have not assembled separate figures on participation by the nobility, but it was low, certainly less than 5 percent. Their names occur, as one might predict, in the upper range of administrative positions, in the military, and again as large landowners. The rural and relatively unmodernized area of Mecklenburg, still top-heavy with remnants of a feudal caste system, showed the greatest participation by persons with the tell-tale *von* in their names.[12] Membership was also low at the other end of the social scale. This was a source of consternation to the bourgeois majority, since it stood in contradiction to what they saw as the generally inclusive idea of Heimat: "It pains us every time," confessed the chairman of a Thuringian Heimatschutz group, "[to see] that our Heimat association has no members from the so-called fourth estate" (Engelhardt 1913–14, 243). It is a rare occurence to find a manual laborer or factory worker actually listed as such in the membership rolls. Some artisans did participate, especially in the smaller towns, but in terms of the overall movement their presence (2.5 percent of the total) was relatively insignificant. In the main, then, it appears that Werner Hartung's observations on the Heimatbewegung in Lower Saxony can be applied to the Heimatschutz as a whole:

TABLE 2. Occupational Overview

Baden	Hesse	Mecklbg.	Lower Saxony		Rhine/Westfalia		Saxony	Schles-Holstein		Thuringia/Anhalt		
Badische Heimat ($N = 1{,}503$)	Hessicher Bund für Heimatschutz ($N = 161$)	Heimatbund Mecklenburg ($N = 1{,}615$)	Niedersächsische Heimatbewegung ($N = 1452$) (cf. Hartung 1991)	Verein für Heimatschutz und Heimatgeschichte Leer ($N = 810$)	Rheinischer Verein für Denkmalpflege und Heimatschutz ($N = 1{,}355$)	Verein für Heimatkunde und Heimatschutz im Siegerlande ($N = 987$)	Landesverein Sächsischer Heimatschutz ($N = 1{,}886$)	Verein Heimatschutz im Hamburger Staatsgebiet ($N = 380$)	Ortsgruppe Heimatschutz für den Kreis Herzogtum Lauenburg ($N = 257$)	Bund Heimatschutz für das Eisenacher Land ($N = 139$)	Verein für Heimatkunde und Heimatschutz von Artern ($N = 111$)	Total
I. Civil Service, Law, Military[a]												
313	61	439	391	116	394	93	484	40	57	30	19	2,437
20.8	*37.9*	*26.9*	*14.3*	*29.1*	*9.4*	*33.8*	*25.7*	*10.5*	*22.2*	*21.6*	*17.1*	*22.9%*
II. Education, Religion, Culture[b]												
502	53	375	301	136	163	208	134	121	65	37	17	2,112
33.4	*32.9*	*23.2*	*20.7*	*16.8*	*12.0*	*21.1*	*7.1*	*31.8*	*25.3*	*26.6*	*15.3*	*19.8%*
III. Commerce and Manufacture[c]												
369	15	249	306	252	460	307	984	127	55	42	43	3,209
24.6	*9.3*	*15.4*	*21.1*	*31.1*	*33.9*	*31.1*	*52.2*	*33.4*	*21.4*	*30.2*	*37.8*	*30.1%*
IV. Artists and Journalists												
31	7	32	87	17	17	7	44	20	7	3	3	275
2.1	*4.3*	*2.0*	*6.0*	*2.1*	*1.2*	*0.7*	*2.3*	*5.3*	*2.7*	*2.2*	*2.7*	*2.6%*

V. Health												
81	8	103	72	42	51	31	61	9	14	11	3	486
20.8	*5.0*	*6.4*	*5.0*	*5.2*	*3.8*	*3.1*	*3.2*	*2.4*	*5.4*	*7.9*	*2.7*	*4.6%*
VI. Agriculture												
54	11	203	63	52	39	21	73	15	15	3	6	555
3.6	*6.8*	*12.6*	*4.3*	*6.4*	*2.9*	*2.1*	*3.9*	*3.9*	*5.8*	*2.2*	*5.4*	*5.2%*
VII. Women[d]												
39	1	78	72	68	73	48	31	39	22	5	6	482
2.6	*0.6*	*4.8*	*5.0*	*8.4*	*5.4*	*4.9*	*1.6*	*10.3*	*8.6*	*3.6*	*5.4*	*4.5%*
VIII. Miscellaneous, Unidentified												
114	5	136	160	127	158	272	75	9	22	8	14	1,100
7.6	*3.1*	*8.4*	*11.0*	*15.7*	*11.7*	*27.6*	*4.0*	*2.4*	*8.6*	*5.8*	*12.6*	*10.3%*
Total												
1503	161	1615	1452	810	1355	987	1886	380	257	139	111	10,656

[a]Includes political titles such as Stadtrat.
[b]Includes academic honor titles such as Prof., Prof. Dr.
[c]Includes architects, clerks and *Handwerker*.
[d]If no listed occupation.

> The Heimat groups clearly did not reach the base of the hierarchical societal pyramid, the dependent artisans and the workers in industry or agriculture. Their capacity for integration ended at the threshold of proletarian consciousness, at the master craftsman and the small farmer who socially considered themselves part of the local Establishment. (Hartung 1991, 119)

One significant exception to this rule may exist, however. The Association for Heimat Study and Heimatschutz in the Siegerland, an organization founded in 1911, listed an extraordinary number of non-noble members without occupation: 247, or fully 25.0 percent of the entire membership. Because of the general nature of a bourgeois association—the kind of gathering where one was likely to give as clear an indication as possible that one was earnest and/or socially acceptable—there is good reason to believe that those who declined or neglected to give their occupation belonged to the lower end of the social prestige spectrum. Siegen was the center of an industrial and mining area as well, and it probably was home to more low-prestige factory workers and miners than some other areas. In view of these circumstances it seems highly likely that the unidentified members of the Siegerland association were indeed factory workers or those employed in similar manual labor. This conclusion is supported by the comments of a man writing in the *Sauerländischer Gebirgsbote* who was trying to start a Heimatschutz asssociation in his neighboring area. He noted that the Siegerland association had grown quickly and could pride itself on participation from every segment of society; in particular, he wrote, "it is highly gratifying [to see] how many workers, farmers, and artisans it counts in its ranks" ("Ist keiner da?" 1912, 253). With a participation by such occupational groups that exceeded 25 percent, the Siegen-based association is an important indication that support for the ideas of Heimat and Heimatschutz was not limited to the bourgeois segments of the German population.[13]

While a rough analysis of the type shown in table 2 provides important information, it does not fully describe where Heimatschutz members came from in terms of their standing in Wilhelmine society. The category of "civil servant, military, law" for example, would cover anything from a small-town clerk to the most influential minister; similarly, "commerce" lumps together occupations as disparate as a longshoreman and a bank president. In order to assess the membership more accurately I adapted a more precise occupational breakdown outlined by Roger Chickering in his study of the roughly contemporaneous Pan-German League (Chickering 1984). Table 3 shows the forty-six occupational areas included in this scale and the percentage of members employed for each of my eleven

Heimatschutz organizations and as an aggregate average. Table 4 shows a compilation of the total percentages related to the civil servants, educators, and others who comprised the movement's strong support in the *Bildungsbürgertum.* The reader might notice that the figure of 46.8 percent given here is a few percentage points lower than the one obtained in table 2. This is because these data do not include the area of Lower Saxony, and because a more restrictive definition was applied that excludes members of the military.

But the occupational descriptions that Heimatschutz organizations dutifully recorded can also be used to make a more precise estimation of each individual's socioeconomic standing. Following Chickering, I broke down the social spectrum into four very rough categories: the lowest-ranked jobs were given a value of 1, while the highest-ranked positions received a value of 4. The resulting socioeconomic matrix is reproduced in table 7, the Occupation Key. The entire process is necessarily very schematic, of course. But in the Wilhelmine era social prestige was tied fairly tightly to occupation, and by computing the average value for a group of several hundred or a thousand members we can get a rough idea of the movement's overall standing in society, as well as a fairly good indication of differences between the various groups. These results are shown in tables 5 and 6.

These tables show clearly that the Heimatschutz was anchored in the middle levels of society. In table 3 we see that the average value for social prestige comes out as 2.25 if one excludes unidentified occupations, and 2.16 if these are included with an assumed low prestige. We can thus say that the movement as a whole had a decidedly solid middle-class profile with a tendency in the direction of high prestige (the value 3 denoting someone with a university education or a businessman with a significant amount of economic and social clout, such as a *Kommerzienrat*).[14] Table 6 indicates one of the main sectors where this tendency toward high-prestige occupations was realized: namely in the civil service, where almost half of the members so employed occupied positions with high prestige. A number of the top-rank individuals in the Heimatschutz were civil servants as well; other "top" prestige individuals were either members of the military or large landowners designated as *Rittergutsbesitzer.* To a certain extent, of course, all of the Heimatschutz associations preselected for better-situated citizens of the Empire. Not only was their entire format decisively bourgeois, but there were important material barriers to be overcome. Membership in a Heimatschutz organization was not terribly expensive, but it did require an expenditure of at least 1 to 5 marks of disposable income per year. To wealthy Germans this would have been a pittance, but to poorer citizens it would have represented a correspondingly greater

TABLE 3. **Participation by Occupational Sector**

	Baden	Hesse	Mecklbg.	Lower Sax.	Rhine/Westfalia		Saxony	Schles-Holstein		Thuringia/Anhalt		
	Badische Heimat (*N* = 1,503)	Hessicher Bund für Heimatschutz (*N* = 161)	Heimatbund Mecklenburg (*N* = 1,615)	Verein für Heimatschutz und Heimat-geschichte Leer (*N* = 810)	Rheinischer Verein für Denkmal-pflege und Heimatschutz (*N* = 1,355)	Verein für Heimatkunde und Heimat-schutz im Siegerlande (*N* = 987)	Landesverein Sächsischer Heimatschutz (*N* = 1,886)	Verein Heimatschutz im Hamburger Staatsgebiet (*N* = 380)	Ortsgruppe Heimatschutz für den Kreis Herzogtum Lauenburg (*N* = 257)	Bund Heimat schutz für das Eisenacher Land (*N* = 139)	Verein für Heimatkunde und Heimat-schutz von Artern (*N* = 111)	Proportion of Overall Movement for selected sectors (*N* = 9,204)
A. Agriculture/forestry												
1. Agriculture	2.8	1.2	*7.1*	6.2	2.4	1.6	3.0	3.9	4.3	1.4	4.5	3.8%
2. Forestry	.8	5.6	*5.4*	.2	.5	.5	.8	—	1.6	.7	.9	1.6%
B. Industry												
1. Mining	.1	—	—	—	.9	1.3	.1	—	—	.7	.9	
2. Metal processing	—	—	—	—	.4	.2	—	—	—	—	—	
3. Mach./engineer	.3	—	.9	1.2	.8	4.5	1.3	3.4	—	—	—	
4. Chemicals	.1	—	—	—	.3	.1	.2	—	—	—	—	
5. Textiles/paper	.1	—	—	—	.3	.1	.4	—	—	—	—	
6. Construction	.5	—	.2	1.4	1.0	1.3	*22.9*	.5	1.6	.7	.9	5.3
7. Polygraphic	.9	1.2	.7	.6	.5	.3	.7	1.8	1.6	.7	1.8	
8. Artistic	1.3	1.2	.5	.1	5.2	1.0	*11.6*	6.1	1.9	5.7	—	4.0
9. Other, unknown	4.0	2.5	.7	2.0	*7.9*	*6.3*	4.2	.8	.8	2.2	4.5	3.8
C. Commerce												
1. Gen. Commerce	*6.9*	3.1	4.1	*15.9*	*11.3*	*9.6*	3.8	*8.7*	2.3	9.3	*9.9*	*7.5*
2. Insurance	.1	—	.1	.2	.1	—	2.0	—	.8	—	—	
3. Transportation	—	—	.2	.9	.1	.4	—	.3	—	—	—	
4. Hotel/Restaurant	2.4	—	.5	3.2	.7	.9	.2	2.6	1.6	2.9	—	
5. Bank/Retail/Other	3.6	.6	2.2	3.1	4.0	1.7	2.6	1.8	.4	5.0	4.5	2.8
D. Dom./Handwerker	2.8	.6	2.8	2.0	.4	2.5	1.9	4.5	*10.1*	2.9	*15.3*	2.5
E. Service & Professions												
1. Military	1.8	.6	3.7	.2	2.1	.8	1.7	—	2.7	2.2	.9	1.8
2. Public Admin. & Law												
a. Law	2.6	4.3	1.9	1.0	2.1	.5	2.1	1.6	2.3	2.2	—	1.9
b. General Admin.	6.2	*11.8*	*6.9*	1.8	*9.5*	2.2	*6.7*	1.3	*8.2*	5.8	3.6	*6.0*
c. Justice	2.1	3.7	4.7	2.1	5.8	1.0	.8	2.4	1.6	9.3	1.8	2.8
d. Finance/ Taxation	1.7	3.7	1.2	.5	.4	.1	1.0	.3	.4	.7	1.8	.9

e. Post/Telegraph	1.3	—	2.1	4.7	.5	1.7	.3	.5	1.2	.7	1.8	1.4
f. Rail/roads/ports	.4	—	.9	.6	.2	.1	.3	.3	1.2	—	1.8	.5
g. Customs	.2	—	.6	.4	.2	.1	—	1.6	—	—	—	.3
h. Industr. regulation	.1	—	—	.1	—	.1	—	.8	—	—	—	—
i. Construction	1.3	*7.5*	2.1	.9	6.2	1.4	*7.7*	1.6	1.2	2.2	1.8	3.6
j. Land/forest/ mines	.5	3.1	.4	.7	.5	.5	1.9	.3	.4	—	1.8	.8
3. Religion	*12.2*	*7.5*	5.0	3.4	4.4	5.3	.8	1.6	*8.2*	6.5	4.5	5.1
4. Education												
a. Volks-, Hauptlehrer	*6.7*	5.0	*7.4*	*8.9*	.8	*9.5*	1.7	*20.0*	*9.7*	5.0	*6.3*	*6.0*
b. Ober-, Gym. lehrer	.9	2.5	4.7	1.7	.7	2.0	.6	2.9	1.9	9.3	—	1.9
c. Dozent, Professor	1.9	5.6	.1	—	.2	—	.8	—	—	.7	.9	.6
d. Rektor, Archives	3.9	*6.8*	1.8	1.1	3.2	2.4	.7	*6.8*	3.1	2.2	3.6	2.5
5. Health												
a. Doctor, Sanitätsrat	4.8	4.3	4.8	3.6	2.7	2.2	2.4	1.3	4.7	6.5	.9	3.5
b. Apotheker	.4	.6	1.4	1.5	1.0	.8	.7	.3	.8	.7	.9	
c. Administration	.2	—	.1	.1	.1	.1	—	.8	—	.7	.9	
6. Journalist, priv. schol.	.7	1.2	.5	.6	.4	.2	.5	1.8	.8	.7	.9	
7. Stenography, clerk	1.5	—	2.4	.5	—	.8	.1	2.9	.4	—	.9	1.0
8. Artistic	1.4	3.1	1.5	1.5	.9	.5	1.8	3.4	1.9	1.4	1.8	1.5
9. Other												
a. Academic title	*7.8*	5.6	4.3	1.4	2.6	1.8	2.3	.5	2.3	5.0	—	3.5
b. Political/other title	3.8	3.1	2.6	1.5	1.0	.8	3.1	—	3.1	.7	1.8	2.0
c. Unidentified Nobility	.1	1.9	1.5	—	.4	.3	.4	—	2.7	.7	1.8	
F. Other												
1. Rentner, Privat	2.3	.6	1.3	.9	1.8	1.1	.6	1.6	1.2	3.6	*5.4*	1.4
2. Women	2.6	.6	4.8	*8.4*	5.4	4.8	1.6	*10.3*	*8.6*	3.6	*5.4*	4.5
3. Students	1.9	—	1.5	6.7	.2	1.1	.2	.3	.4	—	.9	1.4
4. Not Identified	2.3	.6	4.0	*7.9*	*9.7*	*25.0*	2.8	.5	4.3	.7	4.5	*6.7*
												93.5%

Note: For ease in reading, the four largest occupational groups within any given Heimatschutz organization have been set italic.

investment. Time was another consideration. Active members of an association might attend several meetings per year; the most active members, usually officers, might easily triple or quadruple that number. With few exceptions all work was unpaid, and much travel was uncompensated.[15] These high demands on a person's time and financial resources help explain why the professional middle class dominated in Heimatschutz associations.

Until now we have been considering the social composition of the Heimatschutz movement as an aggregate whole. But a number of the regional groups had their own distinctive features. The data in table 5 show four groups with social prestige quotients that are noticeably higher than average. Three of the groups can easily be explained as variations on the standard pattern. The Hessian League for Heimatschutz (*Hessischer Bund für Heimatschutz*) was founded relatively late, in the waning weeks of 1913. Data on its small membership were taken from a public appeal formulated in this early phase of the organization's existence and thus reflect a kind of upscale "movers and shakers" social profile filled with ranking civil servants. It seems likely that this would have become broader as time went on. The Heimatschutz group in Eisenach similarly counted a number of relatively high-ranking local government officials among its members; as its overall size was quite small, the average social prestige went up disproportionately. And although Saxony's Heimatschutz organization was far larger, much the same reasoning applies to it as well. For here the state administration strongly supported the goals of the Heimatschutz and succeeded in encouraging a large number of "high"-level civil servants to join the nominally private group. This combined with an unusually strong participation by a few "mid"-level occupations to produce a higher-than-normal average social prestige.

TABLE 4. Membership of "Bildungs- und Beamtenbürgertum"

Occupation	Percent
1. Civil servants (E2, b.–j.)	16.4
2. Education (E4, a.–d.)	11.0
3. Religion (E3)	5.1
4. Academic Title (E9, a.)	3.5
5. Doctors (E5, a.)	3.5
6. Artists and journalists (E6, E8)	2.0
7. Lawyers (E2, a.)	1.9
Total	43.4
Total including political titles (E9, b.) and students (F3)	46.8

TABLE 5. Social Prestige

	Baden	Hesse	Mecklbg.	Lower Sax.	Rhine/Westfalia		Saxony	Schles-Holstein		Thuringia/Anhalt		Average for all
Average value:	2.20	2.63	2.20	1.83	2.40	1.82	2.41	1.94	2.08	2.38	1.94	2.16
Avg. w/o unidentified	(2.22)	(2.64)	(2.25)	(1.90)	(2.56)	(2.09)	(2.45)	(1.95)	(2.13)	(2.39)	(1.98)	(2.25)
Percent of members in social prestige group (including "Unidentified" as low prestige)												
Low	16.2	5.0	21.0	32.5	15.9	38.7	11.0	24.5	23.0	12.9	32.4	20.2
Mid	49.9	34.2	41.2	52.6	32.0	41.1	55.5	56.8	47.5	37.4	43.2	45.8
High	32.1	53.4	34.0	14.3	47.4	19.9	29.7	18.4	28.4	48.2	22.5	31.1
Top	1.8	7.5	3.7	0.6	4.6	0.3	3.8	0.3	1.2	1.4	1.8	2.7

Note: 1 = Low, 2 = Mid, 3 = High, 4 = Top (see Table 7 and accompanying notes).

These explanations do not, however, hold for the fourth group, which requires somewhat more analysis. The Rhenish Association for the Care of Monuments and Heimatschutz (*Rheinischer Verein für Denkmalpflege und Heimatschutz*) had a significantly higher social prestige than most other Heimatschutz groups in spite of its large size and established character. If we were to calculate the distribution of prestige for the ten groups other than the Rhenish Association, the "mid"-level prestige group would comprise 48.2 percent, and the percentage of members belonging to the "high" prestige group would come in at 28.3 percent. The Rhenish Association presents us with a near reversal of this distribution: in it, 47.4 percent of the members carried high social prestige, while 32.0 percent were from the mid-level prestige category. These figures reflect the unusually large number of *Kommerzienräte, Geheime Kommerzienräte,* and manufacturers who were members along with high administration officials. If we were to bracket out the Hessian Heimatschutz group for the anomalies surrounding its founding, the Rhenish Association would also have the highest percentage of top-level members of any of the studied organizations.

This unusual concentration toward the high end can be traced, I believe, to the fact that the Rhine group devoted almost all of its efforts to a program of historical preservation. It was the only group studied that placed the word *Denkmalpflege* in its title, and it is characteristic that no article appeared in the group's journal on any aspect of nature protection until four long years had passed. Much of the preservation interest was focused on structures that represented traditional systems of authority, namely castles and churches. In the extraordinary clustering of Commercial Councillors around this program of representative preservation we can thus see a vivid illustration of a well-known social compromise that was struck in the Second Empire. The financially strong commercial and industrial bourgeoisie bought into the established system of values, preserving, in this case quite literally, the old structures of "throne and altar." This is not to say that the typical Heimatschutz organization was not inter-

TABLE 6. Prestige Distribution—Civil Servants versus Others (in percentages)

	Civil Servants (N = 1,506)	All Others (N = 7,698)	Overall (N = 9,204)
Low	15.7	21.1	20.2
Middle	35.2	47.9	45.8
High	46.3	28.2	31.1
Top	2.8	2.7	2.7
Total	100.0	99.9	99.8

ested in castles or churches. It valued them as important components of a visual ensemble and often got embroiled in controversies about their renovation or restoration. What distinguishes the reform program of the Heimatschutz from the preservation program that one finds at least partially embodied in the Rhenish Association is the former's emphasis on the complete ensemble as opposed to a single monument, which often was a monument to past authority. The Heimatschutz paid closer attention to the "average" and "unexceptional" buildings, the structures that gave a city or region its own recognizable background "atmosphere."[16]

The state association that handled Heimatschutz affairs for Saxony was perhaps the best-run Heimatschutz organization in the German Empire. One could call the State Association Saxon Heimatschutz (*Landesverein Sächsischer Heimatschutz*) a technocratic organization, for it was heavily populated with architects, builders, and public officials involved in the regulation of construction. The State Association was also remarkable for its very close cooperation with the Kingdom of Saxony. This support and the heavy orientation toward building matters antedated the founding of the State Association and stemmed from its precursor, the Saxon Committee for the Fostering of Heimat-Style Art and Building. The Committee was convened at the suggestion of Saxon officials and integrated a number of high officials into its membership. When the State Association was formed in 1907 (by changing the name of the Committee) it was a state official, the Finance and Building Advisor L. F. Karl Schmidt, who assumed the chairmanship and held it for the next ten years. Dozens of State Association members were bureaucrats in either the Saxon Ministry of the Interior or the Finance Ministry, the branches of government most heavily involved in construction and city-planning matters. These civil servants were often explicitly encouraged to attend Heimatschutz lectures, to the point of having their transportation and per diem costs reimbursed (StA Dresden MdI 17519/81). The Saxon government also supplied generous financial assistance that was the envy of Heimatschutz organizations everywhere, reaching 15,000 marks in the year 1910, 37,000 marks in 1912, and peaking at 42,500 marks in 1914.[17] At every level the State Association and the state seem to have cooperated with the utmost unanimity, so much so that outside observers suspected undue or improper influence. A representative of the building products industry wrote a memo in 1913 complaining about "paternalistic" and "police-power" influence of the Heimatschutz in Saxony's public offices:

> I find the explanation for this in the fact that, in Saxony, civil service and the representatives of the Heimatschutz seem to be essentially identical. This is perhaps not surprising, given the position of leading

TABLE 7. Occupation Key

Occupational Area	Occupational Description and Approximate Social Prestige[a]			
	Low	Mid	High	Top
A. Agricultural/forestry				
1. Agriculture	Gärtner Gutsjäger Landbaumeister Wiesenbautechniker	Domänenpächter Gärtnereibesitzer Hofpächter Landwirt Mineralbrunnenbesitzer Molkereiverwalter Rittergutspächter Tierarzt Wiesenbauschullehrer	Gutsbesitzer Obstgutsbezitzer Vorwerksbesitzer	Rittergutsbesitzer
2. Forestry	Förster	Forstmeister Forstinspektor Forstrendant, -taxator Oberförster Oberforstmeister	Forstassessor Landforstmeister	Oberlandforstmeister
B. Industry				
1. Mining	Schichtmeister Bergmann	Bergschullehrer	Bergassessor Berwerkdirektor Grubendirektor Markscheider Zechendirektor	Berghauptmann Grubenbesitzer
2. Metal processing			Hüttendirektor Walzwerkbesitzer	
3. Mach./instr./engineer	Techniker Modelleur	Ingenieur Oberingenieur	Fabrikant	
4. Chemicals			Chemiker	
5. Textiles/paper		Färbereibesitzer	Fabrikbesitzer	
6. Construction	Bautechniker Stukkateur	Baumeister Bauunternehmer Ziegeleibesitzer	Dir. des Elektrizitätswerks Gas- und Wasserwerks direktor	
7. Polygraphic		Buchdrückereibesitzer	Verleger	

		Werkstätte für Grabmalkunst		
9. Other, unknown branch	Arbeitersekretär		Direktor	
	Fabrikarbeiter		Fabrikant	
	Werkmeister		Fabrikbesitzer	
C. Commerce				
1. Commerce	Auktionator	Kaufmann	Kommerzienrat	
	Handlungsgehilfe	Prokurist	Kommissionsrat	
	Volontär	Verwalter	Konsul	
			Oekonomierat	
2. Insurance	Brand-Versicherungs-Assistent	Agent, Generalagent		
		Brand-Ver.-Inspektor		
3. Transportation	Fährmann	Spediteur	Kapitän	
	Lagerist			
	Stauer			
4. Hotels/restaurants	Gasthofsbesitzer	Hotelbesitzer		
	Gastwirt	Hotelier		
5. Other: Banking, retail	Bankbeamter	Buchhalter	Bankier	
	various Händler	Bürochef	Bankdirektor	
	Kassierer		Bankvorsteher	
	Konditor			
		Sparkassenrendant		
D. Domestic/Handwerker	Bäckermeister	Mühlenbesitzer		
	Juwelier	Sekretär des Innungsverbandes		
	Mauremeister	Direktor der Handwerker-Zentralgenossenschaft		
	Uhrmacher			
	Zimmermeister			
E. Service and professions				
1. Military	Materialienverwalter	Einjähriger	Fähnrich bis Oberstleutnant	Oberst bis General
2. Public administration and law				
a. Law			Dr. jur	
			Notar	
			Rechtsanwalt	
b. General administration	Amtsverwalter	Amtmann	Amtsassessor	Minister
			Amtshauptmann	
	Kreisassistent		Bürgermeister (city)	Ministerialdirektor

(continued)

TABLE 7—*Continued*

Occupational Area	Occupational Description and Approximate Social Prestige[a]			
	Low	Mid	High	Top
	Ratschreiber	Amtsvorsteher	Dr. Bürgermeister	Reg-Präs.
		Bürgermeister (small town)	Referendar	Ober-Präs.
		Drost	Regierungsassessor	
		Kreissekretär	Regierungsrat	
		Regierungsamtmann	GehReg.rat	
			Oberreg.rat	
c. Justice	Amtsgerichtsaktuar	Landgerichtssekretär	Amtsrichter	Landgerichtspräsident
	Gerichtsamtsschreiber	Polizeiinspektor	Gerichtsassessor	Senatspräsident
	Gerichtsvollzieher	Polizeisenator	Gerichtsrat	
			Justizrat	
			Landgerichtsdir.	
d. Finance	Steuereinnehmer	Amtskassenrendant	Finanzrat	
		Revisionsrendant	Rechnungsrat	
		Stadtkassenberechner	Revisionsrat	
e. Post/telegraph	Oberpostasst.	Postbauführer	Postdirektor	
	Postassistent	Postsekretär	Postrat	
	Postmeister	Postverwalter		
		Telegrapheninspektor		
f. Railroads/roads/ports	Eisenbahnassistent	Bahnhofsvorsteher	Strassenbahndirektor	Eisenbahndirektions-Präsident
	Straßenwärter	Eisenbahnbauinspektor		
		Seevogt		
		Sielrichter		
g. Customs		Handelsrichter		
h. Industrial regulation			Gewerberat	
i. Construction	Bauamtsassistent	Bauamtmann	Baurat	
		Bauinspektor	GehBaurat	
		Bausekretär		
		Landesbauinspektor		
		Regierungsbauführer		
		Regierungsbaumeister		
j. Land/forestry/mining		Geometer	Bergrat	
		[illegible]	Stadtvermessungsdirektor	

3. Religion	Hilfsprediger Konventualin Küster	Pastor Pfarrer	Kaplan Konsistorialrat Oberkirchenrat Superintendent	Bischof Oberkirchenratspräsident
4. Education				
a. Volks-, Hauptschullehrer		Lehrer Hauptschullehrer		
b. Ober-, Gymnasiallehrer			Gymnasial-Professor Oberlehrer	
c. Dozent, Professor			Dozent	Univ-Prof.
d. Rektor, Archiv, Museum, Verwaltung		Archivassistent Bibliothekar	DrBibliothekar Gymndirektor Kreisschulinspektor Schuldirektor Schulrat Seminardirektor	
5. Health				
a. Ärzte, Med.- & Sanitätsrat		cand. med.	Arzt, Dr. med Medizinalrat Sanitätsrat	
b. Apotheker		Apotheker	DrApotheker	
c. Verwaltung	Hospitalmeister Bandagist		Blindenanstaltsdirektor Heilanstaltsdirektor	
6. Private scholarship, journalism		Schriftsteller Redakteur	DrRedakteur Chefredakteur Privatgelehrter	
7. Stenography, clerk	Buchhalter Registrator Schreiber Sekretär	Administrator Bureauvorsteher Inspektor Verwalter	Kammerassessor v. Oberinspektor	
8. Artistic		Bildhauer Kantor Kapellmeister Maler	DrMaler	

(continued)

TABLE 7—*Continued*

Occupational Area	Occupational Description and Approximate Social Prestige[a]			
	Low	Mid	High	Top
		Musikdirektor		
		Zeichenlehrer		
9. Other professional				
a. Academic honor		Prof.	Dr; Prof. Dr.	
b. Political and other title		Beigeordneter	DrBeigeordneter	
		Gemeinderat	Geheimer Rat	
		Gemindevorsteher	Hofrat	
		Schulze	Konsul	
			Landtagsabgeordneter	
			Stadtrat	
c. Unidentified nobility			Kammerherr	Fürst
F. Other				
1. *Rentner, Privat*	Rentner	Privat	DrPrivat	
2. Women	Frau			
	Fräulein			
3. Students	Schüler	Student	DrPrakt.	
4. No identification	(No identification)			

[a]Follows general method outlined by Chickering (1984); an important source for corroboration on individual occupations is the extensive list compiled by Reinhard Schüren (1989, 314ff). The basic divisions are as follows.

Low-level: includes peasants, artisans, manual and unskilled laborers, low-level employees, also low-level public-sector *Beamte*

Mid-level: includes public employees and teachers *without* academic education, mid-level white-collar employees, most who describe themselves as *Kaufmann*

High-level: includes public officials, teachers and professionals *with* an academic education, military officers below the rank of colonel, substantial businessmen, managers, and landowners

Top-level: includes high military officers, top-ranking public officials, the commercial and industrial elite, high nobles, *Rittergutsbesitzer,* university professors.

The problems inherent in this kind of classifying are many, of course. Some of the most serious are connected with the ambiguous titular practicie of the Wilhelmine era, according to which private citizens could be given rank designations also used for civil servants: *Baurat,* for example, could mean a man who worked for the government, or one who had achieved a certain solidity as a skilled private architect or builder. In this particular case I chose to err on the side of the civil servant. *Rektor* is another term with different meanings, designating an official who is in charge of an institution—or a church. But a fundamental source of inaccuracy remains the lack of individual research; only this would tell us if, for instance, an otherwise unremarkable "manufacturer" was indeed very rich and well-connected, or if certain *Rentiers* were wealthy investors, etc.

> personalities in the Heimatschutz . . . No official in his right mind will oppose the Heimatschutz when he knows that his superior is a leader in this same Heimatschutz movement. (StA Dresden MdI 17513/152)

Heimatschutz defenders responded to this accusation by noting that a large number of architects and representatives of the private building industry were involved in the State Association. Government and Heimatschutz alike went to great lengths to explain their position to industry, and in some instances they seem to have succeeded in awakening an understanding for environmental reform. An example is the "extraordinary generosity" shown by the League of Saxon Industrialists to Dresden's Hygiene Exhibition (1911) after receiving a Heimatschutz letter supporting the project (StA Dresden MdI 17519/145, 157). A key figure in these good relations was Georg Marwitz, a Saxon manufacturer in the pragmatically progressive tradition of Ernst Abbé. When a group of cement and advertising interests who were allied with the Progressive People's Party (*Fortschrittliche Volkspartei*) attacked Heimatschutz laws at the end of 1913, Marwitz's support backed up the government and allowed it to neutralize the complaints as narrow and self-interested (cf. StA Dresden MdI 17531/5; *Mitteilungen des Landesvereins Sächsischer Heimatschutz* 1914–15, 160ff.).

While the greatest part of the character of the State Association was determined by public employees and members in construction-related branches, it is worth mentioning the popular-education efforts of several of its academic members. Seven men gave talks in the fall of 1908 for the Dresden *Volkshochschulkurse,* or People's College Courses. Among these were the botanist Oskar Drude and the well-known art historian Cornelius Gurlitt. Their participation in the People's College movement demonstrates once again the proximity of the Heimatschutz to better-known progressive reform movements in the Wilhelmine era.

The State Association continued its attempts to reach as many people as possible with a slide-accompanied lecture series in Dresden's Künstlerhalle in the years between 1911 and 1914. For 3 and 2.45 marks, respectively, nonmembers could hear 11 or 7 separate "popular Heimatschutz lectures" over a period of two months (cf. *Mitteilungen des Landesvereins Sächsischer Heimatschutz* 1914–15, 172). Heimatschutz groups in other parts of Germany made similar efforts to reach out to the public: the Pomeranian association had a series of slide shows on hand, whereas starting in 1911, the Brandenburg group put on a series of eight lectures on "applied Heimatschutz" for students at the Technische Hochschule in Berlin-Charlottenburg (cf. *Pommersche Heimat* February 1916, 8; "Vorlesungen über angewandten Heimatschutz" 1911; Mielke 1914b, 59). But the lecture system was still perhaps strongest in Saxony, where the State Association made a habit of sending free tickets to appreciative civil servants, thereby cultivating their already strong support for the movement; nevertheless it is apparent that the lectures were also widely received, often being given in a "completely filled" hall and covered extensively as the first cultural item of the next day's *Dresdener Anzeiger* (cf. numerous communications in StA Dresden MdI 17519/177; *Dresdener Anzeiger* 28 February 1914).

If the organization *Badische Heimat* presents us with a different picture, that can probably be traced to its divergent social makeup. The university and administrative centers Freiburg, Heidelberg, and Karlsruhe provided a large number of academics and bureaucrats, whereas the thinly settled areas in between yielded another major constituency of small-town pastors and elementary school teachers. This social dichotomy can further be explained by the history of the organization, which in full carried the title of Badische Heimat, Association for Folklore, Rural Welfare and Heimatschutz. It arose out of the joining of two separate organizations devoted to the first two areas, on top of which the area of Heimatschutz was added. Socially it seems probable that the two areas of folklore and Heimatschutz attracted a majority of the well-educated bureaucrats and academics, while the rural, conservative program of Rural Welfare appealed above all to the large group of clerics (12.2 percent of the entire total, more than twice as much as even Lower Saxony) and elementary school teachers (6.7 percent of the total, equalled only by Lower Saxony) (on Rural Welfare see Hartung 1991, 163f.). This same constituency played a determining role in the large Bavarian Heimatschutz organization, which drew largely on the "educated rural elite" of teachers and priests (cf. Knaut 1993, 201).

Baden was not the only southwest German Heimatschutz association to institutionalize a partnership with Heinrich Sohnrey's Rural Welfare. But while in Baden it seems that the Heimatschutz component was grafted onto the other organization's preexisting structure, in Württemberg and Hohenzollern it appears that something like the reverse took place. There the organizers of the Württemberg League for Heimatschutz at first rejected the idea of pursuing their agenda through the already existing groups devoted to Rural Welfare (cf. Fuchs 1909a). The independent Heimatschutz association that they subsequently fashioned proved to be extremely strong, attracting over 600 members in the first year (1909–10). From this established position, but seeing in the Rural Welfare a valuable ally with good rural organization, the Heimatschutz now proposed a cartel relationship that was quickly accepted. Elsewhere, however, Heimatschutz groups did not necessarily go along with the program of

Rural Welfare. In early 1908 the Heimat League of Mecklenburg fended off a cartel proposal from the Mecklenburg State Association for Rural Welfare, citing the fact that the two groups' goals lay "in part quite far from another" (cf. *Zeitschrift des Heimatbundes Mecklenburg* 1908, 39).

If in the preceding sections we have described the Heimatschutz, it is now time to begin to evaluate this bourgeois movement in terms of its social and political valences. A useful foil for this undertaking would be to remind ourselves how the prevailing view of the Wilhelmine bourgeoisie has entered into assessments of the Heimatschutz. In the introduction it was noted that almost all of the scholarly interpretations to date have been profoundly underwritten by modernist assumptions about progress. And while references to it are often only implicit, these assumptions are ultimately grounded in a modernization theory of history, one that sees the Anglo-American pattern of industrialization and mass democratic society as a norm. Geoff Eley has described how this pattern assumed tremendous significance for postwar historiography on nineteenth- and twentieth-century Germany, a country whose catastrophic political history seemed to require explanation as a profound deviation from the model at hand. The result was the so-called *Sonderweg* thesis, shared by many of the most influential historians and sociologists of the Federal Republic (Eley 1984; important examples of the thesis are Dahrendorf 1968; Lepsius 1973). Applied to the nineteenth century this theory focused attention on the "weak" German middle class, one that failed to assume the mantle of progressive democratic-industrial change in 1848 and instead collaborated with the "pre-industrial Prussian elites" who held much of the power (proponents of the view are Plessner 1959; Wehler 1985; see also Evans 1987 for a concise summary).

This assumption of a thwarted "natural" development was reinforced by critical theories of culture, and in particular a Marxist understanding of legitimating ideologies. Since the ruling classes had outlived their "objective" usefulness as defined by an economic progress indexed to Anglo-American conditions, virtually any value that they put forward could be decoded as an "ideology," which is to say, a self-serving lie. As the major political discourse of the nineteenth century, nationalism became the main example, the master lie that subsumed scores of smaller ones. It was now demasked as an "ideology of integration," a weapon cynically used by the collaborationist German bourgeoisie in order to bind the unruly proletariat and even itself to the old order.

As early as 1965 scholars began to denounce "Heimat ideology" in much the same fashion: with its origins in class warfare and a historically obsolete vision of the world, Heimat was nothing more than a "rear-guard action of old feudal remnants combined with the struggle of the bourgeois defenders against the forward-striving proletariat" (Schmidt 1965, 35). Klaus Bergmann's 1970 categorization of the Heimatschutz as a "pre-industrial social idyll" spoke the same language of middle-class deception and betrayal (Bergmann 1970, 30). Very quickly this became the conventional wisdom on Heimat. More than two decades later critics are still assuring us that this was not only antimodern daydreaming—a "projection of the educated middle class . . . only distantly related to socio-economic reality"—but also a manipulative "instrument" of German nationalism whose "cementing of pre-industrial values" reveals it as a transparent "ideology of bourgeois consolidation and integration" (Hartung 1991, 51, 148).

Fortunately, over the last fifteen years or so David Blackbourn and Geoff Eley have soundly criticized the theoretical foundations on which this general view rests (Blackbourn and Eley 1984; Blackbourn 1987, "Introduction"). They concede that the German bourgeoisie may not have achieved the same palette of democratic reforms as did its cousin in England; nevertheless, they argue, it was able to promote its agenda across the board in key sectors of civil society such as education and jurisprudence, where bourgeois professionals were increasingly able to guarantee principles of liberality. The idea that the middle class was not just a passive puppet of calcified East Elbian elites has actually gained much ground recently among historians; along with the British triumvirate of Blackbourn, Eley, and Richard Evans, it is a point of view that was given eloquent expression by the late Thomas Nipperdey (Evans 1987; Nipperdey 1986; Nipperdey 1990). The picture of the bourgeoisie that we get from these latest writings is that of a critical, politically active, and socially not unaware class—one that tried very hard, in many different and sometimes contradictory ways, to come to terms with the general process of modernization. This assessment appears to fit with the educated members of the bourgeoisie in particular, with their high participation in reform movements. As Klaus Vondung argued already in 1976, this group "by no means accepted the existing order" and indeed has to be credited with the creation of important "countercultures" (*Gegenkulturen*) (Vondung 1976, 32).

The notion of a counterculture is particularly productive because it implies an element of opposition in a group primarily constituted through cultural, rhetorical, or aesthetic commonalities. With their aesthetic concern for land- and townscapes the Heimatschützer gave a high approximation to this type of organization. Of course, in an age of industrialization and rampant "progress" they found that their aesthetic sensibilities were simply not shared by many of their contemporaries. They were, as a result, an elite, and were unabashed about contrasting their "expert" opinion with that of the "average person," particularly when it came to occupying the seats of the newly established Building Advisory Councils (cf. *Mitteilungen des Bundes Heimatschutz* 1907, 50). A number of regulations brought the latent conflict out into the open, though, with their requirement that construction had to constitute a "readily apparent eyesore" (*offenbare Unzierde*) to be actionable—"readily apparent," of course, to the proverbial man in the street. Cornelius Gurlitt complained that this amounted to making a "non-expert in aesthetic matters into an expert" (Gurlitt 1910–11, 11). Despite this readiness to accept an elite status for themselves, though, most Heimatschützer rejected the idea that they were "elitist." At most they admitted to being part of a historical elite, a small group whose insights had yet to spread. Using this perspective, Paul Schultze-Naumburg compared environmental sensibilities with the history of appreciation for Gothic architecture:

> We live today in a time in which the greater part of humanity is no longer able to see the beauty in nature. There was also a time in which humanity was unable to perceive the beauty of the Cologne Cathedral. The judgment of the majority is here, as in most things, not the same thing as the truth. (Schultze-Naumburg 1906, 90)

Given this situation the Heimatschützer saw their role primarily as that of educators and developers of that elusive quality, taste. As they well knew, though, their "aesthetic education of the people," with its method of pointing out the good, would only function if there was also a way to "push back the ugly," at least freeing up a space in which the new form could take hold (Gurlitt 1911–12, 48, 50). The disfigurement laws were seen largely as fulfilling this preparatory negative function.

These legal measures, though, were not the point of the matter. What the Heimatschutz wanted to encourage was a sensible, considerate approach to the whole aesthetic environment. It saw that no imposed law could possibly provide this motivation. Germany's background of absolutism played a not insignificant role here: as a result of this political history, many members of the middle class, the Heimatschützer included, were highly ambivalent about another form of state intervention. But more generally the Heimatschutz wanted people to be considerate of their neighbors and the environment out of conviction and not, as the general manager of the *Bund Heimatschutz* put it, because of "police enforcement" (Koch 1909, 5).

This long-term cultural strategy is one reason why the Heimatschutz tried to avoid direct attacks on individuals. Robert Mielke described the cautious "negative" activity of his Brandenburg Heimatschutz association in the following manner: After receiving notice of a situation, the first step was to gather information. The next step was a discreet, personalized visit to the offending party in order to discuss things without the pressure of publicity. This policy of "restraint," the Heimatschutz found, was often the most effective, even if it did mean that its "quiet work" behind the scenes did not achieve great publicity (Mielke 1913a, 54). Fritz Koch, Mielke's successor as general manager of the *Bund Heimatschutz,* also stressed the fact that interventions could often not be publicized (letter to Saxon Ducal State Ministry, StA Coburg Min D 18, 126). He characterized the Heimatschutz's primary strategy of conflict resolution as that of "coming to an understanding in private"; only after this level of appeal had been exhausted was there a "flight into the public sphere," something that might involve a series of letters to the local newspaper or the retention of an expert (Koch 1911; similar Swiss strategy in Leisi 1932).

The go-slow, personalized, advisory quality of Heimatschutz activities was a hallmark of its relationship to governmental offices as well. In some areas such as Saxony, where bureaucrats were often either members or at least exposed to Heimatschutz concerns, this method was sufficient to produce results from the very beginning. Similarly, the national *Bund Heimatschutz* owed much of its effectiveness to the "considerable personal influence of a number of its leaders" (Menzel 1910–11, 11). Where Heimatschutz ideas were foreign to the bureaucratic apparatus, though, supporters of the movement soon learned to use an avowedly nonconfrontational strategy. In Schleswig-Holstein the chairman of the regional organization told his fellow members that where active solicitation (*Aufdrängen*) had been attempted, the group had failed miserably: annoyed officials had told the Heimatschutz that they wouldn't let themselves be pushed around ("Schleswig-Holsteinischer" 1909).

But if the Heimatschutz "counterculture" was willing to leave the conversion of the majority to the long term, it nonetheless realized that a key actor in the short term was the public sector. Arguing that its reforms were in the public interest, the Heimatschutz hoped to be able to use government projects as examples of its new vision; in addition, as it realized, the government controlled important sites of education and reproduction. From the very beginning a central part of the Heimatschutz strategy was thus the "influencing of state, church, and municipal organs" (Mielke 1904–5, 67). As we have seen, the attempt often paid off: Heimatschützer managed to gain advisory roles vis-à-vis government bodies, for instance, and what had more impact in the long run was probably their slow infiltration of the schools. Given the tremendous amount of leverage that the government could provide, it is perhaps no wonder that Heimatschützer often praised the "state-preserving love of Heimat" that their reforms would evoke (Schmidt 1913, 50).

It would be easy to blow this relationship to the state out of proportion, though. Rhetoric such as that above was par for the course in the Wilhelmine era and should not be misunderstood as egregious authoritarianism. Heimatschützer were also quite clear about wanting to maintain a popular presence outside the bureaucracies. In purely pragmatic terms they saw that this would make their work more acceptable to the ordinary citizen. A second reason, however, was that the private association had more freedom to press for fundamental change, whereas bureaucracies were limited to what existing laws and procedures would allow (Koch 1909). Carl Johannes Fuchs went even further with his skeptical pronouncement that Heimatschutz might in some cases have to be instituted "*against* the government" (Fuchs 1909b, 6). Andreas Knaut is thus quite correct in saying that the Heimatschutz's relationship to the state was "ambivalent" (Knaut 1993, 237). If in practice the Heimatschutz often aimed at cooperation and collegiality, this may just have been the result of so many civil servants being members. Their individual contributions, as we shall see, constituted an important mode of transmission for Heimatschutz ideas.

It redounded to the movement's benefit that it wanted to make *cultural* distinctions that were often atmospheric and not specified in law or other explicit guidelines. In the area of architecture, for example, certain functional demands and occasionally some rough height limits were often the only things explicitly restricting the fantasy of a building's designer. In practice, of course, the force of political and cultural tradition was considerable, so that the largest and most representative projects in Wilhelmine Germany were not allowed to deviate too much from the classicistic-historistic "norm" that had developed since about 1870 (cf. Lane 1985). Nevertheless it is important to realize that this norm was largely a cultural phenomenon, unsupported by any specific decrees, and open to interpretation.

This was one area where the civil servant could be extremely useful, particularly one with influence over the technical side of administration. Especially in small, run-of-the-mill projects there was less pressure to conform rigidly to the classicistic-historistic norm. This did not mean that a city architect could begin to build the local school out of glass and steel, of course: there were limits to what was tolerated, and it is anyhow unlikely that a wildly inventive artist would have ended up in the relatively staid position of a technical bureaucrat. But with the fairly high degree of authority delegated to public administration there was a certain amount of free space that the Heimatschutz seems to have exploited very effectively. The obituary of a 47-year-old civil servant named Krielke indicates that his ability to promote the movement lay precisely in an ability to color in the gray areas of bureaucratic procedure:

> Everywhere he tried to improve things unobtrusively. During the eight years in his position he provided most of the preliminary sketches for school buildings, for example, probably for about fifty of them. All in complete selflessness, with the sacrifice of much time. According to the law he was only supposed to see that the requirements of the building and school codes were met. If he was unable to convince the community council, the law did not cover him; the effort he had spent was then for naught . . . Fortunately such cases occurred only rarely. Almost everywhere he was able to achieve something good through face-to-face negotiations, and to prevent tasteless developments. (Koch 1912b)

The passage underscores the importance that we must attribute to the recruitment of civil servants by the Heimatschutz, who in many cases were inspired to innovate independently in directions suggested by the movement. The environmental historian Klaus-Georg Wey saw here only a "certain limited resonance," but in fact the response was much stronger (Wey 1982, 134). This was certainly how some of the Heimatschutz's enemies saw the situation, who often complained that "Prussian civil servants in the area of building," for example, were giving disfigurement "an interpretation that goes far beyond the goals and intentions of the law" (PGStA I.HA Rep. 93B 1546). And as we shall see in chapter 3, directives issued by top civil servants also applied Heimatschutz thinking to the area of agriculture in ways not demanded by any legislation. It is in this extra margin of administrative *engagement*—"over and above the office norm and work rule" that we see one of the most powerful resonances of the Heimatschutz idea (Stein 1910, 462).

I think it is accurate to regard the quick uptake and relatively ambitious application of Heimatschutz ideas as a noteworthy achievement for the Wilhelmine era. If in other respects the political and administrative system was calcified and unresponsive, here it showed itself to be on top of modern developments; likewise, the active middle-class Heimatschützer give the lie to that image of "political apathy" that has seemed to characterize the Wilhelminian bourgeoisie. It is true, of course, that much of the educated middle class became frustrated with the inflexible mass political system in which it was losing ground; nevertheless it did not drop out of public life entirely, but rather found other fields of work in which to engage itself.[18] The bourgeois association offered itself as a splendid alternative structure.

Sociologists and historians have often commented on the strong role the association played in Germany's eighteenth- and nineteenth-century bourgeois emancipation; although technically a "private" institution, the bourgeois association was an important site of egalitarian social practice and universal thinking (Nipperdey 1976). These associations, it has been said, were a key part of the process of modernization, inspiring an eminently proto-political form of discourse that took as its reference point the needs of the social whole (Tenbruck and Ruopp 1983). In this respect it has been argued recently that Heimat organizations, too, were a key part of the "process of shaping a new public realm" (Applegate 1990, 107).

If we can conclude that associationism was one of the preferred ways for relatively well-educated bourgeois citizens to express their public *engagement,* it seems that the specifically regional or state form may also have been related to this social grouping. Much research into the Heimat movement presumes that the geographical boundaries were loaded with significance and that the Heimatschützer, too, were deeply involved in "constructing" and "anchoring" regional identities. To a certain extent this is true, of course, but on the other hand Heimatschützer may have merely retained a style of organization that was already embedded in their social habits. Professionals such as lawyers and doctors and a wide range of civil servants often situated their organizations at this level in order to conform to the varying laws and the patchwork political topography of Germany; municipal officials in Brandenburg, for example, gathered to discuss questions of city planning and administration at an annual "Cities' Conference" for Brandenburg. Since Heimatschutz concerns were tied into this very same level of state or regional guidelines, its corresponding geographic form can be seen as one conditioned to a considerable degree on simple expedience, not necessarily as the expression of an inner political ideal.[19]

Instead it is perhaps the associational form itself to which we should look for significance. Like a meeting within a profession, the Heimatschutz offered a unique opportunity for individuals who might otherwise work far apart from one another to consider their common goals and reflect upon the wider significance of their specialty for the population at large. The fact that there was no one profession that could claim to represent the broad vision of Heimat probably helped to achieve an even more open atmosphere in which specialized knowledge was perhaps less important than a common bourgeois cultural background and a certain degree of civic altruism. By uniting civil servants with a cross section of the interested middle class the Heimatschutz *Verein* deployed "that specifically *bildungsbürgerlich*-governmental communications milieu with extension into parts of the propertied bourgeoisie . . . that can be considered a decisive segment of the political culture of the *Kaiserreich*" (Bruch 1985, 85).

As we have seen, this *Bildungs- und Beamtenbürgertum* was deeply implicated in the process of modernization. Large numbers of architects and technical administrators swelled the ranks of the regional organizations; well-informed general administrators met here with concerned local teachers, and doctors came into contact with representatives of commerce. The association was a vitally important site for all of this: it constituted a uniquely flexible, adaptable, informal, permeable, and finally relatively effective forum for discussion. The large quantity of work that it managed to "influence" is in part a tribute to the highly effective transmittal of ideas and suggestions through a bureaucracy that, if it wanted to, could make use of its considerable administrative powers to carry out cultural change. The cooperation here between civil service and the interested middle class is a reminder that, in many respects, the bourgeoisie of the Wilhelmine period was better than its reputation. In an era of colonial tensions and manufactured election passions, the Heimatschutz revealed a sober, methodical, and yet highly idealistic segment of the German middle class that took the initiative in trying to adapt their system to the changing needs of industrial society.

One of the quickest ways to confirm this is to look at Heimatschutz efforts in the built environment. The Heimatschützer were critical observers of the cities they largely lived in; of that there can be no doubt. But as Andrew Lees argued already in 1975, "the positive proposals offered by these aesthetic critics . . . demonstrated their conviction that the big city could be improved. Indeed the objective pursued by many of them was to harmonize the practical imperatives arising from urban growth with the ideals of urban beauty which they had inherited from the past" (Lees 1975, 47; see also Otto 1983, 150).

In Saxony the Heimatschutz organization was deeply involved in questions of city planning and architecture. A huge number of architects were members, and connections to Saxony's governmental offices ensured that they could quickly go to work on public infrastructure jobs. Already in 1908 the association could say that it had "artistically influenced" "an extremely large number of buildings" (cf. *Heimatschutz* 1908, 92). But in Saxony the Heimatschutz involvement in city planning went far beyond an architectural, stylistic level to efforts that addressed the wider urban environment and the needs of the working class. With participation by almost all of Saxony's leading architects and a number of enlightened factory owners, cultural reformers, and top civil servants, Heimatschutz here was, in effect, one of the central forums for socially aware housing reform. The journal of the State Association carried a large number of articles devoted to all aspects of the housing question: Garden City ideas, building cooperatives, and municipal building inspections were topics touched upon in just one of the pieces ("Wohnungsfürsorge" 1911–12). With the help of the Saxon government the State Association soon established a separate office to help with these matters. The socially oriented planning concerns reflected in this "Central Office for Housing Provision" went so deep that shortly after the war the State Association was accused of "one-sided agitation for the cooperative building movement and land reform" (cf. L. F. Karl Schmidt, 1919; *Mitteilungen des Landesvereins Sächsischer Heimatschutz* 1919, endpaper; *Deutsche Bauzeitung* 1919, no. 57).

Elsewhere, too, Heimatschutz organizations were deeply involved in matters of city planning and construction. The Schleswig-Holsteinischer State Association in Kiel was neither the biggest nor the richest Heimatschutz organization, but according to a 1909 report it was able to exercise a considerable impact in its very first year of existence. In cooperation with the League of German Architects it supported the compilation and publication of a "Building Catechism" that showed how to avoid certain frequent aesthetic errors; more than 5,000 of the pamphlets had been distributed to public offices and private citizens by the end of the year. It also helped a community formulate local statutes against disfigurement, helped another town with the design of an administration building, and gave further construction advice to a large number of private citizens and public bodies. It obtained a hearing from the province's Minister of Public Works and from the Prussian railroad agency, extracting from both the promise to take Heimatschutz opinions into consideration in the construction of their buildings. Privately owned small-gauge railroads and the Imperial canal and pier administrations were equally happy to take Heimatschutz design advice. In the city of Kiel the Heimatschutz association was able to convince technical and planning officials to lay out new streets differently. The result was not only better conformity to the terrain, but also the preservation of small areas of landscape beauty and wooded gullies. As if all of this were not enough, at the end of its busy year the association was in the midst of plans to establish a permanent advisory body for architectural questions, a so-called Building Advisory Office (Schleswig-Holsteinischer, 1909).

The multiplicity of its individual projects has sometimes tended to hide the fact that the Heimatschutz was at the forefront of comprehensive planning. This meant general restraint in the development of industry as well as specific zoning controls. One observer noted the fruitlessness of a more limited, merely architectural kind of reform if one did not address the larger question simultaneously:

> Why should we build pleasant houses, if they will soon stand there blackened by smoke? How can there be joy at the building of charming, nice apartments as long as we must always have to anticipate the possibility that our new home will be joined by a blackish-red monster

> factory as its frame or background? Who will seriously think about caring for nature when he does not know what fate will threaten his well cared-for little place tomorrow? (Mugler 1911, 27)

As was typical for the Heimatschützer, this man did not put modern industry radically into question, saying only that it needed to be limited to specific areas. Once the population had been given the security of zoning controls, he said, "let industry spread out uninhibited in its given place, and the sight of huge smoking plants will not disturb us, we shall like it" (Mugler 1911, 28). The Heimatschutz association in Württemberg used a similarly pragmatic conception of zoning in requesting that a planned locomotive depot not be located down in the city of Stuttgart, but on an elevated section of land where the smoke would be more likely to dissipate. In this latter article a more ambitious environmental vision was mentioned: it called for "the complete elimination of gaseous pollution" as the "task of the future" ("Aus der Tätigkeit" 1910, 16). But the road to such a future started with immediate, pragmatic decisions. No doubt with an eye to the complete lack of an environmental consciousness in their surrounding society, Heimatschützer who wanted to influence planning decisions restrained themselves to what seemed possible. Given the overall situation at the time, it is clear that the ideas of zoning that they promoted were extremely perceptive.[20] Such ideas were realized in the 1915 city extension plan for Soest, the work of two Heimatschutz-oriented architects. The design by Paul Schmitthenner and Gustav Langen has been praised as a sensitive accommodation of different functions that was far ahead of its time; it reveals how the Heimatschutz concern for maintaining historic ambience could well combine with principles of zoning and Garden City–type housing (Ringbeck 1991, 231f.).

Zoning was, as we know, only one instrument in an array of means for dealing with the problems of growth. It was for one thing oriented more toward steering new growth than to changing or preserving preexisting distributions. Particularly in these early years of planning it was not well adapted to application to older and mixed-use neighborhoods. Related to this was the fact that city planners and engineers had traditionally not handled the problems of growth and change in an established setting very well. The importance of traffic flow had been a dogma of planners ever since Haussmann's mid-century work on Paris; many an old city center was ruthlessly ventilated with huge traffic arteries and grids as a result. Along with Camillo Sitte and the "aesthetic" planners of the 1890s, though, the Heimatschutz fought against the "expressionless monotony of modern city layouts" and for a more respectful brand of urban renewal (Wölfflin 1907). Their demands were not the "aesthetic foibles of a few antiquities buffs," as Cornelius Gurlitt explained, but rather concerned "an issue that has far-reaching implications for the contentment of all citizens of the cities." One of the key lessons to be learned, he continued, was "how to prevent . . . street straightenings and widenings from robbing our old cities of any more of their original charm than is absolutely necessary" (Gurlitt 1909–10, 12). Comments such as this show that the Heimatschutz was not about "big-city hating," but rather approached cities from the perspective of trying to retain the best of what old urban environments had to offer—*Altstadtfreundschaft* as opposed to *Großstadtfeindschaft*.

The practical result of the Heimatschutz's attention to ambience was a highly reflective approach to planning. A good example is provided by the Lustnauer Allee in Tübingen (for the following account see Lange 1911). This was a tree-lined boulevard near the old university that had begun to attract more and more development around the turn of the century. Municipal officials happily rubber-stamped a proposal to put lanes for vehicle traffic on either side of the old road, leaving the middle for pedestrians and equestrian pleasure riding. It was at this point that the Heimatschutz stepped in to challenge the scheme. It was bad enough, the group pointed out, that the island in the middle of traffic would put "pedestrians in the unpleasant position of having to breathe in dust from both sides." The even bigger problem was that lanes would have to be laid so close to the maple trees that severe pruning and probable death would result. This was also a fault with a quickly thrown-together modification of the plan, which wanted to use the middle as a higher-speed limited-access road.

Against all of these plans the Heimatschutz made the basic observation that the traffic on the Lustnauer Allee was hardly so heavy as to justify these huge increases in infrastructure. It was not the case that the Heimatschützer were stubbornly refusing all change. It was simply that they wanted to proceed in a reasoned manner, considering all of the alternative means. If this was "antimodernism," it was one that was perhaps more in tune with the technological possibilities than the knee-jerk modernizers of the day, for again it was the Heimatschutz that had to ask why none of the plans had "any consideration of an electric tram." Undergirding all of these arguments was the fundamental point of debate in Heimatschutz affairs: were the proposed changes in the interest of the majority? Lange argued that the Lustnauer Allee was a favorite site for many strollers and that the nearby university would also be served best if the neighborhood were to retain its peaceful residential character. Willy-nilly expansionism, on the other hand, was only in the interest of a few businessmen.[21]

Some of the most rancorous Heimatschutz battles had to with appropriate materials. In Württemberg the Heimatschutz association sent an appeal to the Ministry of the Interior in an effort to limit the spread of imitation quoin-stones pressed out of sheet metal. It was no doubt overblown to refer to the "great general significance of this question," but the Heimatschützer did claim to represent the vast majority of opinion in condemning the pieces. It was not the material in itself that disturbed them, they wrote; rather, it was the "fraudulent imitation of other building materials" upon which the "picturesque beauty of our villages" had been based (Bund für Heimatschutz 1910).

Aesthetic arguments such as these did not sit well with the industries whose products were being blacklisted. Their first response was to belittle the movement's "exaggerated sentimentality," but when laws and enforcement started taking effect the manufacturers began to fight back against the "dangerous aesthetics" of the Heimatschutz and its "ministerial accomplices" ("In Sachen" 1908; "Gefährliche Ästhetik" 1909). Industry clearly felt betrayed by a normally pliable government; "Saxon and Prussian building regulators in particular" were accused of "sympathizing with the most extreme Heimatschutz efforts" ("Gefährliche Ästhetik" 1909; "Heimatschutz, Baufach" 1910). In a 1910 counteroffensive the "Association of German Facing-Stone and Terracotta Manufacturers" slyly hinted that government was hurting its tax base by violating the unspoken free-market rule: never "interfere with the development and tax-paying power of industry" (brochure from March 1910, StA Dresden MdI 17526/142; see also Schimank 1983, 17f., on the "expropriation of the state"). But fearing, perhaps, that this argument might be too subtle, the businessmen soon cut to the immense economic impact that they claimed the Heimatschutz was having. Because the Heimatschutz laws had limited the use of certain of their products, they asserted,

> countless factories have had to stop their operations; those which are still running are doing so with losses; *many millions in investment capital have been lost, thousands of workers and foremen have been deprived of their daily bread.* (brochure from March 1910, StA Dresden MdI 17526/142; cf. also letter of 28 June 1910 to Prussian Ministry of Public Works in PGStA I.HA Rep. 93B 1546)

In a similar instance the tarpaper industry had come under pressure due to local Heimatschutz statutes enacted in Opladen. In a letter of May 1912 to the Prussian Ministry of Public Works the Association of German Tarpaper Manufacturers complained that the Heimatschutz was thus "hindering the economic development opportunities of a whole series of building materials industries" and even insisted that "whole industries" had already been "destroyed" (PGStA I.HA Rep. 93B 1546).

Of course it was easy to see through such exaggerations of immense impact: a government official defended Prussia's regulations in parliament by pointing out that the worth of certain controversial concrete products was "not significant in the least . . . when compared to the masses of cement" being used in huge projects all over Germany (HdA 21st legislative period 4th session 1911, Petitionsbericht no. 342). If an industry had actually lost money, though, the Heimatschützer still had a counterargument. Only in a narrow sense would this be of concern, they insisted; when the economy as a whole was considered, it was clear that what one industry did not do, another one would. With this macroeconomic perspective the Heimatschützer denied that they were fundamentally opposed to industry and its main goal of making money. Germans' entrepreneurial drive would survive the Heimatschutz, they observed dryly: "Soon enough it will learn to suck up honey from our flowers, to exploit our ideals commercially" (Hedemann-Heespen 1910, 88).

The Heimatschutz continued to have its critics, however. In determining which materials and forms were desirable it inevitably labeled other materials and forms undesirable; opponents were quick to denounce this as "police aesthetics" (*Kunstgendarmentum*) (Seeßelberg 1911, 61). Somewhat hypocritically, even industrial lobbyists complained of a "bureaucratization of our taste" (StA Dresden MdI 17513/153). Differences on this issue of artistic freedom led to a medium-sized cultural flap with the conservative *Werdandibund* and its leader, Berlin professor Friedrich Seeßelberg. In 1911 Seeßelberg accused the Heimatschutz of stultifying development; instead of artistic creation, he said, it wanted mere preservation and replication. Unfortunately for his credibility, his views were printed in a pamphlet that pretended to speak for the large League of German Architects, but that in reality, as the Heimatschutz found out, was unauthorized and paid for by a "Commission for the Elimination of the Excesses of Heimatschutz Efforts." This was nothing but a front for assorted industrial and cement interests (cf. PGStA Rep. 84a 10747; Verein deutscher 1912). Seeßelberg's warnings of "endangered building materials" rang a little hollow after this (Gurlitt 1911b).

The *Werdandibund* continued to press at the Heimatschutz, though: in 1912, supported by the tar industry, it arranged a competition for designs utilizing a flat roof, a style that the Heimatschutz often dismissed because it did not match well with its environment (cf. Mielke 1912; Pommer 1983). And in 1913 the *Werdandibund,* again relying on strong industry support, erected its modernist Werdandihalle, designed by Max Taut, at the Building Trade Exhibition in Leipzig.[22] They and others still complained that the Heimatschützer were totally absorbed in the past and were not open to new designs. It was such a common accusation that Heimatschützer were constantly having to preface their talks by debunking the impression that "the Heimatschutz aims unconditionally at the preservation of every old pile of junk" (Gaebel 1914, 8). Robert Mielke allowed that the critics had a point, noting that some Heimatschützer were schematically opposed to new ideas; he denounced this tendency, though, saying that in the end it was dishonest and counterproductive. Mielke went on to sum up the position of the Heimatschutz vis-à-vis new architecture in two points. The first requirement was that it fit well into its environment, whether that be a country landscape or a historic ensemble. This did not mean copying every detail, but it did entail an artistic taking-into-consideration of the building's surroundings. The second requirement was that the new work display "an honest sense of workmanship" (*eine aufrichtige Baugesinnung*). In the direct tradition of Ruskin this meant that form and function should be straightforwardly related and that the materials should be both solid and unselfconsciously worked—"quality" production in the sense of the German Work League (*Deutscher Werkbund*) (Mielke 1913; see also Frank 1992).

All of these instances make it clear that the Heimatschutz was deeply involved in city planning in the years before World War I.[23] But perhaps more important than any one intervention into current affairs was its more diffuse effect of increasing the public's level of interest. By 1911 Hermann Muthesius could not help noting the positive impact that the movement had had on architectural matters: it was if a spell had been broken, he said, and "the magic word that dissolved the apathy is called Heimatschutz" (cited in Frank 1992, 115). The Heimatschutz's accessible portrayal of issues was similarly seen as an important impetus behind Berlin's new regional planning council and the parallel rise in popular support for planning (Mielke 1913a, 54). In Saxony there was a strong feeling that the young discipline of city planning had been given a great boost by the Heimatschutz:

> While city planning and the other branches of architecture in their respective developmental phases used to be foreign to the layperson, it remained for the Heimatschutz to introduce these ideas to broad segments of the population, and thus to bring about a work of culture that will only later be able to be appreciated. (Hager 1913, 206)

The obvious success of the program of Heimatschutz in this regard was closely related to its aesthetic emphasis. Unlike the far-off, abstract, and administered world of the planners, a world that could be aesthetically perceived was direct, personal, and above all admitted of an uncomplicated scheme of evaluation. It allowed individuals to place the changes made to their environment in a context that said that it was their right to make sure the changes were tolerable and even pleasant. With this Heimatschutz attitude the average citizen was suddenly empowered to call for more parks or less invasive road and street construction; it was now his or her right to get involved in advertising and architecture, too, and even water and air pollution were open to a straightforward aesthetic challenge (examples are Wedding 1910; Rieß 1911–12; Haupt 1911–12).[24]

Part and parcel of this newly self-assertive Heimat attitude was a critical relationship to the state. In theory, of course, many Heimatschützer expected the state to watch out for the interests of the public; in practice they soon learned to turn a skeptical eye toward the centers of administrative power. Land, in the German administrative system of the time, was typically under the control of a public body's finance department; it was thus not only managed for maximum profit, but also vulnerable to being privatized in order to ease short-term cash flow problems (on cities see Ladd 1990, 200). Prussia's Finance Ministry was particularly notorious on this score. In a 1911 speech Carl Johannes Fuchs took the ministry to task for selling off approximately fourteen percent of its forests closest to Berlin in a space of just ten years. It was hypocritical, he implied, that a government could do this on the one hand, and on the other profess its commitment to housing reform and access to green space (Fuchs 1917, 117–18). In another speech Fuchs suggested that governments were often giving lip service and minor support to the Heimatschutz movement as an alibi for their continued environmental neglect (Fuchs 1912b, 104). Perhaps the best example of the Heimatschutz attitude, though, comes from the case of a new railroad line to be laid out on the west bank of the Havel river near Berlin. The Brandenburg Heimatschutz association argued for a placement of the railroad some distance away from the scenic banks and the simultaneous presentation of a long-term development plan that would assure the preservation of some wooded areas. Noting the government's predilection for "secrecy" in such matters, the Heimatschutz demanded that it "submit the designs for bridges, the building maps, etc. *to the public for comment and criticism before they are finalized*" (emphasis in original; "Schutz den Havelufern!" 1910, 220). With this emphatic defense of the environment as a "public issue" the Heimatschutz gave decisive legitimation to the idea of popular input, and thus deserves to be seen as one of the first steps in Germany toward a democratization of modern community and regional planning ("Schutz den Havelufern!" 1910, 220).

The above examination of the Heimatschutz associational structure as well as its views on architecture and planning has demonstrated that the movement was thoroughly integrated into the social and technological life of turn-of-the-century Germany. Far from standing sullenly apart while the modern world floated on by in the stream of its developments, the Heimatschützer collectively jumped in and attempted to help negotiate a path between the rocks they saw looming on all sides. This is an extremely basic point, but given years of scholarly misunderstanding, it bears repeating. Fortunately the last few years have brought about a growing acceptance of the movement's essential modernity. Andreas Knaut's Munich dissertation, one of the last supervised by Nipperdey, is a fair and balanced portrait of the movement that refuses to pigeonhole the Heimatschutz in the "ghetto of a merely authoritarian organization" and that correctly acknowledges it to have been "an act of emancipation of the bourgeoisie, a sign of its distancing and its independence" from the hierarchical state (Knaut 1993, 436–37). Knaut's sympathy with the environmental aims of the movement is also apparent. Nevertheless, at the level of theory it is obvious that he tends to want to fit the Heimatschutz into Nipperdey's category of a "moderate modernity" (Nipperdey 1988, 75). This is not as gross a distortion as saying that the movement was reactionary, but it effectively assimilates the Heimatschutz into another camp in which it does not belong, that of modernity. According to this view the Heimatschutz constituted one of the era's typically hybrid phenomena: the Heimatschützer were simply part of an incompletely modernized bourgeoisie, one that was still struggling to emerge from its compromises with preindustrial feudalism. At the very least this model suggests that the movement would have done better if it had only been more "modern." More seriously, though, it can leave the impression that the movement's environmentalism was essentially connected with an outmoded historical position.

The tendency to take refuge in the old certainties of modernization theory emerges much more clearly in work done on Heimatschutz architecture by Hartmut Frank (1992) and Matthew Jefferies (1995). To their credit, they, too, are willing to dispense with a number of prejudices about the movement and give sympathetic accounts of its efforts. Nonetheless Frank's words give one pause when one reads that the Heimatschutz was "working toward technical progress" and "modernization" and was clearly involved in the "project of modernity" (Frank 1992, 117, 116, 119). Jefferies exhibits the same tendency even more strongly, dismissing "rabid anti-modernists," "völkish mysticism," and "turn-of-the-century prophets of doom" in favor of "practical, pragmatic" "modernity" (Jefferies 1995, 60, 98, 99). What seems to be going on in this very latest round of critical literature is a conscious effort to make the Heimatschutz a part of our familiar "modern" life and thus to deprive it of its original critical edge against the system of modernity.

If we return for a moment to the important work by Blackbourn and Eley we can see the potential for a different interpretation. The bourgeois agenda that they identify was neither limited to nor necessarily dependent on certain liberal rights or classically democratic political structures; instead, what they see as having been centrally important was the bourgeois domination of the economic realm. What was truly "modern" about the *Kaiserreich* was its free trade, the way that unrestricted disposition of property was instituted, and the fact that even the most stolid *Junker* could not avoid the growing influence of bourgeois manufacturing, processing, and banking ventures. Indeed, as Eley argues, if we look closely at Wilhelmine Germany we will see much the same sort of bourgeois "hegemony" that held sway in other European countries, one that thrived despite the continuing political dominance of Germany's nobility (Eley 1984, 135).

Accepting this notion of a subpolitical realm of bourgeois hegemony helps throw a decisive new light on the activities of the Heimatschutz. This is so for two main reasons. For one, it becomes necessary to revaluate the entire array of anticapitalist and antimodernist sentiments expressed by the organization. These were dismissed all too easily by earlier critics as a kind of quixotic maladjustment to the times, or a woolly-headed attempt to "defy the contemporary developmental tendencies" (Bergmann 1970, 134). Even newer work remains indebted to the same idea of a partial modernity, whereby it is the modern and "realistic" part of the Heimatschutz that is valorized. But the Heimatschutz was not trying to stave off the arrival of capitalism: capitalism had long since arrived in Germany and was a fully entrenched force. Nor does it go to the heart of the matter to see the Heimatschützer as trying to play catch-up with the modern world, heroically, if only partially, casting off their allegiance to the old order and embracing the brave new world of bourgeois dominance. For as these critical members of the middle class saw it, the fundamental problem was not *not enough Bürgerlichkeit,* but decidedly *too much.* Everywhere they looked they found that private, laissez-faire economic activity determined the "realities" and "technical necessities" that functioned to legitimate environmental transformations. Their resistance against "the excesses of modern capitalist development" came from within (Fuchs 1905, 6). This knowledgeable and reflective resistance is thus one that can never be understood adequately or fairly within the terms of a theory that measures a group's "progress" only in terms of its conformity to the ideals of modernization.

The notion of bourgeois economic hegemony further helps illuminate the particular organizational and tactical dimensions of the Heimatschutz as it tried to institute its insights. What it allows us to see above all is that the proto-environmentalist protection of aesthetic ensembles in the Wilhelmine era required a consistent process of cultural and political self-overcoming from its bourgeois proponents who, as a group, increasingly had to leave the comfortable fold of their class. Many Heimatschützer probably did not realize the extent of the alienation involved. But to some at least it was apparent that the idea of Heimat would take German society toward a truly fundamental break with the obtaining Wilhelminian-capitalist order. To study the Heimatschutz in relation to bourgeois hegemony is to see that this journey was difficult, and that environmentalism required considerable effort in the sphere of cultural politics. In order to articulate and popularize their cause the Heimatschützer ultimately had to resort to alternative discourses, ones either neglected or trivialized by the society at large. With respect to the environment it was the aesthetic discourse that allowed them to communicate their new priorities best, a strategy that will be examined in greater depth in chapter 3.

In order to legitimate and explicate its particular social vision, meanwhile, the Heimatschutz looked to ideas of the nation. It must be warned at the outset that critics have often been misled by the reliance on national imagery and, misunderstanding its intent, have all too schematically plugged it into the familiar picture of a rabidly militarist German society. It does not conflict with the present interpretation to grant the partial accuracy of that picture, and, with the Wehler school, to be concerned about the uses to which nationalism was put. Because of this presumption, in the following we will take pains to distinguish the precise meaning of Heimatschutz national references, which almost always were used to summon up the idea of an ethical community of the sort necessary to ensure social and environmental harmony. In the long run, of course, German environmentalism might have developed more smoothly and with fewer political derailments if the Heimatschutz had not partaken in the endemic national rhetoric. One thing that the idea of bourgeois hegemony lets us see, though, is that the particular historical constellation left middle-class Heimatschützer almost nowhere else to turn.

The idea of Heimat, after all, required some of sort of recognition on the part of an individual that he or she was part of a collective: like the environment itself, aesthetic ensembles can only be maintained or changed in a socially acceptable fashion if the individual can be persuaded to give up his or her right to absolute self-determination. The highly individualistic system of capitalism offered very little in the way of such a rationale. Socialism would perhaps have offered a congenial framework for such a project, but most bourgeois Heimatschützer rejected this option due to ingrained political and social habits. The Heimatschutz also needed to tap into a system of value to which a large number of people could immediately relate: something that was familiar and even powerful, but that could also be given a critical thrust against bourgeois economic hegemony. All of these criteria were filled by the idea of the nation.

But as we saw in the discussion of Rudorff, not just any idea of the nation would do; in particular it would not be the mindless and opportunistic nationalism of a Diedrich Heßling. "Here is not national phrase, here is national deed," as one journalist close to the movement put it (Meißner 1908, 317). But of course it is the direction and meaning of this sort of applied "nationalism" that counts. In the following we shall see that practical Heimatschutz work was embedded in a fairly specific deployment of the idea of the nation, one whose central function was to serve as a counterweight to the highly individualistic and remarkably short-term ethics of nineteenth-century capitalism. Not surprisingly, one of the most important and productive applications of the national-communitarian ideal came in the area of economics.

As is true of the Heimatschutz movement at many different levels, it was the aesthetic dimension that started things off and provided a basis for a more deeply probing consideration of this area. As the discussions around the 1907 Disfigurement Law showed, the demand for control of a town- or landscape resulted in an immediate outcry by property owners. Some opponents of the measure were no doubt dismayed at the prospect of having to forgo calculated profits; still others, however, saw that a principle was at stake. For them, having anything less than complete control was "an intervention into private property such as our legal system has never before known nor attempted" ("Die Verunstaltung" 1907). The Heimatschützer Cornelius Gurlitt tried to take the sting out of this idea by portraying it all as a question of better protecting individual rights. Disfigurement laws would prevent one from being subjected to the "aesthetic compulsion" of someone else's incompetence, he argued; all they would do would be to guarantee "that our eye be protected from insult" (Gurlitt 1910–11, 14, 13). By codifying the "right of a neighbor to protection of his property . . . from aesthetic damage," such Heimatschutz measures would actually strengthen the idea of private property (Gurlitt 1909, 158).

But the finely tuned senses of the property owners' associations were probably more accurate than Gurlitt's own. The "Heimatschutz Law," as it became widely known, did indeed deprive them of a right that they had earlier enjoyed, and moreover it constituted the thin end of a wedge that would ultimately split apart the absolute conception of property and of economic activity. In the eighteenth century Kant had given one of this idea's classic formulations in advising the "monarch" to let his subjects do "the best they can to define and promote their salvation" (Kant 1970, 58). Indeed, the German laws regulating commerce came to approximate this injunction to a high degree. But by the beginning of the twentieth century many were beginning to realize that the principle of noninterference simply did not work. And it was in aesthetic matters that the principle found its first widespread negation. Here a new sense of community rights to the environment would be built up and eventually enforced against builders who ruined the historical character of cities, manufacturers who polluted rivers, advertisers who sullied landscapes, and all the others who, inadvertently or not, contributed to a decline of living quality for their neighbors in pursuit of their own particular version of salvation. Almost any private activity had an array of community impacts, the Heimatschutz emphasized, and thus any large project in particular could require a "very careful weighing of the conflicting interests" (Fuchs 1905, 19).

The formulation stems from a man who understood the manifold effects of private economic activity on the environment perhaps better than any other observer of the time. Carl Johannes Fuchs was an economist at the University of Tübingen and one of the most prominent and stalwart adherents of the *Bund Heimatschutz.* In and of itself the participation of an economist is worth noting, since many "ideological critiques" of the Heimatschutz continue to depict it as a movement full of muddle-headed aesthetes. Paul Schultze-Naumburg had to combat similar impressions all the way back in 1905, and it is interesting to see that he, too, turned to Fuchs for help. Putting together a petition against the Laufenburg dam, Schultze-Naumburg asked Fuchs for a signature. His letter continued with a small request: "please, put your 'Professor of Economics' under your name so that people see that we're not just a bunch of dreamy-eyed romantics" (UB Tübingen, Nachlaß Fuchs, MD 876/39).

Fuchs's participation helps locate Heimatschutz concern with community impacts in a much larger context of national thinking, namely that of the "Historical-Ethical School" of German economics. As with most "schools" the unity of this group is often questioned, but it generally includes an 1840s generation around Wilhelm Roscher and Karl Knies along with a younger, 1870s generation that featured Gustav Schmoller, Adolph Wagner, and Lujo Brentano—who in turn was the mentor of Fuchs. What linked all of these men into one "school" was their rejection of the atomizing and pauperizing tendencies inherent in free-market, Manchesterian capitalism. Instead of looking at economics as a science of prices and individual profits, these German *National-Ökonomen* preferred to approach it as an area of national social policy, one where the individual's desire for profit had to be balanced against the historically formed collective's interest in stability and continuity. This was anything but a value-free undertaking, of course, and it comes as no surprise that these tenured civil servants of the German state were at pains not to seriously challenge the existing political structure with their suggestions for reform.

That said, it cannot be denied that the school's strong emphasis on social ethics had decisively beneficial effects. Its proponents Schmoller, Wagner, and Brentano all figured prominently in the 1872 founding of the Association for Social Policy (*Verein für Sozialpolitik*). Especially in its early years the Association marked the leftmost bourgeois position on social matters, particularly with its fundamental demand that workers be given a bigger piece of the societal pie (Bruch 1985). A hostile press immediately branded the professors "Socialists of the Chair" (*Kathedersozialisten*), reminding us that this was already an era in which free-market theories enjoyed an unbroken hegemony. Brentano later recalled that laws regulating factory conditions, for instance, "were seen as an outrageous surrender of a citizen's freedom to arbitrary police powers of the absolutist order" (Brentano 1931, 73). Against vulgar capitalist convictions of this sort the economists of the Historical-Ethical school set their distinctive brand of socially reformed nationalism: all citizens deserved the foundations of a satisfying material life, they argued, and workers in particular needed to be protected from structural inequalities in the capitalist labor market. Despite its statist thrust, Bismarck's accident and old age insurance can be considered an early victory for this line of national-ethical thinking. Over the course of the next forty years members of the Association for Social Policy managed to occupy several key university chairs; in combination with their direct public influence, their remarkably well-attended lectures created an important pool of support for this specifically social nationalism precisely among young civil servants, many of whom were required to study economics.[25]

One innovation of this younger generation was to begin to apply national-ethical principles in a series of reform projects that would eventually extend to the environment itself. Like his teacher Brentano, Carl Johannes Fuchs was a member of the Association for Social Policy and made his early reputation as a proponent of "sweeping reforms" in the area of housing (Fuchs 1904, ix; on Fuchs's comparatively liberal position in the field see Müßiggang 1968; Krüger 1988). But already in the first days of the *Bund Heimatschutz* Fuchs showed how national-communitarian economic thinking could also be applied to the environment.

The case involved a proposed dam on the Rhine near Laufenburg in Baden. The dam promised to and ultimately did destroy a highly scenic series of rapids, "one of the most beautiful landscape scenes in Germany, or indeed the world," according to one of the Heimatschutz petitions

("Laufenburg" 1904–5). Although a series of decisions in favor of the dam had been made by Baden state officials prior to the founding of the *Bund Heimatschutz* in 1904, the fledgling organization nevertheless put tremendous effort into its attempts to stop the project. It was a textbook case of a "plan by Big Capital" clashing with advocates of natural preservation or restraint ("Die Laufenburger Stromschnellen" 1904–5). The economist Fuchs became deeply involved with the dispute, which was taking place not far from his home in Tübingen. He made it clear that he took the claims being made for the dam seriously: that it could potentially inspire industry to locate in the economically weak area, that it could provide jobs, and so on. Nevertheless these gains had to be measured against the overall gain or loss of the wider community. In the case of Laufenburg, Fuchs argued that the dam would hurt aesthetic and economic benefits associated with the area's natural beauty and would also have economic and social impacts because of its destruction of salmon fishing trade.

Fuchs thus explicitly indicated several different variables in setting up a kind of cost-benefit equation for the project. A major factor was of course the irreplaceable and essentially unquantifiable aesthetic worth of the rapids. But given wider society's focus on short-term economic issues, Fuchs also argued the project on this level, pointing out that it was not at all clear whether it would result in a net economic gain. In order to analyze this properly, he said, one must weigh the promised benefits of the dam against the value of non-dam tourism and the current fisheries. Tourism alone promised "a slower but more secure upswing" in the region's economy (Fuchs 1905, 10). Fuchs also implied that the state and the municipalities concerned were vastly underestimating the value of their fisheries, since they were ready to accept cash settlements equivalent to only about ten or twenty years' worth of licensing revenues.[26]

The dam's promised employment opportunities also had to be weighed against its effect of destroying other employment. This was not just a matter of marks and pfennigs. Fuchs pointed out that the local government's acceptance of a monetary settlement would ruin the social and economic position of the fishermen:

> For them it is the destruction of what until now has been their employment opportunity, and it is no replacement for those affected that a bigger new opportunity will be created in new industries in which they either will not or cannot work without sinking socially. (Fuchs 1905, 20)

Fuchs was referring to the real phenomenon of social *déclassement* that threatened displaced independent workers in Wilhelmine Germany. Far from being a question of revenue alone, the dam now had to be seen in terms of its social-psychological impact on the indigenous population. As in the area of housing, Fuchs found "this psychological side" to be "of the very greatest importance" (Fuchs 1904, 14).

According to the Heimatschutz, then, any truly responsible decision would take many factors—financial, aesthetic, social, and psychological—into account and find an optimum balance of the competing interests. In contrast to the dictates of laissez-faire economic theory, this optimum balance might well necessitate a limit to the dam's profit-making potential. Once again the basic precepts of the historical-ethical school came through, for in planning as in "national economics . . . the task is to set boundaries for the pursuit of private economic interest" (Fuchs 1912b, 99). Unfortunately, the boundary-setting and plan-weighing instance was the state of Baden, the same state that had already approved much of the preliminary dam proposal. Somewhat naively in this case, perhaps, the Heimatschutz saw its role as that of forcing the state of Baden to carry out its ideal "duty" of examining alternatives. The *Bund Heimatschutz* thus went to great effort and engaged a leading expert on dams and tunnel construction, who testified that an alternative solution involving no damming of the rapids was possible (cf. "Gutachten" 1905–6; Linse 1988).

But this advice was ignored. A regional advisory board and then the Baden government approved the dam project in short order and let the rapids be obliterated. It was a betrayal of the ethical principles and the hopes that the Heimatschutz had invested in the state, and as such, perhaps, a healthy lesson. But even given the go-ahead for the project, Fuchs had one more argument. Along with the contemporary land reform movement, he demanded that the conversion or harnessing of a socially held resource should benefit only "the general public, not a minority of private persons, of stockholders" (Fuchs 1905, 21). Fuchs's demand reveals that there was a potentially powerful sense of egalitarianism at the base of the Heimatschutz thinking; it is perhaps no accident that it found its farthest-reaching expression precisely in relation to nature and the landscape.

All in all, if we overlook the concrete dimensions of the defeat at Laufenburg we can see that some good came of it, particularly as the incubator of a more critical brand of economic thinking. If in the succeeding years people still accused the Heimatschutz of being "against economic progress," the Heimatschützer responded at a fundamental level by noting that "economic" and "progress" were terms that still needed to be defined (e.g., Mielke 1913b, 76). Taking up the central concern of the national-ethical school, they argued that a fundamental distinction needed to be made between private and public understandings of what is "economic." In an important speech defending Heimatschutz laws, Saxony's Count

Vitzthum von Eckstedt saw this single word as the key to Heimatschutz thinking. An industrialist who complains that environmental protection cannot be afforded, he said,

> is trying to place the conceptual distinction between economic and uneconomic in a place other than the one at which the necessities of modern life force us to put it. From the narrowest egoistic perspective he considers economic to be that which brings the individual an immediate monetary profit, even when that is accompanied by a much greater harm for others and for succeeding generations. ("Industrie und Heimatschutz" 1914–15, 7)

Heimat economics were thus fundamentally predicated on looking beyond the individual, the present, and monetary. They addressed all those factors that had traditionally (and conveniently) been considered "external" in the field of economics; now, it seemed, it was questionable whether there was anything "outside" the economic process that we could afford to ignore.

By calling into question the prevailing division of the economy into private and public spheres, the Heimatschutz mounted a highly important attack on one of the fundamental legitimating discourses of modern capitalism. The centrality of the issue can be appreciated in the fact that this is where many Heimatschützer chose to roll in the heaviest cultural weapon at their disposal: the nation. As we have seen, an invocation of the national community against narrow individual interests was part of the original strategy of the Historical-Ethical School, and showed itself to be a valuable rhetorical tool in limiting exploitation of the environment. A related appeal to national thought appeared in 1910, penned by Heimatschutz member Richard Haupt, an upper form teacher in Schleswig-Holstein and also that province's Conservator. His pamphlet "On the Preservation of Footpaths and the Right of the Nation to Its Forest" portrayed the public trails question as the dividing line between two entire philosophies of law and community. On the one hand he saw "Germanic law," which granted extensive and fundamental rights only to the community, and on the other "Roman law," a system in which only individuals and "selfish principles" counted. Two different ways of treating property in land in particular are contrasted. Ancient Germanic law, according to Haupt, provided for the unrestricted passage of individuals over land, every piece of which was "inalienable property of the whole nation" and that enjoyed unique status as the result of being an irreplaceable resource (Haupt 1910, 4).[27] Under this system the individual had only temporary usage privileges and was held to community standards in his treatment of the land. But with the dawning of the modern world

Roman law took hold, allowing an individual to completely own and dispose of property at will. This development culminated, for Haupt, in the Prussian Forest and Hunting Law of 1880, in which the traditional legal right of citizens to walk and gather in privately owned wooded areas was limited for the first time. Ever since then, land had been increasingly removed from social use and chopped up into little parcels of private domain, inducing in Haupt a kind of claustrophobia:

> That people just put up with it is a remarkable feature of a Volk that praises Arminius as the Liberator and wallows in pride with respect to its *Bildung.* It is fantastic indeed, they even have the water blocked off. There are actually lakes where even driving up to them is against the rules; the idea of some fellow bathing or wading is "strictly prohibited." The goal is the fencing-off of our entire fatherland. (Haupt 1910, 12)

The concrete problem that exercised Haupt was the ongoing restriction of traditional footpaths. In his native province of Schleswig-Holstein he documented 377 paths that were, in bureaucratese, "moved" (*verlegt*) between the years of 1869 and 1905, that is, destroyed. In addition to these there were 476 "changes" listed, which, Haupt asserted, often amounted to the same thing. Adjoining property owners were typically responsible for the actions, wishing to secure their boundaries and their sense of property. Footpaths, to them, were "a soiling of the pure thought of the inviolability and exclusivity of individual property."

Haupt attempted to show how wrong they were by reconstructing the historical sequence. "The paths are much older than the individual ownership of fields," he argued, since only later did the individual properties come into being. Thus the path did not, as was commonly thought, cross anyone's field; rather, it would be more accurate to think of the field having obstructed the path at the time of some acquisition (Haupt 1910, 13). While one can quibble with Haupt's construct of an old Germanic system of paths (surely not all of them were that old!), it is clear that the underlying thought shares much with the national economic ideals discussed above. In fact it represents the concrete geographic embodiment of those ideals: a network of socially held resources (paths and public ways) forms the framework within which private economic activity (a house lot or productive acreage) is to be allowed. The model also helps demonstrate the origin of the evils of the present capitalist system: namely, in the illegitimate extension of private economic concepts onto public domain. Haupt also noted that the damage done here was not abstract, but rather impacted precisely those members of society with the least opportunity to avail themselves of recourse: "the users . . . are in the main poor people who do not even dare to open their mouths" (Haupt 1910, 12).

Haupt's appeal to Old Germanic law may strike us as anachronistic or even eccentric, but the important fact of the matter remains that he was applying it in service of a well-intentioned and closely reasoned attack on the prevailing habits of landowners. Paul Weber, a professor of art history at the University of Jena, took up the same set of distinctions in arguing that the present reliance on "Roman" law was a drawback precisely because it allowed the individual an "unlimited right of disposal over his property" (Weber 1910–11, 226). As he saw it, though, the Heimatschutz was on the leading edge of a great cultural-legal movement that would restore Germanic principles to the economic realm. What he meant by this was "the new, social belief"

> that feels it as an injustice to the general public when an individual mercilessly destroys the beauty of a previously harmonious streetscape, a square, or an entire town, or when the town's surroundings are ruined by egoistic exploitation. (Weber 1910–11, 226)

These references to a Germanic tradition are interesting, not so much because they conclusively prove anything about the past, but rather because they show how hard some Heimatschützer were looking for alternative discourses on the environment, points of cultural purchase that might allow them to challenge the hegemonic bourgeois order. We also see how some middle-class reformers were drawn toward instrumentalizing contrastive national ideas in particular, though. These could have beneficial effects, of course; nevertheless a contrastive celebration of "Germanic" qualities could all too easily reinforce chauvinistic stereotypes. Here, perhaps more than anywhere else, it is therefore necessary to take into account the use to which national rhetoric was put.

We are fortunate in that we can refer to a much wider context of national rhetoric within the Heimatschutz movement in order to understand the intent of such remarks. The overwhelming tendency here was to summon up the unifying spirit of nationhood in order to combat concrete environmental and social drawbacks of capitalism. The economist and Heimatschützer Werner Sombart backed up the sense of urgency with his warning that

> Society cannot withstand capitalism without protective measures. The power that resides in capitalist forms of organization and that is called forth by the uninhibited drive to acquire is one that is so mon-

> strous, that it pulverizes land and people, culture and morality, everything around it, wherever it operates freely. (Sombart 1904, 118)

This description of capitalism's destructive powers reminds one immediately of the very similar analysis contained in the *Communist Manifesto;* unlike many Socialists of the time, however, Heimatschutz supporters placed no special hope in the eventual curative effect of system-immanent or even politically revolutionary forces.

More than anything else that separated the two groups, this difference was probably based on the bourgeois economists' profound disgust at capitalism's liquidation of a many-dimensioned historical inheritance. This led them methodologically away from explanations based primarily on wages and class interests, factors that did not seem to capture adequately the senselessness of the rampage nor its uncritical acceptance by so many. Sombart theorized that the capitalist "system" had managed to anchor itself in people's practical consciousness; this "economic rationality" functioned "like an invisible spirit" to drive the ever more intensive cycle of profit and production to levels that no longer bore any relation to the real, material requirements of "the living person" (Sombart 1920, 447, 423). In the entrepreneur this false version of rationality culminated in a pathological state of destructiveness. Sombart saw the profit-driven capitalist as a loose cannon on the deck of society, an individual of impoverished abstraction who has lost touch with the concrete social and environmental dimensions of his home: "For the entrepreneur, the Heimat becomes like a foreign country (*die Fremde*). Nature, art, literature, government, friends: everything disappears into a mysterious void for him" (Sombart 1920, 454).

This description from 1913 tells us much about the interrelated ideas of community and economy that drove the Heimatschutz movement. Heimat is associated with a broad range of noneconomic ties to the wider human and natural environment: to have feelings for one's homeland means to pay attention to the art and literature around one, to be aware of politics, to not ignore the surrounding nature. Critics might argue, of course, that behind all of these more or less positive values there lurks a strategy of "negative integration": Heimat is opposed to *die Fremde* and thus seems to imply a chauvinistic attitude.

Such objections lose their justification, though, if we look at the specific function of this opposition. It is the same one that we find in the opening lines of the Call for the Founding of a *Bund Heimatschutz,* the 1903 document that started the movement. Playing on the contrast of Heimat and foreignness, Ernst Rudorff opened this first public appeal with the following words:

> Homeland protection is our demand!—Foreign invaders are not what we fear; all the more so our native vandals. Since the founding of the new German Empire it has become more common than ever before to speak of "German interests," "patriotic undertakings," and similar slogans; but the homeland itself, our German land, the fertile ground of all our customs—it can be brazenly degraded, raped, disfigured. (59)

With exceptional clarity these lines claim national feeling for the program of Heimat at the same time that they attempt to redefine what patriotism is really all about: namely, it is implied, a respectful stewardship of land and customs. Nevertheless this positive project does go hand in hand with the naming of an enemy; thus we do see a strategy of "negative integration." Decisively, though, this is not the negative integration about which most narratives of Wilhelmine history are worried. The enemy is not the working class, for instance, nor is it another race. And as is explicitly stated, neither is it some foreign power. All the fuss about a bigger navy was misplaced, according to the Heimatschützer. In their eyes Germany was threatened not by an invading army, but by "native vandals"—a homegrown class of capitalists who had become more thoroughly indifferent to it than any distant culture could.

Capitalism as danger: this basic thought resounded through many of the more critical reform movements of the turn of the century. Many thematized capitalism's effects on the working class, but it remained for the Heimatschutz to point out the additional connection between capitalism and the environment. No one argued the point better than the economist Fuchs, who saw that workers and nature alike needed protection from laissez-faire policies:

> We have to get rid of a leftover from the period of economic liberalism: the view, namely, that capitalism has to and ought to take over everywhere without limit. Just as we have used worker-protection laws for some time now to set limits to its exploitation of human productive forces, of *man,* so, too, must we accept the principle of setting limits to capitalism's exploitation of *nature.* (Fuchs 1906a, 93)

Werner Sombart agreed with Fuchs on the necessity of limits and envisioned a comprehensive package of "protective measures . . . in the form of worker-protection laws, Heimatschutz laws and so forth" (Sombart 1920, 462).

While these quotations emphasize the parallel nature of the problems in two distinctly different areas, the reformers who gathered around the banner of Heimatschutz saw an increasingly tight connection between social and environmental realms. Heimatschutz was "a piece of social policy," they insisted, noting that it was "in the social interest of the whole that every person possesses a right to a full enjoyment of nature and the landscape, a right that may not arbitrarily be ignored or taken away" (Wirminghaus 1913, 179).

Their standpoint here was virtually identical with that of smaller bourgeois reform groups such as the Garden City and "People's Parks" (*Volkspark*) movements; although each organization concentrated on a different area, all sought to supply workers needed compensation in the form of aesthetic and recreational opportunities in nature. The land reform (*Bodenreform*) movement, too, combined anticapitalist thinking with a desire to promote contact with the land; finally Germany's growing cadre of professional garden architects should be mentioned, a group that worked closely with the Heimatschutz in an attempt to guarantee access to green spaces across the face of the nation.

The overlaps between these groups' respective work areas and that of the Heimatschutz are confirmed by multiply overlapping memberships: Fuchs and Sombart were both members of the German Garden City Society, for example; Karl Keller shared contacts to the Garden City movement and was active in the land reform movement along with other Heimatschützer such as Heinrich Erman; Arthur Glogau was an important link between the Heimatschutz and professional garden architects; his colleague Ludwig Lesser, finally, was not only a moving force behind the People's Parks, but was also involved in Garden City building and Heimatschutz issues. The overriding "green reform" conception behind all of these movements was one that the Heimatschutz had pioneered: a recognition that a worker's immediate environment had a profound impact on his or her sense of well-being, and a genuine commitment to an egalitarian minimum standard of living. Fuchs put it best with the simplest of words: the job was to provide everyone in Germany with a "home" (*Heim*) (cf. Fuchs 1908; Fuchs 1911–12).

In terms of its overall class implications this effort reflects that sort of bourgeois social reform theory that Paul Boyer, looking at the American context, has labeled "positive environmentalism" (Boyer 1978, 220f.).[28] In the main what is meant here is the Schillerian strategy outlined in chapter 1: assuming a decisive impact of a person's environment on his or her character, it recommends the thoroughgoing improvement of daily living and working conditions as the most practical path to overall societal betterment. But with upper classes deciding what is "good" for an individual of the lower class, this kind of environmental reform has long had the reputation of being an insidious type of social engineering. It cannot be denied that the Heimatschutz occasionally revealed a similarly manipulative ambition: one representative of the movement in Lower Saxony, for instance, promoted it out of the perceived necessity of "spiritually chaining our Volk to the Heimat again" (Kettler 1906, 32). Even Carl Johannes Fuchs, in his early days, subscribed to a similar strategy of keeping workers in their place. In an 1899 assessment of building cooperatives he saw the projects' value precisely in the fact that

> Workers will learn to appreciate the limits of their own abilities, will recognize and respect the superiority not only of capital, but also of the higher intelligentsia and of education, and will grasp their societal necessity. (Fuchs 1904, 13)

But by 1911 Fuchs had begun to emphasize a much broader and more egalitarian program of Heimatschutz, one that would not be realized until each and every worker, no matter how poor, could enjoy all the benefits of a humane and rewarding physical existence. This kind of Heimatschutz was far removed from an elitist and merely aesthetic appreciation of land- and cityscapes:

> The Heimatschutz movement must not lead to a situation where the beauties of the Heimat, cultural and natural beauties, are protected by it for only a minority of aesthetically enjoying persons; rather it must be seen in much broader terms . . . Its mission is to provide the masses of the population with a housing scheme that is more worthy of humans, living quarters that are more worthy of humans, adjusted to what remains of the old conditions, standing in harmony with their surroundings. The best Heimatschutz is to create for the Volk a home again! (Fuchs 1917, 116)

It cannot be denied that there are still strong paternalistic elements in the idea of "creating a home for the Volk"; nevertheless it is significant that Fuchs increasingly rejected any Heimatschutz that would be the "privilege of a small rich minority" (Fuchs 1917, 115). In looking at the "green reform" program of the Heimatschutz I think it is therefore important not to underestimate the degree of sincerity that was present and to be aware of the growing potential for solidarity across class boundaries.

A quick look at the decade in which the Heimatschutz rose to prominence shows us that such a development would have been quite possible. Many portrayals of the Wilhelmine era convey the impression that members of the bourgeoisie were riven by a deathly fear of the proletariat and could not approach the working class in anything but a cynically manipulative spirit. What newer historical work emphasizes, though, is that by the last years before the war the German bourgeoisie as a whole, and educated, reform-minded *Bürger* in particular, tended slowly but surely toward an understanding with the Social Democrats; indeed, it can be said that the outlines of Weimar's Social-Liberal coalition were "already emerging in the last years before the outbreak of World War I" (Langewiesche 1990, 230; Nipperdey 1986, 183).

Heimatschutz efforts to protect the local environment deserve to be seen against this background of a real rapprochement. These were first of all not fear-laden "rear-guard actions" against the proletarian enemy, but rather products of an atmosphere in which bourgeois Heimatschützer saw a "profound reduction in political tensions" (Hedemann-Heespen 1910, 88; similar optimistic assessment of lower class in Naumann 1909–10, 425). Moreover, Heimatschützer often revealed an acute sensitivity to the tangible inequalities of life in Wilhelmine society. One of the clearest statements in this direction was provided by an upper form teacher and Heimatschutz member in the city of Siegen. Heimatschutz, for him, went beyond a question of moral education or internal refinement and was founded in the basic "human right" to health and material equality. What kind of social ethics do we have, he asked,

> when whole areas of our Fatherland are transformed into smokestack forests, thereby robbing the inhabitants of pleasant surroundings, air, light, nature—the things that people need most desperately, the best that they have? when we have to stand by and see how not only the development of emotional life is inhibited, but also that the external elements of life are provided in so meager a measure that the basic conditions for health are lacking? It only adds insult to injury that those who have grown rich off their industrial enterprises usually move away and set up housekeeping in beautiful nature under a clear sky unspoiled by smoke—leaving it to the less well endowed, who are doomed to stay there, to enjoy the view of black mills, and to digest clumps of soot. Again, the better-off can compensate for the continual deprivation by going on a trip—while it is the simple citizen who is tied to the spot, and who consciously suffers under the wretched conditions of his homeland. (Mugler 1911, 26)

Although the speaker of these lines waxed more eloquent than many of his colleagues on the subject of industry, I believe that he was nonetheless representative in his genuine desire to eliminate or at least attenuate social inequalities by working to protect the entire vernacular environment.

In terms of the overall social constellation of Wilhelmine Germany the program of Heimatschutz thus stood for the tendency toward a genuine reconciliation between bourgeois and proletarian elements. With its concern for landscape and environment it closely resembled contemporary schemes of "municipal socialism," where technical, public health, and infrastructure problems served as a key site of newly egalitarian class cooperation. In the area of Heimatschutz, too, it appears that local Social Democrats increasingly accepted and reinforced progressive middle-class initiatives.[29] Social Democrats in Saxony's lower house, for example, were instrumental in supporting the government's Heimatschutz laws; environmental protection was seen here as being unproblematically "in the interest of the whole" (cf. Second Chamber, 59th session 11 March 1912; also Dominick 1992, 62f.). In the case of the Heimatschutz, then, it may be that the politically illiberal Wilhelmine system sheltered "a policy of reform that did not conform" to the bourgeois economic hegemony (Langewiesche 1990, 231). Germany's die-hard free-market liberals were understandably irked at being outflanked in this manner: government-sponsored Heimatschutz measures, they now said, revealed a dangerous new "tendency toward Communism" (cf. record of Saxony's Second Chamber, 77th session 24 April 1914). Whether this was true or not is almost beside the point. What is important is that Heimatschutz probably helped usher in a new era of understanding between Social Democratic and bourgeois segments of the German population, and as such it can even be seen as a precursor of the more egalitarian atmosphere that was to characterize the Weimar Republic.

Is the nation an important environmental category? The ecophilosopher Klaus Michael Meyer-Abich has suggested that it is. As an ethical bond between people, the idea of the nation limits an individual's exploitation of a given environment when and where other citizens are adversely affected (Meyer-Abich 1990, 64f.). It is an anthropocentric standard, to be sure, but Meyer-Abich sees it as a key stage in the gradual widening of people's ethical community, one whose practical significance for the environment can be enormous.

As valid as the abstract philosophical formulation seems, though, its construct of an ethical nation leaves much to be desired as a guide to actual historical processes. Nineteenth-century nationalism did not automatically and evenly lead to an altruistic sharing of the environment, just as it did not inevitably result in universal suffrage or material equality. If people did manage to constitute themselves as an ethical nation and began respecting each other as citizens, this required conscious and continuous intervention, the creation of cultural practices, and the slow buildup of institutions that were not necessarily identical with those of the political state. From an ethical point of view Wilhelmine Germany in particular was a nation still profoundly in the making, and not a nation made.

The aesthetic environmental demands of the Heimatschutz interacted with this process of nation-building. On the one hand it made use of preformed aesthetic sentiments along with preexisting discourses on national unity; these latter discourses, as we saw, gave it valuable ammunition in its efforts to limit capitalism. On the other hand, by showing how such sentiments could be applied to the concrete benefit of the whole, it helped further define and develop social solidarity within the German nation, to break down rigid class boundaries, and to free up a critical spirit vis-à-vis states and administrations. Here the Heimatschutz was far more than just "modern." It was in important and valuable respects antimodern, opposed to the fragmentations, the marginalizations, and the conveniently limited, "private" notions of responsibility on which bourgeois capitalism relied. This was as true for the social sphere as it was for the environmental one. As we shall see in the next chapter, though, the task would be even more difficult with respect to the environment. Here, more than in the social sphere, it would be necessary to address fundamental problems of perception, and to bridge a gap whose origins lay even more deeply rooted in the history of Western civilization.

CHAPTER 3

Envisioning the Heimat Landscape

"Learning to see, teaching to see: that is the goal of the Heimatschutz!" ("Leuchtende Stunden" 1911–12). If protecting nature was, as many saw it, "the main objective of all Heimatschutz," then the simple line cited first contained a penetrating insight into the origins of the problem at hand (A. Fischer 1914, 332). For nowhere was there a more fundamental blockage of Germans' perception at the turn of the century than in the area of the environment. Under a hegemony of capitalist-utilitarian thinking, people assessed the world surrounding them largely in the one-dimensional terms of finance. Powerful discourses of modernity helped sever any remaining ties to the land: if a favorite shade tree was to be cut down or a brook straightened, this was all in the service of a greater "rationality" and a scientifically legitimated "progress." Given this dominant mode of rationality, a strongly aesthetic attitude toward the land was in some ways the most radical and effective form that environmentalism in Wilhelmine Germany could take.

In chapter 1 we saw some important precedents for a specific connection between aesthetic perception and antihegemonic reform. In the area of housing, for instance, bourgeois citizens convinced themselves through the imagery of "light and air" of the necessity of reforms that they could not justify in terms of the dominant, individualistic rationality. Much the same process can be seen in efforts promoting environmental reform. The powerful aesthetic fascination with nature that was integral to the educational tradition of Germany's middle class was an all-important precondition for the organizational success of the Heimatschutz; moreover, as I will argue, their aesthetic discourse was not just any way of valuing the environment, but rather fed directly into specifically ecological thinking with its emphasis on a concrete, "synoptic" perception of the integrated ensemble.

The stage for this particular developmental pathway of environmentalism was set by a common cultural framework. In nineteenth-century Germany many of the most popular and vivid conceptions of nature, the kind to which a writer or speaker could readily refer when trying to communicate a point to his or her audience, came out of the storehouse of aesthetic tradition. German Romanticism was especially familiar due to its highly accessible Late Romantic and *Biedermeier* forms. Paintings by Moritz von Schwind and Ludwig Richter were widely reproduced and became icons of a rural idyll; lines from nature poems by Joseph von Eichendorff and Nikolaus Lenau sprang to people's lips when confronted with a pleasing landscape. Popular culture was nothing if not eclectic, though, and so it easily included Friedrich Schiller's encomiums to the mountains and Goethe's more difficult, plant-thronged poetry in the same catalog of nature conceptions.

It is clear that the commonalities here were directly related to relatively uniform school curricula: Schiller and the Romantics were so highly quotable because they were the kinds of authors students were required to read. In the selection of certain favorite lines and poems we can see the additional influence of the German tradition of the *Lied* and the many singing associations (*Gesangvereine*) that were an important component of social life across the class spectrum. Given the intertwining of songs and nature imagery it comes as no surprise to read that the Heimatschutz organization in Württemberg, for instance, was able to rely on two different *Gesangvereine* in the town of Nürtingen to help host its annual gathering in 1914 (Lörcher 1915, 18).

This broad-based cultural tradition was reinforced by institutional factors that helped cement "nature" to an aesthetic mode of contemplation for the educated bourgeoisie in particular. As Jürgen Kocka has written, the *Bildungsbürgertum* was held together precisely by its adherence to a common body of knowledge developed "above all through through schools"; it was, moreover, a knowledge "shot through with aesthetic judgments" (Kocka 1989, 18). To no subject did this apply more clearly than to the environment. Through the 1890s German *Gymnasien* were guided by strongly antiscience, "humanist" curriculum requirements. Fearing the insidious effects of Darwinism, authorities saw to it that biology courses in particular were more or less banned from this level; only in the universities could students have access to such a touchy subject (cf. Trommer 1990, 198ff.). But even here the path toward exact biological science was hindered, this time by a different mechanism: the prestigious and influential positions in civil service, namely, those in the so-called general administration, went to those university students who studied law and political matters, not science. Students of the more applied sciences were not even eligible to attend the same schools as their humanistically oriented companions: they were educated at the polytechnical universities, which until 1890 were not allowed to grant the prestigious university diploma. This configuration of the German educational institutions went a long way toward ensuring that the common environmental language of the educated and governmental elite was that of their school days' aestheticism.

Thus even though the natural sciences and technology were continually breaking new ground and driving an enormous German economic expansion at the turn of the century, there existed on the cultural front and among the educated elites an unmistakable habit or tradition of dealing with the natural world in nonscientific terms. In this tradition "nature" was primarily an aesthetic or psychological experience of the individual subject, one that was triggered by or grew out of the actual immersion in natural surroundings. With the Romantics and their national-vernacular emphasis these surroundings were also almost always the surroundings of home: German oaks, the Rhine, the Brocken, or just the area outside one's own city walls. It is worth noting that this understanding of where and how to find nature dovetailed rather nicely with German middle-class leisure habits. As we mentioned in the beginning of chapter 2, the bourgeoisie was beginning to develop a characteristically close-to-home pattern of tourism and relaxation. Whereas the nobility and even the *haute bourgeoisie* continued to pursue the "grand tour" tradition of travel to foreign capitals, time and money constraints forced the average middle-class vacationer to set his or her sights considerably lower and nearer. But a virtue was quickly made out of this necessity, and an entirely new taste for the various regional cultures and landscapes of Germany began to flourish. This specifically middle-class phenomenon meant that clear recreational value supplemented an already strong aesthetic appreciation of landscapes, and it helps explain why the Heimatschutz drew so much of its support from this segment of the population.

In summary we can say that a number of cultural and institutional factors determined that environmental concern in Wilhelmine Germany would be communicated within a highly aesthetic vocabulary, and perhaps most compellingly in the visual concentration on landscape. The unfortunate tendency of much recent scholarship is to call this bottle half-empty, though, when in reality it should be seen as half-full. Klaus-Georg Wey, for instance, laments the ineffectiveness of this early phase of environmentalism, for "as long as protecting landscapes was seen to be necessary only in aesthetic and not in rational scientific categories, the seemingly so rational economic criteria could easily assert themselves" (Wey 1982, 133; cf. also Andersen 1987). While true, the implications of the statement nevertheless make it historically unfair. Rational economic criteria enjoyed hegemonic dominance at the time; to have challenged that dominance and put some serious reservations into people's minds should be seen as an achievement, and not as a failing. A further implication is that things would have gone better for the environment if only groups like the Heimatschutz had dropped their aesthetic pretentions and applied themselves to scientific research. Andersen points to some contemporary efforts in Germany such as the journal *Rauch und Staub* and the International Association against the Pollution of Rivers, Soil and the Air (*Internationaler Verein gegen Verunreinigung der Flüsse, des Bodens und der Luft*), both of which were working on practical remedies to pollution problems. From this perspective it is argued that an aesthetically based landscape concern was, by 1900, unnecessary, or that it even constituted an objective hindrance to the full development of applied scientific ecology.

The problem with this argument is that it refuses to confront the enormous gap that even today separates scientific fact from environmentalist value. In the historical situation of Wilhelmine Germany, though, it not only underestimates the significance of a cultural component, it also grossly overestimates the potential power of a few scientific findings to influence environmental policy decisions in the *Kaiserreich.* For one thing, it ignores the very real political and institutional limits that were imposed on scientific reformers. Even the most obvious kinds of regulatory reforms were frequently stymied by the conservative intransigence that was the hallmark of Wilhelm II's political style. A good example of the regime's stubborn unwillingness to listen to reason is provided by the protracted struggle around the housing surveys (*Wohnungsenquêten*). For years, scientists, bourgeois social reformers, and more progressive members of the Reichstag bombarded the government with demands that it do something in the area of housing reform. When it refused even to take the first step of carefully studying the problem, the forerunner of the AOK insurance group took matters into its own hands and commissioned studies of housing conditions by experienced social scientists and medical personnel. Even this was considered too provocative. In 1907 a consortium of landlords successfully petitioned Prussian authorities to prevent the use of state-subsidized insurance monies for the allegedly "misleading" surveys. By 1910, as the AOK persisted in publishing its bleak summary of lower-class housing conditions in the empire, those responsible were officially reprimanded for printing such agitational materials (cf. Asmus 1984; Bollerey and Hartmann 1978).

The implicit assumption that a little more science would have turned things around is thus belied by the experience of the housing reform proponents, who had plenty of scientific evidence but, in the end, very little to show for it. The assumption becomes doubly improbable when we consider the very large differential in institutional prestige we are dealing with. Supporters of the housing surveys included some of the best-known individual scientists of the era and representatives of prestigious and well-established academic disciplines such as medicine and economics. What could have been expected of a corresponding "environmental survey"?

One has to recall that "ecology" was at this time an obscure term at best, unknown to all but a tiny minority of biological scientists. As a branch of scientific inquiry it was hampered not only by fairly crude empirical methods, but also by the fact that it was only beginning to build up a theoretical base (Worster 1977, 198; McIntosh 1985, 15ff.). And especially in the hidebound world of German academia it had achieved none of the intellectual standing or institutional clout that could have carried it, qua science, over innumerable obstacles to wider societal application. It is of course true that one can protect the environment without having obtained a degree labeled "ecology," and that we should perhaps not make too much out of a name. Certainly there were biologists, botanists, and chemists who worked diligently on various aspects of pollution as well as foresters and agriculture experts who concerned themselves with interrelationships on a given piece of land. But their individual studies were typically just that, individual bits split up among different disciplines, different publications, and different overall interests. These obstacles were built into the entire system of education, academic production, and technical administration and could not have been simply dissolved, for example, by the wishes of an outside citizens' movement. Even among these advanced proponents of science what was often lacking was a basic, unifying impetus, an awareness that they were all working on the same urgent collective problem and toward the same vitally necessary goal.

If we look at things in a sober light, we must see that there were too many obstacles to be overcome, too many scientific developments that had to run their course before ecology would be able to assert its presence in German society. In turn-of-the-century Germany, scattered scientific findings, processed through accepted channels of publication and mulled over in dusty seminar rooms, stood little chance of overturning the immense cultural consensus in favor of industrial expansionism. What was needed and what eventually did the job was something else: the idea of Heimat, an overarching philosophical model that was able to foreground environmental damage for the broader population. And in the relative absence of scientific ecology, this kind of environmentalism took place within a discourse of aestheticism.

As I noted above, it would be wrong to lament this development. That would be the half-empty interpretation, which continually decries the fact that the Heimatschutz was not Greenpeace. Instead it seems much more accurate to take the half-full position, which not only sees the Heimatschutz as a powerful environmental consciousness-raiser in its own right, but also acknowledges that science and Heimatschutz aestheticism were not mutually exclusive. As we saw in chapter 2, the Heimatschutz movement in Saxony included connections to scientific organizations from the very beginning. In other regions as well, Heimatschutz ideas on protecting the environment soon penetrated to scientific organizations: some of these were the Natural Sciences Association in Karlsruhe, in Kassel, in the Frankfurt/Oder region, and in Bremen respectively, the Nature-Researching Society in Görlitz, and Württemberg's Association for Patriotic Nature Study (cf. Focke 1905; Hennings 1908–9; Schaefer 1909–12; Graner 1912). In Bremen the Natural Sciences Association even became an affiliated member of two regional Heimatschutz groups. At the national level a similar degree of understanding was evidenced by an early membership in the Bund Heimatschutz on the part of the Kosmos-Gesellschaft, a national organization of popular science interests that helped introduce conservation ideas to a broad cross-section of the population through its journal and a widely read monograph series. In Berlin the *Urania* society was also a piece of common ground between Heimatschutz and popular science education; Franz Goerke, the director of the *Urania,* was for many years the treasurer of the *Bund Heimatschutz* and booked a large number of lectures on conservation topics (cf. Günther 1913, 499).

These contacts suggest that the aesthetic message of the Heimatschutz had some success in reaching a broad cross-section of the scientifically interested public. More significant, though, is the fact that the Heimatschutz enjoyed strong support from a large number of professionals in specifically environment-oriented occupations. Foresters comprised about 1.6 percent of the movement, a small but significant number that included professors at forestry academies such as Franz Mammen in Tharandt and Karl Eckstein in Eberswalde. As we shall see below, garden architects were deeply influenced by the Heimatschutz, and a number can be identified as dues-paying members. Botanists and zoologists were also active in the movement. Aside from Hugo Conwentz these included Ferdinand Pax and Theodor Schube of Breslau and a remarkable pair of Dresdeners: Bernhard Schorler and Oskar Drude. Schorler was a coeditor of one of Germany's most-used plant atlases; Drude was the author of a seminal work on plant ecology. Their participation gives added legitimacy to a view of the Heimatschutz as a key site of an emerging ecological consciousness. It should be noted that aestheticism, in these men's eyes, was not a drawback or embarrassment: instead they saw it as a way to engage and inform the broader public. Drude in particular seized on the Heimatschutz as an opportunity for "educating the population . . . to the point that it becomes conscious of what it sees" (Drude 1909, 20).

Drude's program of education confronted an era in which Germany was fully committed to the program of industrial growth. Lingering doubts from the 1890s debate on the "agricultural state" had largely been swept away by a tide of arguments in favor of industry. It promised something to virtually everyone: workers could look forward to more and better employment, domestic consumption could increase on the strength of exports, the middle and investing classes could ride a rising tide of manufacturing wealth. As Rolf-Peter Sieferle has noted, the industrial consensus was therefore fundamental, crossing party and class lines and embracing an overwhelming portion of the population. This consensus extended even to the Social Democrats, who in spite of their opposition to the manufacturing and owning classes remained focused on industrial expansion for both pragmatic and more philosophical reasons (Sieferle 1984, 141).[1]

The immediate past history of Wilhelmine Germany probably helped to cement the remarkable unanimity behind the idea of expansion. After a boom period in the early 1870s Germans from all walks of life had had to live with the existential uncertainties of a sputtering economy and a long and debilitating period of structural change in which many of the smaller manufacturers and shop owners were eliminated. Financial security, even in the form of a meager income, was no small matter. Captains of industry were able to exploit these real fears whenever there was the danger that their considerable freedoms might be infringed; moreover, the industrial consensus allowed them to portray their own interests as those of the masses. Given this mindset a reform program such as that of the Heimatschutz could be tarred with the brush not only of economic amateurism and pie-in-the-sky idealism, but also of dangerous and even revolutionary destabilization. As one industry representative named Hermann Meissinger put it:

> To the broad mass of the people their daily earned bread with a healthy piece of meat is understandably more attractive than the most charming landscape. We would not want to see the representatives of the Heimatschutz with all their ideals one day having to experience how dangerous it is to try to bring happiness to a people that does not feel any need for their happiness! ("Industrie und Heimatschutz" 1914–15, 5)

The passage is telling for the alliance it rhetorically consummates between two otherwise opposing classes: it paints a picture of industrial entrepreneurs coming to the rescue of their workers and defending their solid, meat-on-the-table jobs against the "dangerous" ideas of environmentalists. Not all segments of the German population were so aggressive in their rejection of the Heimatschutz, of course; this almost allergic reaction could only have come from an industry whose profits were at stake. Still, memories of the economic depression lingered, and the primacy of employment could be activated almost at will as an argument against conservation. More subtle commentators from the liberal-industrial sector thus contended that Heimat was "more than a beautiful forest and a charming lake" in an "age of railroads and the automobile"; instead, they insisted, most Germans were more interested in the home-building value of "work" ("Erworbene Heimatliebe" 1911). Given this kind of a false choice between "Iron and Bread" or landscape, particularly in the context of the many other pro-expansionist pressures, there was little debate in turn-of-the-century Germany about which was the better pick.

This unholy bias in favor of unfettered economic activity was codified in a series of laws and understandings whose provisions throw considerable light on the state of official environmental awareness. The government itself, if we take the case of Prussia, largely renounced any fundamental role as environmental policeman. Its only measures against pollution, for instance, were those made possible by the *Gewerbeordnung,* the regulations pertaining to commerce. Here it could require factories to maintain furnaces and waste units "in good working order," but this was an extremely weak provision since it was linked with a static definition of what that order should be. As long as the equipment corresponded to the widespread "state of the art" (*Stand der Technik*), nothing extra could be required: in particular, no new waste technology could be imposed, and neither could any technology that might cost the company more money (Andersen and Brüggemeier 1987). This governmental program, weak as it was, suffered additionally from poor funding and the fact that not nearly enough inspectors were hired.

Civil law provided for suits against a polluter, but here there were other obstacles to overcome. A major problem was that the damage had to be quantified and traced to a specific defendant or defendants. The still relatively crude state of biological and chemical analysis made this legal burden especially onerous in situations where a number of factories or cities dumped their effluents into a river, for here it was often an unsolvably complex scientific problem to isolate each component and reliably connect it to the point of origin. The costs of litigation were also high and tended to discourage action by even the wealthy.

Another legal wrinkle made it almost impossible to combat pollution precisely where it was heaviest. Since these areas were generally occupied by workers and less well-to-do families, the chances that legal action would be taken were small from the very beginning. But the law added insult to injury with its concept of "*ortsübliche Belastung,*" or "normal level of pollution for the area" (Andersen and Brüggemeier 1987, 72). Dirty areas, the law more or less stated, were meant to stay that way. Only in relatively untrammeled sections, without a history of pollution, could action be taken to prevent a new and higher level thereof.

Even this level of protection often failed, though, in the face of the tremendous bias toward private economic activity. A case from the year 1900 is an instructive example. It concerned a cement producer who erected a new facility in a Rhenish village. Soon neighbors began to complain about the clouds of dust the plant generated and sued in court. But their suit against the owner was dismissed: the judge found that cement plants always produce dust, and that because the owner possessed the right to carry out his trade, the neighbors would simply have to live with it (Wey 1982, 111).

The above examples have concentrated on the issue of pollution because, as an odious and dangerous environmental effect, it constituted the most likely source of comment and action. As they show, the Second Empire showed little promise of perceiving pollution as a significant problem. This lack of concern was generally even greater where more subtle types of ecological damage were concerned: in monocultured agriculture, for instance, or in the spread of suburban housing. The legal principles tell a very clear story: this society was extraordinarily unwilling to interfere with property owners, even where their activities tended to reduce the quality of life for others. Observers of public legislative processes rightly attributed the lack of environmental action to a kind of hegemonic saturation: "the universal reluctance of our day vis à vis new legislative interventions into the individual's freedom to act" (Wirminghaus 1913, 176). "Freedom" here was not just economic: it was also the virtually unlimited right of the individual entrepreneur to foul and despoil the surrounding area. These fundamental legal doctrines of the Second Empire, when added to the many empirical shortcomings of its institutional mechanisms, reveal a systematic lack of environmental awareness.

But the story does not end there. For in addition to the economic and legal biases inherent in the system, there was a strong aesthetic or symbolic reinforcement of the ideals of industrial expansion. The great trade exhibitions of the nineteenth century were just the beginning. Soon this same kind of industrial-age advertising had infiltrated into many aspects of daily life and helped orient the public toward values of technological progress. It was readily assisted in this by the nearly obsessive journalistic coverage accorded to technological feats. Looking back at history, it is thus clear that the overall trend toward modernist design and industrial chic did not suddenly arise in the Weimar Republic, for it had already become well established in the Wilhelmine era. Contemporaries of the Heimatschutz were already fascinated with the precision of railroads, inspired to awe by the sight of sleek ocean liners, absorbed by the drama of steel mills, and impressed with the tooling on a long-range Krupp gun. This far-reaching complex of psychocultural commitments was what Thomas Nipperdey had in mind when he identified the turn of the century with a "massive vulgar progessivism" that united Germans in a kind of "secular religion" (Nipperdey 1990, 821).

All of these trends came together and still come together in the way in which people relate to landscape. As is still true today, the linguistic badge of one caught up in the glories of technological might was a coolly detached view of the surrounding world. A typical author would survey capitalism's effect on the environment in the neutral-sounding vocabulary of "changes," "transformations," a "shift" or a "reorganization" (one example is Felix 1883, 290f.). The reader should note the curious ethical effect of such words: they make it sound as if the object at hand is a ball of wax, or a pile of building-blocks with no inherent order. As far back as Francis Bacon the project of modernity in its basic economic sense had encouraged such value-free thinking; what had started out more or less as an *alibi* for tinkering with nature, though, was by now the culturally sanctioned and heavily valorized *necessity* of technological transformation. The brave new landscape of modernism was one where transformation in and of itself was good. Masochistically, almost, people took pride in the economic "realism" of such a perspective. An author from Denmark named Jensen confessed his initial shock at visiting a village he had known thirty years earlier, but quickly suppressed his doubts with a rosary-like chain of arguments:

> The first impression is hideous, but the price for the loss of the old idyll with all of its atmosphere is not too high, given new and better earning opportunities. One cannot live in a museum without having something off of which to live. The transformation was a necessity. Any other point of view is absurd. (Jensen 1914, 131)

The monolithic and thus uncritical tendency of much modernist thought is clearly expressed here: "The transformation" in its entirety was a necessity; "any other point of view" is absurd and ridiculous.

Progress, according to this line of belief, demanded that one adjust one's aesthetic expectations to conform to the given mode of production. After some reflection our author found that "factory smokestacks are not such an ugly growth in the landscape" and went on to praise the "tremendous expression of perspective" created by a string of telephone poles. Anticipating the spread of industrial modes of production to every corner of the earth—he named Russia and China as the next participants in the urban-technological "Great Leap Forward"—Jensen suggested "that we must look for comfort precisely in technical uniformity" (Jensen 1914, 133). While Jensen and other modernists had a right to call for an aesthetic appreciation of new landscape features, the apologetic tendencies in this approach cannot be overlooked. The idea that "technical uniformity"—a uniformity based on a particular historical form of capitalist-industrial rationality—should be taken as a "comfort" is especially disturbing when one considers the importance of diversity for the ecological health of the land.

But with their intense inner drive to accept and authorize transformation, modernist aesthetics cannot make an exception for the land, nor accept any old-fashioned standards of beauty. Speaking of the landscape of the future, Jensen remarked only: "We will see it in a different rhythm than before . . . we will be forced to expand our framework for sensing beauty" (Jensen 1914, 134). This kind of modernist apologetics has to be credited with smoothing the path of industrial expansion all over the world. Less than twenty years later, the arch-modernist Ernst Jünger could look at the changed German environment with full approval: "The landscape is becoming more constructive and more dangerous, colder and more glowing; all the last remnants of the old *Gemütlichkeit* are disappearing from it" (Jünger 1982, 173).

As subsequent history shows, landscape aesthetics is a realm of cultural activity where simple relativism can be dangerous (cf. Shepard 1985, 6f.). Some people find the pockmarked and starkly lit surface of the moon to be interesting. Deserts and windswept dunes are also not without a certain amount of aesthetic appeal. Despite their aesthetic qualities, though, neither of these situations would make sense as a landscape aesthetic toward which one would want to strive as a society. Somewhere cultural freedom has to bump up against the "real characteristics of the landscape" as an ecological (and therefore social) system; as far as the land goes, in other words, some aesthetic configurations and some aesthetic tendencies are better and more reasonable than others (cf. Finke 1986).[2]

Although not without its potential dangers, this kind of naturalistic framework has the advantage that it would be able to take protests more seriously than the opposing modernist view, which holds that the environment, along with our attitudes toward it, is infinitely malleable and historically ephemeral. It would also have less trouble accounting for the intransigence that has crept up repeatedly in industrial societies. Modernist critics are genuinely confounded by the fact that protest after protest has been sparked by people who apparently are not satisfied with the kind of landscape and the kind of environment that capitalist developers have given them. Noting these kinds of "obsolete" preferences, Rolf-Peter Sieferle has worriedly asked: "Has a specific image of landscape thus got fixed in our heads? why is everyone so reluctant to give it up?" (Sieferle 1986, 260).

Sieferle's comments are not an isolated phenomenon, but rather represent a much wider set of related opinions within planning disciplines and cultural geography.[3] J. B. Jackson is a prominent American landscape thinker whose concerns run in the same direction of valorizing vernacular change, an "incessant adaptation and remaking of the landscape," over the dangerously stiff nature of a landscape that is forced to conform to aesthetic or (God forbid) environmental preconceptions (Jackson 1984, xii).[4] Of even more relevance for this study are the points raised by Joachim Wolschke-Bulmahn, who has sharply criticized landscape ideals developed by the German youth movement in the same years as the Heimatschutz movement around the turn of the century. A central concept in his work is "the fear of the new landscape" that characterized members of middle-class youth in particular in that "epoch of rapid societal and spatial processes of change." Against this frightening world of change they set "a static landscape ideal oriented to historic conditions," an "ideal image of preindustrial peasant-dominated cultivated landscapes" that contained no recognition of social conflicts (Wolschke-Bulmahn 1990, 91–92, 228).

With slightly different emphases all three of the named critics set the living forces of change—social and economic—against what they see as the dead hand of aesthetic reification. But this opposition is too simple. An aesthetic imposed from above may properly be described as inflexible and undemocratic; at the same time many of the vaunted changes within society and the landscape are themselves undemocratic, the result of immense capital being brought to bear on corruptible structures and people who are not in a position to resist; indeed, the idea of hegemony shows that we have to be very skeptical about even the most basic processes of social decision making. In this constellation popular aesthetic ideals can be a valid and important avenue for the articulation of real concern, just as they can contribute to the consolidation of a popular movement.[5] Aesthetic conceptions thus often deserve to be looked at more closely and to be brought into a process of explication and discussion. This is quite the reverse of that urge to dismiss those "maladjusted" people whose mental images disagree with the present, as Sieferle apparently does. He finds it an "annoying situation," after all, "that we still have the dream of a landscape for which there is no longer any correlate in the outside world" (Sieferle

1986, 260). Other people find this situation annoying as well, but unlike Sieferle, they tend to find fault with reality, not the dream.

In the end the answer will lie in a compromise. The dream must be an economically feasible and socially acceptable one; meanwhile, the "reality" of modern society, its "rationality" and its "common sense," must be contested in order to aspire to the new and presently only "dreamlike" priorities of a future society. As we shall see, the program of the Heimatschutz was a significant attempt at this kind of a far-sighted compromise. Against an entire age of "massive vulgar progressivism" and its compensatory, romantic dreams, the Heimatschutz aesthetic represented a better—and ecologically more responsible—version of the German landscape.

The key difference between the Heimatschützer's ideal landscape and the present "modern" one lay rooted in the central term "Heimat." Every inch of Germany was "home" to someone, and thus there could be no question of splitting off pieces of it to be despoiled or degraded; whatever economic development was necessary would accordingly have to be compatible with a continued (or regained) high quality of life. With this radically integrative view the idea of Heimat contradicted the geographic fragmentation inherent in modern industrial society. As we shall see again in our examination of the nature protection movement, later in this chapter, this was nowhere more evident than in the specifically modern attempt to appreciate nature. For the vast majority of people around 1900, the kind of nature that deserved admiration and inspired interest was always somewhere else, someplace more exotic, more exceptional, or more, in a certain sense of the word, romantic. It was what Herbert Muschamp has labeled a quintessentially nineteenth-century conception of nature: nature was a realm apart even in those few situations where it was appreciated, "as a scenic backdrop or as a wilderness to be segregated into parklike preserves" (Muschamp 1993). Prussia's 1902 Disfigurement Law codified precisely this pre-Heimatschutz understanding by singling out only canonized views of the Rhine valley and its castles for protection.[6] Economic interests combined here with a widespread lack of concern about the issue to ensure that the provisions would go no further. The Heimatschutz movement formed as an explicit rejection of this kind of environmental cream-skimming. In its turbulent first years it sought new legislation that would drastically widen the 1902 law's scope. Unfortunately, as we have seen, it failed to obtain a satisfactory result. The so-called Heimatschutz Law of 1907 retained the earlier law's narrow standard: protected landscapes would be only those deemed to be "exceptionally beautiful."

It is worth digressing for a moment to examine the particular weaknesses of the 1907 law because the Heimatschutz subsequently defined what it really wanted in opposition to the measure. Section 8 of the law provided the basis for denying permits to construction activity that would "seriously disfigure" the appearance of any areas designated or zoned as "exceptionally scenic" (Bredt 1908). Soon after the fanfare of enactment had died down, though, critics noted that many authorities were remarkably reluctant to zone any areas in which industry had expressed interest; precisely those locations where "construction can most likely be expected" were "left out of consideration" (cf. "Siegerland" 1913; Keller 1910). Since a major feature of the law was that it only *allowed* the promulgation of local regulations, its effectiveness was directly dependent on the ability and willingness of town or county officials to educate themselves and take steps to enact statutes. Already by the end of 1908 the central administration noted a conspicuous lack of activity (letter of 24 December in PGStA I.HA Rep. 93B 1546). By 1913 it was clear even to outsiders that local authorities were taking advantage of the disfigurement law "unfortunately in many instances only with hesitation and dislike"; their impression is confirmed by the remonstrative memo sent to thirteen different *Regierungspräsidenten* in July pointing out that they had made "no or hardly any use" of the law's provisions (Wirminghaus 1913, 178; PGStA I.HA Rep. 93B 1546).

Nevertheless some local *Regierungspräsidenten* were relatively active in placing large swaths of land under protection, although less with respect to the economically more sensitive activity of building than with respect to advertising billboards. They were spurred on by a *Bund Heimatschutz* petition of March 1911 that specifically requested restrictions on the land adjoining rail lines and overland roads (PGStA I.HA Rep. 87B 3133/183–86; letter of 18 April in Rep. 93B 1546). Partly because of this additional push a number of railroads were given billboard-free corridors 300 meters in depth in Brandenburg and Silesia; lakes were given 500- or 600-meter zones; there were even moves to protect the landscape along canals (cf. *Brandenburgia* 1909–10, 82, and 1911–12, 292; *Heimatschutz in Brandenburg* 1910, 197, as well as 1911, 94f.; *Braunschweigischer Heimatschutz* 1912, 117; PGStA I.HA Rep. 87B 3134/26–7).

Despite these innovative applications, though, the disfigurement law's fundamental weaknesses prevented it from having anything like the kind of effect that the Heimatschutz had initially envisioned. The building of structures, for instance, was covered; but quarrying, draining of lakes, and the extermination of plant and animal species were not (Keller 1910, 181). The high threshold of "serious" disfigurement was a further obstacle to fighting subtle landscape changes. Perhaps the weakest link of all was the required designation of an area as "exceptionally" beautiful. This made protection dependent on a slippery game of determining different degrees of beauty. Advertising interests argued that almost no rail corridors met this criterion; far from being exceptional, the land beside the tracks was of a "virtually typical scenic dullness and uniformity" (petition of 1 October 1912, PGStA I.HA Rep. 84a, Akte 1983). In a 1913 court case having to do with advertising in the Berlin suburb of Spandau the defense was able to convince a jury that "truly exceptional landscape beauties" could be found "at best on the Rhine, or in the Harz or Sudeten mountains" ("Reklame" 1913; "Hervorragend schöne" 1912–13; "Was sind" 1913).

The statute was to be battered further in the courts. Already on 23 September 1912 Prussia's First Criminal Senate concluded that an administrator's designation of an area could be subject to judicial review. It was now a question of waiting to see how Prussian judges would interpret the standard. An advertiser from the Silesian town of Hirschberg gave them an opportunity. On appeal the highest Prussian court (*Kammergericht*) agreed that his billboard near Riebnitz could not be prohibited because the area was simply not exceptionally beautiful—something that would require, as the court wrote, an "alternation of forests, water, and mountains." This meant that huge sections in the northern half of Germany could not be covered by the 1907 law; again, as the court wrote, "north German flatland cannot, as a rule, be considered to be scenically exceptional" ("Norddeutsches Flachland" 1914; for very similar decision of 29 June 1915 see "Reklame" 1915).

The court cannot properly be blamed for its decision: outrageous as the result was, it merely reflected the highly restricted sort of environmental concern that had been built into the law from the beginning. But the increasing dissent in the German public is good evidence that the Heimatschützer had successfully intervened against the idea of protection only for limited, special, or exceptional areas. Exceptions, they pointed out, proved the more important rule of no protection. Providing protection only for exceptional areas was also attacked as a fundamentally unjust act of "distinguishing between first- and second-class landscapes" ("Weg mit dem" 1912). Over the course of time it is apparent that this Heimatschutz logic became more and more accepted by other segments of the population: people now observed that there was "no nature that is not 'exceptionally beautiful' for large groups of people" ("Hervorragend schöne" 1912–13). The extent of the swing in public opinion can be judged by the general dismay that greeted the setbacks of 1913 and 1914 and the resultant Reichstag resolution calling for wider designating authority ("Was sind" 1913; Nordhausen 1914). Against the stereotypical demands of the court for "mountains" and a certain "emotive quality," newspapers now argued that "the plains, too, can be pretty (or in legalese: 'scenically exceptional') . . . There are even people who love their vastness and stillness more than the aesthetic tumult of the mountains" ("Reklame" 1915). And whereas in the implementation guidelines of the 1907 law the government had stated that "streams of visitors" were a better indicator of a landscape's quality than the "Heimat feelings of inhabitants," by 1915 an official in the Ministry of Public Works could respond with an emphatic "quite right!" to the above article's praise of plains, thereby revealing a far-reaching conversion to the Heimat principle (cf. *Zentralblatt der Bauverwaltung* 4 September 1907, 33; PGStA I.HA Rep. 93B 1537/177). Thoughts of any significant legislative advances, though, soon perished in the face of the increasingly intensive war effort.

Heimatschützer were quick to point out that discrimination in landscapes was intimately related to a kind of social discrimination. "Exceptional" landscapes were protected for tourists, a Pomeranian wrote during the war, or "so that a painter can paint a picture that disappears in the salon of a war profiteer." But the Heimatschutz law was not there to serve the "accidental visitor" to an area, he emphasized; instead it was clearly meant to benefit *"the population that lives there"* (Reepel 1917; emphasis in original). A similar analysis was heard from the industrial city of Siegen, where a Heimatschutz author complained that the fixation on "exceptional" areas constituted a kind of discrimination against the urban working class:

> It is precisely landscapes like ours that need protection, landscapes that are subject to drastic transformation by means of their rapid industrial development—even if they don't carry the designation "scenically exceptional." Only a minority is so lucky that it can recuperate body and soul in a yearly trip to the beach, or to towering mountaintops, or to some other location that God has blessed with particular beauty. The majority of our citizens stand day in and day out in the noise and the exhaust of factories, in the monotonous driving work of the business world. When the holidays come they must be able to seek out untrammeled pleasures for body and spirit in their own local mountains, valleys, and fields—they therefore have a special right to beauty in the world that surrounds them. ("Siegerland" 1913, 50)

These lines bespeak a clear commitment on the part of Heimatschützer to helping to provide hygienic and recreation facilities for the lower class. Once again one is struck by the way in which an egalitarian social ethic seemed to flow out of the discussion of nature. It confirms Celia Applegate's observation that the nature of Heimat in many ways helped constitute an important "alternative milieu" within Wilhelmine culture, one that "tentatively represented a concept of general citizenship" (Applegate 1990, 77–78).

A 1908 article by Karl Keller shows how the specifically aesthetic evocation of nature helped to develop this concept of general egalitarian citizenship in one of its most radical forms: namely, as a restriction on private property rights. As a member of both the Brandenburg Heimatschutz organization as well as the land reform movement, Keller was appalled by the plague of private homes and vacation cottages that were rapidly colonizing every last inch of shore frontage on lakes around Berlin. To him this demonstrated the fundamental flaws in the "dominant conception of property in land" (Keller 1908, 597). But a frontal assault on the principles of private property would have immediately activated even sympathetic middle-class readers' affective filters; Keller's first task was thus to convince his audience that land could even be seen in something other than the property-oriented terms of money or individual ownership. Significantly, he chose an aesthetic reference in order to set up this alternative perception:

> We all are familiar with the lakes of the Mark, dreamily lost in the forest and mirroring dark firs on their reed-encircled surfaces. These are landscapes like those that the famous painting "A Grunewald Lake" by the recently deceased Leistikow summoned up for us. If the development that has begun in the area surrounding Berlin progresses uninterrupted further, in the not-too-distant future these landscapes will be a thing of the past. (Keller 1910, 597)

The structure of Keller's argument supports one of my main lines of interpretation with regard to the Heimatschutz: that aestheticism, as here applied, was not ineffectually "romantic" or hopelessly esoteric, but rather constituted a central ground of perception and discussion from which real problems could be addressed.

Keller's allusion was a skilled public-relations choice: for one thing, the painting was extremely well-known, and for another, it depicted precisely the kind of sandy Brandenburg lake that was being threatened by Berlin's suburban sprawl. It is also easy to see that the painting lent itself to a discussion of environmental qualities in its formal aspects: the viewer is nearly enveloped by the enormous canvas and is drawn into the harmonious, quiet round of a twilit lakeshore.

Having evoked this entirely new attitude toward the landscape, Keller went on to apply it to the concrete world of the Berlin suburbs:

> Anyone who has observed the change in Berlin's surrounding country with open eyes in the last decades must unfortunately come to the painful conclusion that one magnificent forest lake after another has been closed to the friend of nature. Where you used to be able to stroll freely along the shore, now the barbed-wire fences of a land speculator scream out an energetic "Halt!" to the hiker, and it is only from the garden of a noise-filled restaurant that you can cast a sidelong glance at the beauties of nature that are now locked away. (Keller 1908, 597)

In providing a language with which to talk about the environment, the function of the aesthetic landscape view was in this case more social than directly prescriptive of how an area should look. Keller did not take the preservationist position that development near lakes in and of itself destroyed all natural beauty; rather, he argued primarily that this was an example of unfair distribution, of a fragmentation and an unrightful taking of natural beauty for a small minority. Expensive villas constituted a violation of the principle of equal access to the natural resources of the nation, for neither workers nor even members of the "modest middle class" could possibly afford the housing being built there; it was thus nothing more or less than "the *monopolization* of natural beauties by a small number of the wealthy" (Keller 1908, 597–98).[7] Keller's article demonstrates clearly how the newly recognized aesthetic value of the landscape dovetailed with social egalitarianism.

It was the same underlying conflict between individual rights and those of the larger community that made an apparently superficial issue such as billboards important. Advertisers often pretended that it was merely a pragmatic question, of course, and pointed to the "great value for modern economic life" that their huge signs possessed. In good modernist fashion they would also insist that the public's aesthetic taste had already "modernized itself" far beyond the old-fashioned and implicitly absolutist "police sense of beauty" (Pickardt 1911; "Polizeilicher Schönheitssinn" 1911). But when push came to shove advertisers would always resort to the legal institution of private property, insisting that what they did there was nobody's business but their own. An aesthetic point of view, however, pointed to a more social—and thus more environmental—understanding of the issue. A billboard's visual effects clearly emanated out beyond the strict legal borders of the piece of property. And prominent structures such as lookout towers, monuments—even misplaced cafés—could be said to have similar impacts on their respective environments.

In these instances aestheticians walked in where lawyers feared to tread. If we look at the vocabulary of aesthetic protest and begin to take it seriously we can see that it entailed a new and more community-oriented definition of property along the lines of Ernst Rudorff's suggestive term "ideal co-ownership" (*idealer Mitbesitz*). At least in terms of an "ideal," that is, not directly economic perception, one could begin to view the world as connected, not split up into atoms of individual properties. Indirectly and tentatively, then, a fundamentally new social-environmental ethic arose out of this kind of aesthetic reasoning. Overly bold advertisements and inappropriate structures attained a new significance before this background of paradigmatic change. They were not merely cosmetic problems, but rather important test cases for a deeper principle, namely "the feeling that the collective has a right, a kind of right of co-ownership, not only over monuments of art, but also to a greater degree over nature" (Hennings 1908–9, 120).

This declaration of a formal right of "co-ownership" was not an end in itself, though. The Heimatschutz goal was to infuse all co-ownership with an entirely new spirit of relating to the environment. It is a subtext present wherever the Heimatschutz fought against projects that exploited the landscape as a mere substrate for arbitrary or unmotivated monuments. A case in 1911 in the Thuringian town of Gehren concerned the planned installation of a memorial lookout tower on a scenic exposed ridge, the "Langer Berg." The Heimatschutz objected to this "misguided" bit of development because it seemed to be designed to accommodate people without a "deeper sense for nature" who wanted to take in a panoramic view "just by turning on their heels" ("Gegen ein Denkmal" 1911). Symbolic and cosmetic as it was, the project thus provided an important pretext for bringing up general questions about the desirability and extent of development.

In Jena's Heimatschutz circles there circulated a related anecdote about a supposedly educated man who simply could not enjoy nature without a full complement of infrastructure:

> It is said that he always gave expression to his rapture and joy over a beautiful spot in nature with the seriously intended words: "Here it is wonderfully beautiful, here there ought to be a restaurant!" ("Burschenplatz" 1913)

This anecdotal man had many very real contemporaries. The spread of cafés, lookout towers, monuments, and amusements across the landscape of turn-of-the-century Germany was virtually unstoppable. Although in many cases conditioned by a property-owner's desire to make money, this drive to occupy virtually every corner of Germany with human structures could not be put down to profit mechanisms alone. Instead it reflected the much subtler hegemony of a manipulative cultural attitude toward nature. Ordinary people *expected* a café at every turn of the path, and the hiking groups that in theory stood closest to the mountains were often the most avid supporters of lookout towers.

In such a situation the Heimatschutz focus on aesthetic issues was perhaps the most accessible and productive way to start a discussion. It was not the Heimatschutz position to oppose any and all construction of this sort, just as it was not the movement's goal to stop all industrial expansion. Instead it took a much more flexible approach: one that proposed, as a first step, that people should at least think about the range of possibilities. Since one of the most-ignored possibilities was that of simply leaving an area as it was, Heimatschutz authors tended to emphasize this. But even if this minimalist solution did not appeal to people, the Heimatschutz could still influence the terms of the debate with their aesthetic axiom that all rural construction should fit well with its environment. As Joseph August Lux noted in a pragmatic article critiquing different lookout towers, "a feeling for nature is a feeling for beauty"; in both it was "the relations of the individual to the picture as a whole" (Lux 1906–7a, 167). With its aesthetic arguments the Heimatschutz thus positioned its audience to pursue a deeper awareness of environmental influences.

This strategy owed its particular effectiveness to the fact that Germans' environmental perception at the turn of the century was deeply "aesthetic-sentimental" (Nipperdey 1990, 183). The role of poetry was mentioned above, but ideas about nature and landscape were perhaps even more inextricably linked with pictures. The Heimatschutz itself was no exception here: Paul Schultze-Naumburg, the first chairman of the *Bund Heimatschutz,* actually started his career as a painter of landscapes, and the movement he later led featured a large number of artistically oriented people among its activists (on Schultze-Naumburg see above all Borrmann 1989). Since the Heimatschutz needed to be able to reach large numbers of people, the shared aesthetic cast of environmental perception provided strategic opportunities. As one of the many amateur photographers in the Heimatschutz remarked, "the pictorial arts are among the most powerful allies of the Heimatschutz" (Rieß 1913, 19).

Perhaps the most successful attempt to capitalize on this was the idea of creating picture-postcards with Heimatschutz content. Seeing in them an "educational means of the first order," Fritz Koch pushed his colleagues to embrace this and other populist media:

> We simply have to speak to the broadest segments of the population in the manner they are used to, otherwise our thoughts will never

> become known! We must not consider it underneath our dignity to try our luck with articles in local papers and with picture-postcards. (Koch 1907, 48)

Koch masterminded the first use of postcards in 1906, when the *Bund Heimatschutz* used a picture of the rapids at Laufenburg stamped with the *Bund Heimatschutz* name as a recruiting and publicity tool. This case apparently proved the "extraordinary significance of picture-postcards for the propaganda of the Heimatschutz" (Koch 1913a, 104). At a meeting of regional representatives in 1912 there was uniform recognition that the "Heimatschutz card" was of "tremendous significance for education" and also represented an "advertising mechanism of incalculable value"; the low-price cards could even be seen as a "cultural bacillus" that would help reproduce reform thinking within the public organism ("Von der Berliner" 1912; "Postkarte" 1910). By 1913 Fritz Koch listed twenty-six series of eight to twelve cards each issued by the Foundation for Heimatschutz alone; sixteen other regional Heimatschutz organizations issued cards of their own (Koch 1913a). The content varied: scenic street views of well-preserved villages and cities were probably the most common, but there were many cards with picturesque or record-breaking trees and also some of broader landscapes. Most were without the bold propagandistic captioning of the Laufenburg card, but many of the packaged series included a lengthy introduction to the respective material and information about Heimatschutz activities.

The move to take advantage of the relatively new medium of picture postcards tells us something about the movement. It shows, for one, that the Heimatschutz was serious about trying to reach as many people as it could with its ideas. Other, similar strategies were pursued that confirm this tendency: in Saxony a member recommended that Germany follow the English and Scottish practice of outfitting train compartments with pictures of local natural attractions (Baer 1909–10). Slides were another relatively new and popular technology, and most of the larger regional groups kept a series of canned slide shows on hand for public lectures. And several regional Heimatschutz associations, such as those in Pomerania and Württemberg, also issued posters, both with and without organizational names. But it would be insufficient to merely point to the generically aesthetic mode in which the movement argued; there was also a particular content. Perhaps the most direct way to appreciate this is to turn to some of the simple sketches that often decorated Heimatschutz journals.

A first example is shown in figure 3. In isolation this image may strike the reader as hopelessly naïve or idyllic. And yet this image was a definite part of the serious reform movement discussed in the last chapter. The thing that we must remember is that a simple vignette like this was a symbol, not a blueprint for a new world order. Its purpose was to set a programmatic emphasis within the context of Wilhelmine society and begin or confirm a process of group communication. If we ask what the programmatic emphasis here is on, it would seem to me that it is on a decidedly panoramic view of the countryside, something that is reflected in the broadly stretched format of many pictures. Figure 4 is an example drawn from the journal of the Heimatschutz organization in Saxony. Although the Jugendstil influence results in a different and more consciously ornamental style, the perspective is arguably much the same. This perspective is perhaps the central "ecological" message of Heimatschutz pictures, for without preaching it reinforces the conviction that everything in the environmental ensemble is connected. This includes not only the hills and the fields, the streams and the towns, but implicitly the viewer and the picture, too.

In figure 4 this integration is carried out by the imaginative path traced out by the eye as it follows along the body of water from left to right. The vignette in figure 5 uses the similar device of a road to guide the viewer on a roundabout path out into the countryside. In another picture a cleverly positioned row of sheaves takes on this function (fig. 6). Here there is almost no mediating frame; the viewer is thus almost thrust out into the vibrant agricultural landscape. The winding road, stream, or path is, and was already at the time these images were created, a standard trick of landscape painting. But hackneyed though the technique may be, it nevertheless succeeds: confronting the pictures, one cannot help being drawn into the scene, and one can almost smell the cut grass and feel the give of newly bared earth underfoot. Although limited here to the world of the imagination, this participation of the individual in his or her environment is an important dimension of what Heimat attitudes would eventually strive to achieve in real life.

A common feature of Heimatschutz pictures is that houses and towns blend into their surroundings. In figure 7 we see a woodcut in which the roofs of a small village protrude only slightly above the valley floor; only the picturesque castle asserts itself by filling the left-hand side of the scene. But if we look closely we see that even here nature is equal to the task, for trees crown several of the battlements, and speck-like birds circle above. Figure 8 shows us, once again, Otto Ubbelohde's title page woodcut for the *Bund Heimatschutz* journal. It concentrates into one picture virtually all the elements seen above: thus we see the broad, panoramic sweep of the countryside and an inviting pathway for the eye. But if anything it carries the blending impetus one step further, programmatically reducing houses to mere ripples in the greater sea of vegetation.

Fig. 3. Field scene. (From *Heimat und Welt* 1911–12.)

Of course none of these pictures shows an industrial landscape, and some critics would probably take this as a chance to argue that the Heimatschutz just gave up on the twentieth century and industry altogether, fleeing into an intact rural idyll. I do not think that this is true. We saw a whole series of cases in chapter 2, after all, in which the Heimatschutz promulgated a very serious, hands-on program of managing growth and development. Nor does it go to the heart of the matter to complain that such pictures are "stereotypical" or repetitive.[8] For these pictures were largely meant to be symbolic for a general cast of mind, a perspective. This is why the details of Ubbelohde's scene have been eliminated, reduced to mere squiggles. Friedrich Ratzel offered an interesting analysis of "vagueness" or "indeterminacy" in landscape paintings that indicates the direction one needs to take here: "Precisely the indeterminate quality of the subject of many landscape pictures," he wrote, "brings with it an agreeable distraction, out of which we slowly come to a feeling for the atmosphere of the landscape" (Ratzel 1906, 270). Ratzel perceptively highlighted an antihegemonic characteristic that inheres in such

Fig. 4. Stylized countryside image. (From *Mitteilungen des Landesvereins Sächsischer Heimatschutz* 1908, 59.)

Fig. 5. Typical landscape view as section header. (From *Heimat und Welt* 1910–11.)

Fig. 6. View of fields. (From *Heimat und Welt* 1913.)

sketches: the lack of definite, graspable, contoured objects exercises an important "wiping clean" function for us. Our ingrained utility analyses are, for a short while, disengaged, and we come to a wider and less manipulative understanding of the land's total and interconnected "atmosphere." This makes the all-inclusive perspective and the few recognizable features all the more effective in communicating a simple message: this is a scene in which human settlements and the natural world peacefully coexist.

This perspectival message is significant, and it should not be dismissed. If these vignettes are not exciting enough in a technical sense, if they are too simple, too direct, or too kitschy, then this is where we must part ways with the art historians. For historians of culture the organization of a new and critical movement around such pictures is a sign of something potentially important. As Antonio Gramsci wrote, a truly effective counterhegemonic aesthetic will not take root by being complex and artful, but rather by being almost programmatically uninventive:

Fig. 7. Wide-angle perspective on town and countryside. (From *Mitteilungen des Landesvereins Sächsischer Heimatschutz* 1914–15, 448.)

Fig. 8. *Bund Heimatschutz* icon. (From *Heimatschutz* 1908, 1.)

> It must aim at elaborating that which already is, whether polemically or in some other way does not matter. What does matter, though, is that it sink its roots into the humus of popular culture as it is, with its tastes and tendencies and with its moral and intellectual world, even if it is backward and conventional. (Gramsci 1985b, 102)

"Backward and conventional" may accurately describe the formal qualities of some of the Heimatschutz art, but these stylistic descriptors must not be understood as an indication of "backward" content. Instead it seems clear that readily comprehensible aesthetic images were a key part of the Heimatschutz attempts to introduce their antihegemonic ideas for environmental reform to the wider public.

Figure 9 shows how this kind of aesthetic communication could be applied directly to an understanding of the local environment. It is a full-page set of two photographs that Fritz Koch published in one of his short "Heimatschutz Issues" brochures in 1914. The method of a contrasting pair of photographs was one that Paul Schultze-Naumburg had made popular in his *Cultural Studies,* a series of widely read and richly illustrated works that he had begun as early as 1901. But if Schultze-Naumburg had employed it mainly to criticize the choices ordinary Germans were making in their architecture, Koch now used the method to show the consequences of decisions being made in the wider arena of agricultural land use. The way in which he guided his readers down a path, from aesthetics to a much closer consideration of the environment, can be considered paradigmatic for the Heimatschutz as a whole.

The photographs set up a contrast between a barren agribusiness landscape on the top and a far more diverse set of stream and fields on the bottom. The top caption reads: "The Stream of the Future. To what sort of songs will this one inspire our poets?" To the modern reader this inscription with its invocation of poets and songs might seem corny, and an unnecessarily indirect way to approach the issue. But it may well be that this device was meant to convince the contemporary reader to discard his or her notions about agricultural improvement and to begin to look at the land within the much more pliable aesthetic mode. This was quite important given the wide distribution of Koch's brochure, for among other places it appeared as a Sunday supplement to a newspaper based in Hildburghausen, a predominately rural area where the conventional wisdom of the agriculture extension "experts" no doubt held sway. The caption tries to circumvent this level of opinion by asking the readers a different question. Pretend you are a poetic sort, it posits; what kind of an aesthetic impression does the "improved" stream make? A negative one, the readers would have been forced to admit.

Der Bach der Zukunft. (Aus Südthüringen.)
Zu was für Liedern wird wohl der die Dichter begeistern?

Ein Bach der Vergangenheit.

Fig. 9. Contrastive pair of landscape photos by Fritz Koch. (From *Heimatschutzfragen—Beilage zur Dorfzeitung*1914 (3), 3; photo reproduction courtesy of Forschungs- und Landesbibliothek Gotha.)

Their answer would have been conditioned, of course, by a comparison with the picture below, which represents a more scenic stretch of what was actually the same stream. In that lower picture the assumed aesthetic sensibility and imaginative powers of a poet probably led them to investigate the land with uncommon thoroughness. Aside from the presumably winding course of the hidden natural stream bed, the photograph shows an abundance of bushes and trees. It is a variegated little biotope: the song of birds and the rustle of animals in the underbrush are almost perceptible, and even the contemporary readers could have begun to grasp the ecological health of such a landscape. Comparing this once again to the picture above, they no doubt began to appreciate in some detail the things that were missing. The ruthlessly modernized landscape had no trees, no bushes—not much life at all, other than the monotonous expanse of field.

What songs will this inspire? Implicitly the reader was being asked to connect the achievements of a culture with that culture's environmental base. If the reader was at all educated this reference to songs inspired by nature would have brought to mind verses by Eichendorff, Lenau, or Goethe. But these highlights of German national achievement had clearly not been inspired by the stone-lined gutter. Their inspiration was—and here the reader confronted the caption underneath the lower photograph—"A stream of the past." The presumptive tone of this label was of course a ploy: it sparked resistance and encouraged Germans to start asking questions of the dominant paradigm of utilitarian progress. "Is this stream a thing of the past?" they might have wondered. "Why can't it be retained? Why do we have to have an ugly and lifeless landscape?" Today, these are still powerful questions. Inspired by aesthetics, they nonetheless progress inevitably to a consideration of how economic priorities structure the environment.

With its emphasis on protecting *all* of the German countryside, the program of Heimatschutz was far more radical than that of nature protection, or, as it will be referred to here, Naturschutz. In Germany today this latter term is often used as a synonym for a broad kind of environmentalism, and, indeed, that is in accordance with the spirit of Ernst Rudorff, who is credited with coining the term in 1888 (Buchwald and Engelhardt 1968, 101). But Rudorff obviously sensed the difficulties stemming from the common idea of nature as a primeval condition and soon dropped the word *Naturschutz* in favor of that human-environment symbiosis favored in the word *Heimatschutz.* By the turn of the century the terms designated very different things. Naturschutz in any organizational sense was inextricably linked with the emphases of Hugo Conwentz and his role as the first director of Prussia's State Office for the Care of Natural Monuments, first organized and funded in 1906 as a special section of the Ministry of Culture (cf. Wettengel 1993, 365f.). It was the first state-level agency specifically devoted to any kind of environmental protection in the dominant German state, and thus one whose wider public influence went well beyond the rather narrow and circumscribed advisory role it had been given vis-à-vis other government departments. Unfortunately Conwentz was interested only in pursuing a relatively limited kind of nature protection, one whose primary goal was to assure a complete natural-historical record for scientists. Under his leadership official Naturschutz reserves, for instance, were aimed explicitly at a "preservation of original nature" justified on the basis of the area's "scientific and teaching significance" (Moewes 1926, 60f.).[9] This emphasis was again reflected in the organization's central concern for "natural monuments," formations that was implicitly "special," "unusual," or rare.

This narrow definition at the heart of official Naturschutz meant that in most situations having to do with ordinary environmental destruction there was little the government could or would do. In 1914, for instance, the Heimatschutz association in Brandenburg reported handling twenty-eight different cases involving landscape conservation; in the vast majority of these there was no assistance from the provincial government's Committee for the Care of Nature precisely because they did not have to do with "natural monuments" in the strict sense of the word (Mielke 1914b, 59; also Mielke 1907).[10] As Carl Johannes Fuchs of the Heimatschutz critically observed, this was a minimalistic *Naturschutz:*

> This Naturschutz, as Conwentz understands it is . . . only one part of the entire program of Heimatschutz. The motives and the objectives are fundamentally different: [for Naturschutz] they are purely scientific . . . , in no way artistic and general cultural ones. With Naturschutz it is only a question of preserving the native animal and plant world for the zoologist and the botanist and certain areas and landscapes for the geologist. From this comes the tendency among the former to be satisfied with little "reservations," or in other words a kind of museum—the museum spirit that hangs over the whole enterprise. (Fuchs 1909a, 5)

The "museum spirit" of Naturschutz was paired with an almost painful diffidence in practical questions of protest and preservation. Although there were exceptions, Naturschützer in the Conwentz mold typically avoided any fundamental challenge to the dominant rationality by making only the most minimal of demands.[11] Birds of prey, for instance, were labeled by one author as "unquestionably damaging"; nevertheless, he timidly continued, they should not be eliminated "absolutely completely"

(Peckelhoff 1908). Calls for protecting animal species also frequently demonstrated a unique disconnection between problem and solution. One representative article gave a long and devastating list of man's economic pressure on the environment: regulation, drainage, monocultured forests, quarrying, industrial pollution, species extermination, and other measures were stridently lamented—only to conclude: "We cannot halt the progress of civilization. We cannot change that." The article proceeded to recommend only slightly less relentless application of hunting and harvesting techniques (cf. Knauer 1912). Conwentz himself was so loath to upset property owners and bureaucrats that he often spoke of Naturschutz in the singular, as when he wrote that "a single moor" of a given type was enough (cited in Schoenichen 1954, 221).[12]

The narrowly focused program of governmental Naturschutz was complemented on the private front by the Association for Naturschutz Parks (*Verein Naturschutzpark*). Founded in late 1909 with the support of organizations close to the Heimatschutz such as the *Kosmos* society and the Dürer League, the Association for Naturschutz Parks wanted to establish a chain of smallish nature preserves throughout Germany and Austria, each about 50 to 100 square kilometers in size; this was quickly reduced to plans for only three separate parks. A combination of members' dues, philanthropy, and—most importantly—a government-approved lottery would in the end make the land acquisition financially possible (Korb 1915). The idea caught the public's attention immediately: membership shot up to 16,000 by the end of 1913, and soon the association was able to take over a good-sized piece of the Lüneburg heath as its first nature preserve.

Max Kemmerich invoked for this program the name of Heimatschutz, but this unilateral appropriation of the concepts of the movement did not sit well with many Heimatschützer (Kemmerich 1910). It appears that part of this was due to the fear of a possible drain on its membership, but it went to a difference in principle as well. Ernst Rudorff put his finger on precisely this point when he remarked that a nature-protection park was little better than a "showpiece," a "zoological-botanical garden grown large" that would very possibly detract from the kind of "laborious, noiseless work over years" required to institute total reform (Rudorff 1910).[13] But the *Bund Heimatschutz* did not want to unnecessarily divide the ranks of environmentalists and soon officially accepted the concepts of the Association for Naturschutz Parks as consistent with the larger mission of the Heimatschutz (cf. Heicke 1912, 70). Asked by Prussian authorities to comment on the group's ideas for a preserve in the Lüneburg heath, Ernst Rudorff himself wrote a very supportive letter in favor of the project (PGStA I.HA Rep. 87B 3144/47). That preserves and Heimatschutz were not mutually exclusive can be seen from the further fact that the Heimatschutz supported a number of initiatives to form smaller and more focused preserves in Saxony (Naumann 1909–10; Schorler 1909–10). And Heimatschutz associations all over Germany followed the efforts to establish bird preserves with special interest.[14] Oskar Drude emphasized time and again that the idea of Heimatschutz was the idea of preserving diversity; more limited scientific preserves and the wider aesthetic agenda came together in the common goal of preventing a "desolation of our forest flora, our fields and our waters" (Drude 1909).

Well before the Association for Naturschutz Parks was founded, though, Joseph August Lux expressed the fundamental criticism that the Heimatschutz applied to the idea of setting apart certain protected areas. These constituted for him "a sad sign of the times," for this kind of special treatment constituted "if not an invitation, then at least the tacit permission to deal ruthlessly with the much greater remaining portion of the face of the earth" (Lux 1906–7b, 60). In fact it seems clear that the establishment of nature preserves is often a form of compensation for exploitation and thus helps perpetuate the very behavior that lies at the root of the problem. Contemporary statements that described nature preserves as a "point of honor that modern civilization has taken up vis à vis maltreated nature" confirm the fundamentally dichotomous conception and lack of reformative power that lie at the heart of the preserve idea (Ribbeck 1910–11, 126).[15]

The Heimatschutz position, outlined both in its critique of contemporary Naturschutz and in its arguments against specialized scenic areas, was by contrast radically integrative and universal. As one author put it, the goal was simple:

> The joy of his homeland's appearance should be maintained for each and every person. For this very reason it is not a question of protecting just the old and the old-fashioned, nor a question of protecting just some monuments of nature somewhere, some individual "exceptionally scenic areas": rather, it is a question of protecting the landscape, period. (Wirminghaus 1913, 179)[16]

This engaged and practically oriented Heimatschutz thinking had an immediate and lasting impact on German society by helping to redefine the goals and aesthetics of professional gardeners. Nineteenth-century garden architecture in Germany had been deeply indebted to the landscape garden style. Peter Joseph Lenné and Gustav Meyer were two key figures whose lifespans and teachings formed an almost unbroken line connecting the professional gardeners of the turn of the twentieth century with ideals that had sprung up in the latter half of the eighteenth century. Much in society had changed since then: agricultural and industrial technology, in particular, had altered the face of the land, producing a bleaker, far more sober countryside. Meanwhile, though, in an odd dialectic of change and calcification, garden architects continued to churn out extravagantly "wild" and "romantic" pieces of scenery to be installed in the middle of Germany's growing cities. Berlin's Victoria Park is a good example, where gardeners and construction laborers spent untold hours making a rock-strewn mountain cataract out of a modest sandy hill. Stylistic innovations of the original landscape gardens, such as curving walks, were radicalized and applied with dogmatic fervor: whereas the original curving line was a breakthrough that allowed a path to conform to its environment, nineteenth-century park-goers were confronted with what critics called *Brezelwege,* "pretzel paths" that curved, writhed, and twisted around the enclosed space of the park in a highly improbable and unnatural manner (cf. Richard 1984, 224f.). Professionally designed gardens thus echoed much of the same split personality with regard to the environment that obtained in mass consciousness: nature was something to be had in tiny little preserves or romantically distant corners of the earth and was in any event not to be confused with the pragmatic patterns of ordinary life.

This carefully contrived world of the garden architects was destroyed by a new, pragmatic landscape aesthetic introduced by members of the Heimatschutz. Alfred Lichtwark, Joseph August Lux, Hermann Muthesius, and Paul Schultze-Naumburg—all men who came to the Heimatschutz from the Ruskinian/English tradition of craftsmanship—argued that it was preposterous for gardeners to try to reproduce untouched nature in their city parks (cf. Schultze-Naumburg 1902; *Gartenkunst* 1903, 16–20; *Gartenkunst* 1904, 54–55; Lux 1907; Lichtwark 1909). This was inauthentic, they said: parks should freely reflect the functional requirements put to them and should not be ashamed of their eminently constructed nature. It was a point of view that struck a responsive chord with garden architects throughout the profession, leading a number of them to join various Heimatschutz organizations: these included Arthur Glogau in Lower Saxony; Rudolf Goethe and H. R. Jung in the Rhine Province; Max Bertram, Bouché, and Degenhart in Saxony; and C. O. Berz, who was a member of the steering committee of Württemberg's large Heimatschutz organization. Through their work and that of many obvious sympathizers Heimatschutz ideas helped bring about a long-overdue turn to designing municipal parks "as if people mattered": in the years after approximately 1906 new parks were increasingly oriented toward the everyday recreational needs of children, workers, and sports enthusiasts.

Paths were streamlined to reflect people's natural traffic patterns, facilities were given more room, and people were now even allowed to lie on some of the manicured lawns! Gardens were now more "architectonic": like a house, they were straightforwardly designed in order to fill the real needs of people who lived in them.

This frankly functionalist aspect of the new Heimatschutz approach has been noticed and even praised before. Especially critics with a modernist slant have liked the fact that the Heimatschutz endorsed a pragmatic, no-nonsense style (Richard 1984, 288). But these modernist critics have generally either missed or dismissed the far broader and more important shift in opinion that had to do with the vernacular landscape. It has thus often gone unremarked that the new functionalism within the city was paired with a new, environmentally considerate aestheticism in the surrounding countryside.

This is an important blind spot even in Winfried Richard's book, which is the best critical appraisal of German landscape architecture at the turn of the century (Richard 1984). Richard reserves his highest praise for Friedrich Bauer, a gifted young landscape designer from Magdeburg who boldly applied functionalist principles in the service of socially progressive facilities. But Bauer also identified strongly with the Heimatschutz, as Richard admits (256). He attempts to write this off as a regrettable lack of consistency on Bauer's part, but in fact Bauer had thought the issue through quite clearly and saw no contradiction here whatsoever.

Bauer expressed his reflections in a key 1906 article that condemned the nineteenth-century landscape-style gardens as having been true neither to gardens nor to landscape. As gardens they did not meet people's needs, he said; and as landscape they were fake: "artificially constructed," they "show little consideration for the native landscape that carries them" and "sit in it like foreign body." Indeed, Bauer detected a degree of modernist arrogance in the so-called landscape garden: the area's own naturally developed landscape was seen as "imperfect, unattractive, in need of improvement" (Bauer 1906, 111). This would all change, Bauer wrote, "when people have learned to grasp . . . the unrecognized value of the native landscape, to yearn for it and to take walks far out beyond the city—then many an artificial landscape will show itself to be purposeless and superfluous" (112). What Bauer clearly indicates with these comments is that utilitarianism in limited gardens can be completely consistent with a more protective and less interventionist approach to the wider landscape.

Critics may have disliked this component of the Heimatschutz program because this is where it clearly "betrayed" the project of functionalist modernism; instead of cheering on the advance of rationalized production, the Heimatschützer objected, saying that rationalization had actually gone too far in many parts of the German landscape. Lux discussed the transition in his theoretical treatment of a park that might stretch out into nonurban areas. In this situation the true garden architect, he said, would strive for a "preservation of existing features" as far as possible, and would touch "neither existing old trees or tree groups, nor hedges or overgrown walls, nor old roads and paths, nor the original inventory of meadows and woods" (Lux 1908a, 173). What is important here is Lux's non-"functionalist" recognition that the unique natural and historical features of an area were interesting and rewarding, and thus they were parklike in their own right. The new split paradigm could be embodied within even a small urban park, where small brushy areas could be planned in as not-to-be-disturbed bird habitat (Heick 1910).

As the example of this park shows, the Heimatschutz attitude was not just preservation-oriented: in parts of the working countryside where natural aesthetic diversity had been wiped out, it actually required intervention into the status quo. This was certainly the case in the income-producing forests at many cities' edges, whose "aesthetic and sanitary value" for the industrial population became increasingly apparent (Wilski 1908). Given these new recreational demands Heimatschützer, gardeners, foresters, and social reformers began to try to reform the forests, to bring back some of the natural variety that had been eliminated. The thought of treating forests as aesthetically valuable, functionally parklike areas owed a great debt to Heinrich von Salisch and his 1885 book on *Forest Aesthetics,* which went through several printings, was widely reviewed, and whose impetus was taken up in countless articles (along with Salisch see also Kozesnik 1904 and Felber 1906; some of articles that apply the idea of forest aesthetics are Schier 1904; Lüdtke 1905; Walther 1906; Trip 1908; Hanisch 1909; Verein Deutscher 1909; Schneider 1913). Raoul H. Francé expressed the wide agreement among green reformers with his enthusiastic promotion of a "girdle of woods and meadows" for every German city: "The demands of public health coincide here with those of the schools, and also with the wishes of aesthetics, friends of nature and Heimatschutz" (Francé 1909, 89). Indeed, it was only a small step from this greater-metropolitan point of view to the old ambitions of land beautification, an idea that now regained its strength among garden architects and Heimatschützer as comprehensive regional planning (cf. Schoch 1902; Glogau 1903; Tutenberg 1903; Mielke 1908).

The distinctive contribution of the Heimatschutz to this process was to apply reform ideas to the countryside proper, eventually subjecting common land use practices to the same kind of multidimensioned analysis as municipal forests. Both Lux and Schultze-Naumburg taught that the countryside was at its most beautiful where the forms of human intervention sprang only from necessity. This was not to be confused, though, with the total manipulation inherent in intensive, rationalized schemes for the maximization of profit. Lux saw this exaggerated form of intervention in "State-operated forests in their 'cleaned-up' condition . . . where one and the same species of tree stands in an inexpressibly monotonous rank and file formation that has been pedantically cleared of underbrush" (Lux 1908a, 172). In any quantitative and historically concrete sense, this is where the true significance of the Heimatschutz aesthetic must be looked for: in the confrontation with everyday environmental abuse.

Over the course of the nineteenth century large portions of Germany had been turned into a featureless, monocultured agricultural production platform: huge swaths of the German countryside were thus, as Ferdinand Avenarius said "an instrumental terrain, but no Heimat" (cited here from *Dresdener Anzeiger* 20 March 1914, 8). Landowners tried to maximize the return from their acreage by removing all "superfluous" brush, draining wet spots, and running the plow right up to the straightened borders of brooks and streams. In the largest state, Prussia, over 11,000 kilometers of rivers and streams had been regulated in the fifty years ending 1910; what is perhaps more disturbing is the fact that another 12,000 kilometers were in some stage of preparation or planning. Projected new field drainages were to encompass approximately 750,000 hectares of Prussian land—again, an area equal to that drained in the entire previous half century (cf. overview in PGStA I.HA Rep. 87F 2891). Nor were the forests excepted from this trend: all over Germany, many areas of mixed or hardwood forest had become solidly single-age monocultures of fir or pine. And even in the immediate neighborhood of cities the landscape was often deprived of trees lining roads, framing houses, or gracing town squares.

Against these tendencies of modernity the Heimatschutz pursued a bold program of landscape reform whose cardinal principle was to put natural diversity back into the landscape: hedges, brushy field edges, wooded stream banks, forest undergrowth, and the admixture of deciduous trees into pure conifer stands were just some of the more important features aimed at creating a more beautiful landscape. In looking at the measures that the Heimatschützer recommended, though, we will see that their standard was never *merely* aesthetic. In the tradition of Ruskin and the arts and crafts movement they were convinced that beauty was linked directly with overall functionality: "in all objects of daily use true beauty consists in complete functionality" (Fuchs 1905, 9). The key here was the word *complete.* Beauty was not identical with the functionality of capitalist expedience, but rather with a much expanded version of functionality: one that included beside profit also health, social effects, and ecological effects. As the Heimatschutz never tired of pointing out, an ugly, dangerous, inhumane factory operated on the basis of "merciless exploitation" would be "in the final analysis not a rational operation"; similarly, the monotonous aesthetic poverty of modern agricultural plantations was an indication that they, too, were functionally unsound in the long run (Schneider 1907, 235). When one looks closely at Heimatschutz arguments, then, one sees that aesthetic appreciation and ecological evaluation were explicitly interlocked in a program aimed at rejuvenating a vernacular landscape ravaged by capitalist-mechanist neglect. In terms of the landscape, as one insightful member noted, "Heimatschutz would actually have to be a kind of restoration" (Gradmann 1910, 47).[17]

Nowhere was there a stronger connection between aesthetics and the ecological dimension of the landscape than in the Heimatschutz advocacy for hedgerows. This was no obscure detail for the Heimatschutz: as may be recalled, Ernst Rudorff had defended hedges as far back as 1880, praising them not only for their landscape beauty, but also in their function as bird habitat. By the time the Heimatschutz movement was in its prime it had become *de rigueur* to mention the hedgerow's role in alleviating the "apartment shortage in the bird world" (Greiter 1914, 24; also Mielke 1906; Ehrlich 1917; Robien 1917; Zimmermann 1917). It was also pointed out that the birds themselves were not just aesthetically pleasing but also highly significant as crop protectors (cf. among others "Schonet die Hecken" 1916). In all of the articles authors challenged the narrow agribusiness definition of "rational" utility that had led many German farmers to rip out their hedgerows (cf. 1907 handbill from Gera, in THStA LADH Nr. 756). One noted that it did not make sense even in immediate financial terms: the return on the acreage created would be far less than the costs involved ("Schutz der Heimat" 1913, 15). This argument became even stronger once the ecological benefits were factored in. Readers were reminded, for example, that hedgerows could "influence the climate" of the immediate area: they helped increase soil moisture, allowed snow to accumulate, and prevented winds from drying out crops (Greiter 1914, 24; Koch 1913b; Buddin 1907). Elsewhere hedges and edge habitat were credited with a beneficial effect on the water table; in this capacity they helped prevent "disturbances in the water cycle" over a large land area (Jünger 1913, 2).

In other articles authors rightly emphasized the simple beauty that hedgerows added to the landscape. Figure 10 shows a panoramic photograph of hedged north German countryside that appeared in the journal of the Heimatschutz group in Mecklenburg. An unstated caption beneath it could read—as did one at an exhibition later arranged by Fritz Koch—"Try to imagine that the vegetation was not there: what would remain of the landscape?" (review of an exhibition held in July 1921, in THStA LADH 52/3). Another writer from northern Germany linked this kind of large-scale beauty with a more precise appreciation of the hedgerow's diversity:

> Just take a look at our hedgerow landscape: how much diversion it offers in its spring, summer, autumn and winter clothes! Study it closer and be amazed at how many different shrubs find a Heimat in the hedgerow. The smaller plants have their unique features, too. Throughout the hedge there is life, nowhere in Germany at the same latitude is there a fauna as diverse and abundant. For its sake alone the hedgerow should remain a custom with us. ("Schutz der Heimat" 1913, 16)

As this shows, the beauty of the hedgerow was directly and consciously linked with its function as a refuge and habitat. The reader of Heimatschutz articles thus came to a deeper appreciation of the hedgerow: underlying its aesthetic qualities was its ecological role as an entire "community of life, a cooperative, an indispensable member of the natural life of an entire landscape" (Christiansen 1907, 72).

The Heimatschutz found a certain amount of support for its hedgerow and brushy-margined landscape in government offices. The directive whose impact was felt most widely across Germany was one issued by Prussia's Agricultural Ministry on 4 June 1907. It was distributed widely: to top members of "Settlement Commissions" in Prussia's sparsely populated eastern areas, to the "General Commissions" in charge of overall agricultural and infrastructure planning, further to field rationalization offices and those in charge of moor drainage. The directive was ostensibly inspired by the newly discovered value of Naturschutz "in accordance with the recommendations of Professor Dr. Conwentz," but in fact the author(s) showed the influence of Rudorff's Heimatschutz thinking in going far beyond a narrow scientific concern for primeval nature. A few of the changes in procedure that it suggested were, for instance, to get away from the "straight line as the sole measure of things"; to "leave brooks and waterways in their original position"; to "conform to old field borders with their scenic trees and hedges" as much as possible; indeed, technical officials were now urged to think not only of the "material" purpose of their work, but rather also to understand its larger "ideal purpose" (cf. PGStA I.HA Rep. 87B 3132/72).[18] Not surprisingly, lower agencies and the wider public interpreted the directive in its wider Heimatschutz sense as calling for a general "reinvigoration of the landscape"; indeed, Fritz Koch immediately recognized it as "one of the greatest triumphs of

Fig. 10. Photograph of hedgerow landscape. (From Buddin 1907, 118–19; photo reproduction courtesy of SBPK.)

the Heimatschutz idea" ("Landwirtschaft und Naturdenkmalpflege" 1907; letter of Koch to Ministry of Agriculture 10 Nov. 1907 in PGStA Rep. 87B 3132/102).

Adherence to these new principles would of course be anything but easy for agricultural technicians who were warned in the same directive not to reduce "economic efficiency" at any time; as a reminder sent out in 1910 shows, too, "not everywhere" was the kindler, gentler attitude given the "desired amount of attention" (letter of Agricultural Ministry to General Commissions, in PGStA Rep. 87B 3133/149).[19] Nevertheless the effort probably helped make some converts for the Heimatschutz. Indeed, some local offices' *engagement* for the Heimatschutz may have gone too far for the cautious bureaucrats in Berlin. The president of the General Commission in Münster wrote in 1912 that he had had "much opportunity" to conduct his work "in accordance with the efforts for the care of natural monuments and Heimatschutz over the last several years" and requested permission to attend a meeting of the Heimatschutz; permission was denied because it would be "unseemly" (PGStA Rep. 87B 3134/22).

All over Germany we can assume that Heimatschutz groups made similar attempts to approach parts of the agricultural establishment. As one of its very first actions the Eisenach local section resolved to establish contact with the agency responsible for rationalizing fields; here it would be the goal to get the officials to spare as many small tree groups and bushes as possible. In nearby Gotha the Heimatschutz helped spark administrative interest in practical Naturschutz and even worked together with government bodies in ironing out its regulations pertaining to bird and plant protection (cf. *Heimatschtuzfragen* 1914, 1; 1913, 11). In the small state of Lippe the Heimatschutz group reported that "several agricultural associations" had approached it "on their own initiative" requesting informational lectures (Lippischer Bund 1909, 3). In Silesia the local Heimatschutz association also reported good cooperation with the local agricultural association and was able to advise on field rationalizations as well as participate in forest inventories. The agricultural association had further helped by publishing "a number of larger and smaller Heimatschutz articles" (Schube 1914). In Hesse there appears to have been a similar rapport: a Heimatschutz publication praised "the friends of the Heimatschutz in the special and general commissions" and commended them for having done "much to reconcile the demands of progressive agriculture with those of a feeling for nature" (cf. *Mitteilungen des Vereins für Naturdenkmal- und Heimatschutz in Kurhessen—Werbenummer* 1912, 3).

A possible foothold for this rapprochement may have been the contacts established early on between the *Bund Heimatschutz* and the Raiffeisen agricultural cooperative movement. As we know, it was a representative of the Raiffeisen coops who helped organize Koch's Foundation for Heimatschutz; Carl M. Wuttig saw this as an extension of his farmers' general commitment to raising the conditions of rural life (cf. *Heimatschutz* 1909, 91; 1910, 84f.). In the journal of the Raiffeisen movement the Heimatschutz was accordingly praised "repeatedly" for its "extraordinary cultural significance" (cf. *Raiffeisen-Bote* 1910, 286; also 1911, 302).

The relations between Heimatschutz and organized agriculture were not always so rosy, of course. The same Raiffeisen members who read about the Heimatschutz in their journals could simply flip the page and be told that the "full exploitation of each and every acre" and "the highest possible production" was the point of their work (Schloßmann 1914). As this corresponded closely with everyday financial pressures, many or even most farmers probably agreed that bushes and shrubs "are out of place in advanced agriculture" (cited in *Heimatschutzfragen—Beilage zur Dorfzeitung* 1914 [3], 2). Many, too, were the agricultural extension officers and surveyors who refused to accept the idea of limits to rationalization. The writer Hermann Löns gave a biting portrayal of such a ruthless land surveyor in his poem "Field Rationalization" (cited here from *Braunschweigischer Heimatschutz* 1911, 101):

Es geht ein Mann durch das bunte Land; Die Mißkette hält er in der Hand.	A man walks through the bright gay land; The surveyor's chain is in his hand.
Sieht vor sich hin und sieht sich um; "Hier ist ja alles schief und krumm!"	He looks ahead, then around, and says: "Everything here is a crooked mess!"
Er mißt wohl hin und mißt wohl her; "Hier geht ja alles kreuz und quer!"	He lays out one leg and then an angle; "Everything here is one big tangle!"
Er blickt zum Bach im Tale hin; "Das Buschwerk dort hat keinen Sinn!"	His scrutiny falls on the brook so fair; "They serve no purpose, those bushes there!"
Zum Teiche zeigt er mit der Hand; "Das gibt ein Stück Kartoffelland!"	The pond to his vision has to yield; "From that we'll get a potato field!"

Der Weg macht seinen Augen Pein; "Der muß fortan schnurgerade sein!"	The winding road, it pains his eye; "Straight as an arrow she'll have to lie!"
Die Hecke dünket ihm ein Graus; "Die roden wir natürlich aus!"	The hedge to him is a special horror; "We'll dig it up! We'll start tomorrow!"
Der Wildbirnbaum ist ihm zu krumm; "Den hauen wir als ersten um!"	Too crooked is the wild pear, poor tree; "We'll chop that down first!" says he.
Die Pappel scheint ihm ohne Zweck; "Die muß da selbstverständlich weg!"	Senseless seems the poplars' row; "That bunch there will have to go!"
Und also wird mit vieler Kunst Die Feldmark regelrecht verhunzt.	No trick of the trade did he avoid And so a charming land destroyed.

Even men who were self-declared allies of the Heimat movement could occasionally revert to the rationality of mechanistic agribusiness. A pamphlet by Richard Hempel (1914) titled "Heimat Beauty and Land Cultivation" demonstrates the less-than-smooth path that environmental reform had to travel. Hempel was truly insightful in portions of the pamphlet, for example where he criticized the effects of rationalization and deforestation on the overall water cycle. But in other portions Hempel voiced his troubling "conviction that we need have no worry about using our Heimat earth as thoroughly as possible." Heimatschutz, on this view, would be a pitifully superficial and cosmetic affair in which the landowner was advised to "leave a tree or a bush standing every now and then in an appropriate spot." Hempel dismissed the idea of a more comprehensive landscape reform in characteristic terms as "unrealistic": "The iron demands of our time have already rolled over many an idyll" (quotations from Hempel 1914).

We should not let this kind of unavoidable backsliding color our perception too deeply, though. What is obvious is that the Heimatschutz expressed a whole set of arguments in favor of more hedgerows and bushes in the landscape, and that it was precisely the integration of aesthetic elements with functional, ecological ones that gave its demands a broadly powerful appeal. The valuable qualities of hedges had been recognized here and there before the Heimatschutz came along, but as a garden architect noted, ordinary people began to recognize the value of hedgerows "only after the Heimatschutz efforts pointed it out in writings and in public lectures" (Maaß 1910, 207). When articles in popular media equivalent to the *Farmer's Almanac* began calling for a "Protection of Hedges and Trees in the Fields," this can likewise be traced to the tremendous impact of the Heimatschutz (cited in *Heimatschutzfragen* 1914 [1], 1). With respect to this very concrete element of the working landscape the Heimatschutz really did succeed in opening people's eyes: following the program of "teaching to see," it undergirded its aesthetic with a far-reaching examination of functional ecological matters—thereby giving Germans the tools with which to better appreciate one of their own most immediate environmental formations.

The same kind of concern for aesthetic diversity in the landscape that sparked the defense of hedgerows was often applied to ornamental trees. Heimatschutz associations were continually reacting to news of a threatened stand of poplars or an *Allee,* a double row of trees lining a country road (cf. among others *Heimatschutz in Brandenburg* 1909, 40f.; *Mitteilungen des Schlesischen Bundes für Heimatschutz* June 1913; *Pommersche Heimat* May 1917, 8). Through a highly successful petition of March 1909, "Concerning the Protection of Remarkable Trees and *Alleen*," the *Bund Heimatschutz* managed to awaken similar concern in government agencies all over Germany. In Prussia the petition was picked up first by the Ministry of the Interior and transmitted to its construction overseers as well as to other ministries; although the Ministry of Agriculture was not on this original list, it requested the document soon afterward and promptly distributed more copies to all *Oberpräsidenten* and all of the "Royal General Commissions" responsible for rural improvement schemes (PGStA I.HA Rep. 87B 3133/87–89). In Württemberg, meanwhile, the same petition from the *Bund Heimatschutz* was received and quickly outfitted with official recognition in a directive to lower agencies from the Ministry of the Interior; in Saxe-Coburg the central State Ministry similarly transmitted copies of the petition to regional offices without delay (cf. HStA Stuttgart E151/07 Nr. 570I/107–113; StA Coburg MinD 20/26f.)

Since they obtained apparently widespread agreement on the unique aesthetic value of trees in the open landscape, it is no surprise that Heimatschützer turned to the forest as well. One part of their recommendations concerned individual trees. In Mecklenburg a forester member of the Heimatschutz repeatedly stressed that old and singularly impressive trees should be preserved by local foresters. He also advised foresters to begin a planned campaign of setting aside certain young trees as "natural monuments for the coming generations" (Arnswaldt 1908). But these demands were really just the beginning. The monocultured pine or fir plantation that had dominated German forestry since at least the middle of the nineteenth century was, in the eyes of many Heimatschützer, a practice that had to be changed from the ground up.

Historically the Heimatschützer were not the first Germans to criticize monocultures. Far-sighted foresters such as Emil Adolf Roßmäßler and Karl Gayer expressed deep and principled opposition to this form of forestry already in the nineteenth century, arguing that the mixed forest with limited-area or single-stem harvesting created a far healthier woodland (Pflug 1969, 250). But their warnings about long-term consequences went essentially unheeded in a hegemonic system oriented toward the short-term financial gains that rationalized and clear-cut monocultures made possible. As the forester and Heimat enthusiast Lorenz Wappes perceptively wrote in 1887, interested parties such as landowners showed a disturbing ability to ignore scientific doubts about intensive monocultures, and this was only encouraged by a lack of wider interest in the question:

> We will not put a stop to selfishness and all its pressure by proving its negative consequences for the distant future; how long now have foresters been fighting against this evil that threatens the existence of the forest, and how often have their voices died out unheard, or rather—intentionally—unheeded, because the direct interest of large segments of the population did not stand behind them as support. Popular beliefs and the will of the people are very difficult to combat with scientific deductions. (Wappes 1950, 24)

Wappes did more than just lament the weakness of scientific arguments, however. Having recognized the power in the hegemonic system of "popular beliefs" described above, he seized on aesthetics as a way to get things moving against the grain of short-term profit making:

> We must strive more in the direction of setting up against the destructive party another one that has a direct interest in the forest and its full protection. And for most that quality that is most obvious and is most easily understood is without a doubt beauty, and the effect of the free atmosphere of the woods on soul and body. (Wappes 1950, 25)

This was a substantial insight that effectively anticipated the tactics of the later Heimatschutz movement. In a bid to save the forest from its owners, Wappes called here for a renewed aesthetic and recreational appreciation as a way to build up social pressure against woodland abuse. Instinctively, it seems, he grasped the necessity of a counterhegemonic environmental culture within Wilhelmine Germany.

Wappes's call was implicitly answered two years before it was even made. The response came in the form of a book by a forester and landowner named Heinrich von Salisch, titled *Forest Aesthetics.* Salisch put his new aestheticism explicitly into the context of the traditions of land beautification. He proceeded from the view that the best harvesting or foresting method was one that fulfilled its functional goals of producing wood while maximizing an area's aesthetic potential. For him there was no question that only a mix of deciduous and coniferous trees at different ages, carefully harvested in the single-stem manner, would yield this optimal result. Salisch supported his aesthetic argument by pointing to the social benefits of a more beautiful forest: foresters and lumbermen would lead a more fulfilling life, the woods would inspire more care and respect on the part of the local population, and possible hikers or tourists would reap the benefits of a more scenic land. With these interlocking arguments it is no accident that Salisch's book became something of a bible for making Germany's municipal woodlands into accessible and rewarding recreational facilities.

Given their philosophical congruence it is no surprise that Salisch joined the *Bund Heimatschutz* in its earliest days and that his book on forest aesthetics was cited repeatedly over the course of the next ten years.[20] Forests, however, were an area where hegemonic limits and environmentally progressive aesthetic demands clashed with particular vehemence. Franz Mammen, a member of the faculty at Saxony's forestry school in Tharandt, tried at every opportunity to pull the Heimatschutz away from its radical aesthetic position. According to him, there were "serious reasons of forest-technical and financial nature" that necessitated fir and pine cultures, and thus also clear-cutting (Mammen 1909, 56). No doubt some of Mammen's enthusiasm was due to the fact that the Kingdom of Saxony, his employer, had one of the highest percentages of pure fir forests in the entire German Empire and was deeply involved in the rapid-growth management regimen (Jacobi 1912, 13; Plochmann 1985).

Professional foresters were an integral part of the system of agribusiness and were often portrayed negatively by Heimatschützer as "senior foresters who manage their woods like a turnip field" or "those forest-wrecking forestry officials"; one Heimat advocate (who was himself a forest regulator) referred to professional foresters as "the most damaging insect in the woods" (cf. *Brandenburgia* 1906–7, 275, 277). Full-bore softwood production was the be-all and end-all for these typical members of the guild; at most they would be willing to make a few limited concessions to aesthetics, perhaps by leaving a thin screen of trees around their clear-cut areas. The Heimatschützer took a more radical position: foresting, they said, "must be beautiful, functional, and useful"—no matter what part of the woods one was in (Walther 1906, 8). The old rationality stood to be reformulated as the result of this equation, which accepted that "the ground rules of aesthetics can very often have an enlightening effect" (Mammen and Bruhm 1910–11, 39). A number of articles on the topic of "Heimatschutz in the Forest" applied this principle and showed that the aesthetic dimension of the forest was closely related to its ecological functionality (along with Mammen and Bruhm see "Heimatschutz im Walde 1905; Jacobi 1912; König 1915).

The general strategy in these articles was to contrast the ugly monocultured fir plantations of high-output capitalism with more beautiful and diverse kinds of mixed-species, low-impact management methods. Many authors cleverly anchored the distinction in a linguistic difference made possible by the German language: the beautiful mixed-species area was called a *Wald*—a term rich in positive Romantic and artistic connotations—whereas the production-oriented monoculture was consistently labeled a *Forst,* a more management-intensive term that almost involuntarily gave rise to visions of an "ugly lumber factory" (which was another favorite designation). The Heimatschutz authors pointed out to their readers that the *Forst's* monotonous appearance was due to the eradication of "the unique lower flora of the *Wald*" (Arnswaldt 1908). The "living forest edges that consist of bushes and birches, oaks and maples, ash and beech trees" also stood "in exquisite contrast to the monotony found in the rest" of a plantation ("Heimatschutz und Landwirtschaft" 1910, 63). Animal diversity was similarly eliminated in a monoculture; this had practical disadvantages in that the lack of birds exacerbated insect damage. Pure conifer stands were also criticized because they tended to exhaust the soil; last but not least, the comparatively poor humus production meant that a conifer stand would not retain as much soil moisture (Rasser 1910–11; Schlesischer Bund 1911; Jacobi 1912, 157f.). This last feature only made the effects of other interventions such as stream regulation, the dredging of rivers, and field rationalization worse. Taken together, all of these added up to a "system of impoverishment," an "organized exhaustion of the soil's strength" in which the absorption capacity of whole regions would be drastically reduced. As a result the water table would drop, flooding would become worse, and far-reaching climatic changes that they dubbed "steppification" (*Versteppung*) could take place (Hempel 1914; "Erhaltet die Knicks" 1912; Koch 1913b; Jünger 1913; Dammer 1916).

This far-reaching discussion marks the Heimatschutz as a significant popularizer of ecological concepts. Its aesthetic approach was remarkable not just for the discussion it started, though, but also for the sureness with which it led to a plan for mediation. As one supportive forester noted, a few general rules would go a long way: bushes and thick protective edges should be allowed to grow up around individual stands; unusual plants and animals should be preserved; and all sorts of birds should be encouraged—"in a nutshell *one should combat boring monotony*" (Walther 1906, 8; emphasis in original). As Wappes had predicted, the directness of the Heimatschutz aesthetic appeal succeeded in making the condition of Germany's forests a public issue where generations of scientific arguments had failed. Already by 1908 one insider could say: "The Heimatschutz issue has recently been a much-discussed topic in foresters' associations . . .; the applied forest literature is very occupied with it" (Mammen 1909, 74). Newspapers began calling for "the protection of our hardwood forests" for the sake of "Heimatschutz" (cited in Jacobi 1912, 174). And in Saxony, of all places, officials issued a far-reaching administrative guideline that encouraged the state's foresters to lay more emphasis on preserving natural variety in the public forests; indeed, the "forest-wreckers" were now ordered to "develop a keen interest in Heimatschutz efforts" ("Generalverordnung" 1911–12).

Aesthetic worries could be found by a Heimatschützer no matter where he or she looked. A sense of urgency was given by the prospect that in the not-too-distant future the entire landscape would be ruined: "our mid-size mountains leveled down, our proud rivers made into canals, our idyllic rivers polluted and poisoned" (Nordhausen 1913, 319). The simple beauties of a relatively natural stream appreciated in value immensely before such a dystopian perspective. One author wrote of a particular stream in northern Bohemia, one that gave its broad valley variety and beauty with its thickly grown, meandering banks. "A bitter feeling," though, came over him with the thought "that this little river, too, will sooner or later become a victim of regulation" (Korb 1913, 108).[21] This passage recalls the powerful effects that we saw above in the contrastive photographs of two Thuringian streams. A more didactic display of ecological-aesthetic sense can be seen in a set of captioned photographs that circulated in a publication of the Silesian Association for Heimatschutz (fig. 11). Using the well-known contrastive technique, it shows two streams: on the left the pleasant sight of what the caption identifies as a "natural stream-bed"; on the right, however, we see the same stream in regulated form. The caption here teaches the viewer to associate ecological implications with particular aesthetic impressions, for it reads: "This straightened section of the same stream not only looks boring, but the living conditions for the animal and plant world are worsened." Brief and to the point, this demonstrates very nicely the increasingly self-aware integration of aesthetics and ecology that makes the Heimatschutz such a significant development in environmental history.

Given the obvious interest of the Heimatschutz in limiting stream reg-

Natürliches Bett eines Gebirgsbaches

Fig. 11. Contrastive pair of stream photos. (From Schlesischer Bund 1911; photo reproduction courtesy of SBPK, Kunstbibliothek.)

Begradigter Lauf desselben Baches wirkt nicht nur langweilig, sondern es sind auch für Tier= und Pflanzen=welt die Lebensbedingungen erschwert

ulation, it comes as no surprise that a number of the regional Heimatschutz associations were actively involved. One of these was the state association in Reuß, based in the Thuringian city of Gera. In 1912 the association opposed plans to carry out a no-holds-barred regulation of the Elster; to this end it issued picture postcards of the threatened area and also commissioned a study on an alternative scheme that would leave a scenic bridge and landscape intact ("Von der Berliner" 1912). The Heimatschutz group in Silesia was able to influence procedures through contacts in governmental agencies. It held talks with that province's hydrological bureau and *Oberpräsident* to prevent the typical elimination of bushes and trees on river banks (Schube 1914, 3). In 1912 the imminent regulation of the Alster river became a source of concern for Heimatschützer in the Hamburg area. The matter was brought to the attention of the central *Bund Heimatschutz,* which submitted its standpoint on the matter to Hamburg's Senate. At the same meeting Fritz Koch reported that he was keeping an eye on the regulation of the Werra river ("Von der Berliner" 1912). And in Hesse, Heimatschutz influence was clearly visible in a local administrative guideline on the "Protection of Trees and Bushes along Waterways." Formulated by the Kreisamt Erbach, this general guideline of 14 April 1914 codified the Heimatschutz emphasis on educating people to everyday nature protection. Not only did it denounce the "excessive" cutting of trees and bushes, but it went on to give a detailed explanation of how the much more attractive tree-lined stream was also a far more vibrant little ecosystem, with more fish and more places for agriculturally important birds to nest ("Schutz der Bäume" 1914). Even if a direct Heimatschutz influence on the content of this guideline may be hard to prove, we can still say that the movement made an important contribution with effective distribution. It was reprinted and sent out widely by Koch's Foundation for Heimatschutz, obviously gaining far wider exposure than would otherwise have been possible (cf. StA Coburg MinD, Akte 20/196–98).

In its goal of restraining intervention into streamside ecosystems the Heimatschutz was encouraged by other reformers within the German hydrological engineering branch. One such man was H. Gravelius, the chairman of the Association for Earth Science in Dresden and an early supporter of conservation causes. Like many Heimatschützer, Gravelius expressed a healthy reserve vis-à-vis the "modern" belief in a "lordship of humans over Creation." No complete control was possible, he argued in 1905; in the long run humans could not aim at "overcoming" nature, but rather only at "closely conforming" to it (Gravelius 1905, 35–36). The Heimatschutz reflected this attitude in looking skeptically at technological fixes, particularly as a response to flooding.

A case involving an overeager public works department in Bavaria was cited with obvious interest. After a flood of the sort that might occur only every 50 years or so hit the town of "L." in 1909, its civil engineers carried out a far-reaching program of regulation: they cut down trees on the banks of the river, lined both sides with stone, installed iron weirs to catch floating objects, and further used the opportunity to cover a good section of the river with a massive street overpass. A review by an enlightened Bavarian state administrator named Cassimir leveled heavy criticism at these "technical mistakes." Fundamentally, he argued, the measures were misplaced because of the rarity of extensive flooding. But more specifically they ignored the fact that the best flood protection consisted of two natural features: "a thick cover of woods and a good blanket of plants." The Heimatschutz journal boiled Cassimir's insights down to the basic recognition that "radical cures" would always exact "bitter revenge," and it recommended conforming to the "natural configuration" of the waterway as much as possible ("Wasserbauten" 1911–12). Cassimir had impressed the Heimatschützer so deeply that he was invited to give a presentation at the large biennial meeting in Dresden in 1913; through copies of this speech his low-intervention ideas found their way into a wide range of government offices in Prussia and, most likely, other German states (cf. distribution list 26 February 1914 in Hess. StA Marburg 180, 2603). If in subsequent years more and more civil engineers came to see, with Cassimir, that maintaining natural stream banks was one of the most effective forms of flood protection, it only confirmed that equation between the environmentally "beautiful" and the economically "functional" that the Heimatschützer had always promoted (cf. Weyrauch 1915).[22]

Within the Heimatschutz movement there was a similarly informed opposition to moor drainage. The journals are replete with articles celebrating the protection of individual moors (see for example *Heimatschutzfragen* 1914, 12; *Blätter für Naturschutz* 1 December 1912, 22f.). Even crash cultivation efforts during World War I were recognized as a significant threat to environmental diversity and opposed on those grounds. With the pressure to show patriotism intense, and with a large number of prisoners of war being pressed into service, the motto of the Heimatschutz was caution: "Our saying here is: watch out that there is not too much cultivation going on!" (Geisenheyner 1916; Bödecker 1915; *Pommersche Heimat* April 1916, 13f.). The draining of moors and swamps was opposed in part because of its long-term effects on the water table, already sinking due to stream regulation and field rationalization. As one author wrote, this reckless reliance on a "one-sided principle of utility" would result one day in damage "of unimagined dimensions" (Jünger 1915, 3).

Although the Heimatschutz remained focused on large-scale visual ensembles, its commitment to environmental health and natural beauty led it to also take on the problem of water pollution. Critics in the past have erroneously asserted that the Heimatschutz showed no interest in the problems of pollution (cf. Andersen 1987, 149; Wettengel 1993, 361). In the later years of the movement, as of about 1912, this was certainly not the case.

At this time a medical doctor named Georg Bonne became virtually omnipresent in the Heimatschutz movement. One of his first contributions was a presentation on "The Increasing Pollution of German Rivers" to a gathering of regional leaders in February 1912. Here he described the "economic, health-related, scientific, and aesthetic" impacts of water pollution in such horrifying detail that the Heimatschutz not only vowed to communicate Bonne's findings to all government agencies but also printed the presentation as an article in the journal *Heimatschutz* (Bonne 1912). As a crowning achievement Bonne give a long talk on the subject at the biennial Heimatschutz conference in Dresden in 1913.[23]

Under the impact of Bonne's forceful presentations and the supporting views of other authors, the issue of water pollution soon attained the status of "one of the most important Heimatschutz issues" (cf. *Siegerland* 1913, 95). A typical case came from Gera, where the local Heimatschutz association joined swimming groups in protesting the pollution of the Elster. In the Heimatschutz petition to the river police it was emphasized that "the German people has a right to live by pure waters" and that no pollution should be tolerated that stemmed from the egoistic activities of individuals (Landesverein Reuß 1913, 4). But the aesthetically driven concern with pollution also led to more scientific questions. Given adequate attention to the problem, the Heimatschutz noted, it was "theoretically" possible "to purify all kinds of household and industrial wastewater to any desired degree" (Haupt 1911–12, 457).

But of course this was where the real problems of the Second Empire began. In theory the means and even the ethical issue of responsibility were both cut and dried matters. But as Haupt, the author of the article, noted,

> In practice it is a widespread and in a certain sense understandable tendency of those involved either to shirk these tasks altogether or to provide only inadequate means for treatment, so that this has to remain more or less illusory. (Haupt 1911–12, 458)

A good part of the pollution could be traced to the "dividend sewage" of an "unsocial industry"; by the time of the war Heimatschützer angrily contrasted the desperate need for better antipollution measures with the 25 to 68 percent dividends companies were paying out to stockholders (Bonne 1915; Dießner 1918).

But municipalities were also guilty of spewing tons of untreated waste into Germany's rivers, and a huge legal problem had to be overcome before either kind of polluter could be taken to task effectively. Because rivers flowed from one German municipality or state to the other, an effective and uniform code had to be established at the level of the Empire before anybody would be prepared to commit significant amounts of money to pollution control. Haupt emphasized that this kind of *Reichswassergesetz* was in the works, but that a tremendous political battle would take place before it could ever be enacted. In this situation, he wrote, the most effective thing for the Heimatschutz to do was what it did best: "work towards enlightenment and education!" (Haupt 1911–12, 458).

Haupt meant this in a broad sense. He saw that the pollution problem was not a dry, technical matter, but that its solution would depend upon deep-seated ethical and environmental feelings. Haupt contrasted Germany to England in this regard. Although both countries had experienced a "similar, brilliant industrial development," England was nevertheless further ahead when it came to pollution control precisely because of its more developed popular sense of the environment. Whereas in Germany companies shirked their duties to the populace and shamelessly polluted the countryside, in England, Haupt said, pollution-reduction "burdens are carried willingly." Part of the reason for this was a set of "imponderables" that were rooted in the "love of nature." "The Englishman is perhaps ahead of us in this area," Haupt conceded, and he pointed to the almost universal concern of the English for birds and hedgerows, as well as the popular weekend custom of picnicking and relaxing on the banks of rivers. This kind of popular environmental culture, Haupt contended, was the base upon which effective pollution control had to be built:

> All these people appreciate their relatively clean river . . . They would feel its pollution to be a kind of personal offense against them, and they stand united behind the government, which strictly enforces the laws preventing this pollution . . . The English industrialist, for his part, usually sees it as the duty of a gentleman to make the sacrifices necessary in order to take care of his wastewater . . . [In England] the views with regard to the impermissibility of extensive river pollution have become common property, and precisely therein lies the best guarantee for a true protection of the rivers. (Haupt 1911–12, 459)

One could question whether pollution was in fact so well controlled in England, but what is more important is that Haupt's analysis of that country indicated a clear strategy for the Heimatschutz. Its most important task would be to help create the general cultural conditions under which political solutions to the pollution problem would become possible, "to awaken an understanding according to which the protection of bodies of water is a kind of duty of propriety . . . a duty to the whole . . . and also a duty toward nature" (Haupt 1911–12, 459; see also Haupt 1917).

The Heimatschutz strategy of awakening general ethical and cultural awareness, though, did not exclude practical legal and political maneuvering. On the contrary: in the area of water pollution, as elsewhere, their aesthetic sensibilities guided the Heimatschützer toward an increasingly well-informed and activist brand of environmentalism. Pointing to the pathbreaking waste treatment union that was in the process of being set up along the Emscher river, Haupt himself urged fellow Heimatschützer in Saxony to "work toward the formation of such cooperative structures" (session protocol 25 November 1912, Stadtarchiv Dresden LvSH I/124). This kind of action was already under way further west, in the neighborhood of the heavily industrialized cities of Barmen and Elberfeld (now joined as Wuppertal) on the Wupper river. In spite of abundant complaints that the river had become a reeking cesspool, it was not until 1910 that the two cities built a crude water treatment plant. But it only provided for a mechanical filtering of effluents, and thus downstream users continued to suffer under the stench and the strong chemicals and dyes that the Bayer concern, among others, was allowed to dump into the waste stream. Governmental structures failed to redress the problem, since the responsible commission consisted of even portions of the feuding upstream and downstream parties. The only remaining solution was a civil suit.

In 1911 citizens formed the Bergisch Association for Heimatschutz in the downstream city Solingen. Since one of its "primary" goals was the cleansing and protection of area rivers, from the very beginning it tried to gather the money necessary to sustain a costly civil suit (cf. protocol of founding in Stadtarchiv Solingen S2663). Membership dues were set at the low level of one mark in order to obtain broad support from the population; the association even gathered donations from passersby on bridges. Figure 12 shows a postcard used to promote the cause. Somewhat incongruously, it traced the architectural beauty of a nearby castle to "gifts of the people," a generosity that should now be turned toward making the "Wupper as clear as it used to be." The propaganda's lack of historical acuity and its "romantic" image of a solitary wanderer should not mislead us, though: it was directly tied into a very concrete battle for environmental rights in which a rather broad segment of Solingen's population wanted to "show the government that it had to take care not only of the kings of industry, but also of the health of the little people" (*Solinger Zeitung* 18 September 1911).

By 1913 a suit was under way against the cities of Barmen and Elberfeld (cf. *Gesundheit* 1913, 27). The legal battle was long and involved and was put on hold throughout the war years. When the cities still showed no signs of improving their facilities after the war, the Heimatschutz association tenaciously pursued the suit and simultaneously applied pressure with publicity campaigns and petitions. Through 1920, 1921, and 1922 the association continued its court fight, amassing huge legal bills. Finally, in 1922, it agreed to settle the matter out of court, in part out of the fear that it could be attached with costs. Although the settlement called for a strict timetable in forming a workable regional waste treatment scheme, the same opposing interests and unreliable stuctures managed to drag out the process; it was not until 1936 that the first new plants went into operation (a thorough examination of the legal process is given in Wey 1982, 90ff.).

The intransigence of the cities in this case is a reminder of the kind of power that can inhere in a system of outdated regulations, financial interests, political timidity, and lack of environmental concern. The chilling disregard that officials of Barmen and Elberfeld showed for their fellow citizens and the environment demonstrates once again the very basic ethical and structural barriers that stood in the way of even the most obvious kinds of environmental protection. But if the Wupper case reveals how far the regulatory system and certain segments of the German populace still had to go on the path toward environmental thinking, it also demonstrates that aesthetic concerns were a tremendously important force in helping to change that situation. It shows us the Heimatschutz as an organization that gathered around a highly aesthetic vision of the landscape, but one that never remained on a superficial, cosmetic level: increasingly, as we have seen, it was an aesthetically inspired concern that was moving in the direction of an explicit, concrete, and engaged kind of ecological activism.

It would be impossible to measure the impact that Heimatschutz ideas for landscape reform had on Wilhelmine society with any accuracy. We saw how it achieved a not inconsiderable amount of governmental support for different parts of its program, but probably much more important in the long run was its less documentable, overall cultural impact. A recent study accorded the Heimatschutz a "prominent role" in environmental affairs in Wilhelmine Germany—an assessment that is all the more notable for being based on an assumed membership many times smaller than the actual one (Dominick 1986, 276).[24] Another study has shown that the Heimatschutz was the prime source of inspiration for the *Wandervogel*'s interest in conservation (Morris-Keitel 1991, 131). In the field of gar-

Fig. 12. Postcard appeal for assistance in cleaning the Wupper River, around 1916. (From StA Solingen, Ve 005–004.)

den architecture Heimatschutz ideas had a similarly definitive impact, as we saw above. No doubt part of this far-reaching influence can be attributed to the simple organizational strength of the Heimatschutz movement. But it was also rooted in the quality of the Heimatschutz ideas. Within the broad spectrum of environmentally interested groups at the turn of the century the Heimatschutz stood out as an organization with perhaps the widest and most detailed vision of Germany's environmental future.

This breadth of perception allowed the Heimatschutz not only to tackle problems at hand with good effectiveness, but also to anticipate and criticize weaknesses within the loose network of environmentally interested groups. We saw above how the limits of Naturschutz were recognized early on; the Heimatschutz was similarly aware of shortcomings of the Garden City movement. Already in 1912 Carl Johannes Fuchs anticipated the danger of suburban sprawl, warning that the excessive "expansion of the area required for housing schemes of the Garden City type" was a potential problem, and that the result would be in many cases "a not very pleasant bastard between city and rural area" (Fuchs 1911–12, 316; Fuchs 1911, 148). A similar critique was brought to bear on the land reform project of "inner colonisation," in which settlers would take possession of so-called empty land—areas such as heaths, moors, and marshes. Along with various Naturschützer, the Heimatschutz distanced itself from this idea of draining and plowing under what little remained of Germany's wild or at least nonproducing acreage.

As these examples show, the Heimatschutz constituted an important ecological corrective to a number of more narrowly oriented reform programs. The effectiveness of its position has to be traced back to its emphasis on a wide, even panoramic view of the land such as that seen in a few of the vignettes in this chapter. The Heimatschutz perspective did not allow a one-sided pursuit of this or that element of the countryside, but rather visually argued that the whole was the precondition for the parts. The ecological implications of this unique emphasis on an entire landscape can be seen when one compares it to the program of many bird protection organizations. In the few histories of environmentalism that touch on Germany's early developments, the bird protection organizations are usually admired as some of the most active "Green" groups (Barthelmeß 1981, 133f.; Dominick 1992, 53f.). They were some of the largest organizations, it is true, but for many years the bird lovers were singularly unable to see that the bird population decline was due to the effects of rationalization and habitat destruction; a few of the strategies they developed were thus such decisive measures as a tax on household cats and admonishments to nest-destroying "rampaging youth" (Schwarz 1903, 14; "Vogelschutz im Sauerland" 1909). Another supposed cause of the decline were the Italians, whom many German bird enthusiasts accused of capturing and eating unconscionably large numbers of their feathered friends. Only slowly did it occur to the bird groups that habitat destruction might be at the bottom of it all:

> Yes, it is true that the Italians trap masses of our songbirds in the autumn; but it was like that centuries ago . . . Only the complete disappearance of good nesting places led to the rapid decline in numbers of our songbirds. All the hedges and field borders in our fields are disappearing . . . no little corner remains undisturbed anymore for the songbirds. (Grimm 1908, 180)

Unfortunately, even this insight did not necessarily lead to ecologically decisive action. Instead of analyzing the root causes of habitat destruction, much of the bird-protection establishment signed on to a program of providing artificial nesting boxes designed by Hans von Berlepsch (cf. Kaeflein 1902; Schwabe 1915).[25] This kind of monofunctional box was of course not the equivalent of the ecologically complex hedgerows and underbrush, which as far as most of the bird protectors were concerned could continue to be cut down and hauled away. In this situation it required the "interests of a comprehensive Heimatschutz" to correct what was rightly identified as "a one-sided kind of bird protection" ("Kalender" 1914, 100).

This very concrete effect of the Heimatschutz's landscape aesthetic in turn shows how far the movement had brought the discussion of nature. In the short span of ten or so years the Heimatschutz managed to carry out a far-reaching deconstruction, if one will, of the rigid borders separating humankind and nature that had been established by mechanist capitalism. Eighteenth- and nineteenth-century enthusiasms for nature or landscape had tended to view nature as the absolute Other, as wilderness or as an exceptionally impressive landscape feature. Thus right up until the arrival of the Heimatschutz, people typically saw "the 'real' landscape . . . in some other place, in an ever more remote distance, out in the countryside, in the mountains, on the coast," and explicitly not in their own industrialized and manipulated back yard (Sieferle 1986, 252). This purist definition of nature or landscape probably persisted for so long precisely because it did not seriously challenge the hegemony of environmental exploitation. The movements to establish nature parks or animal preserves were similarly innocuous. These and other "green islands" can be easily tolerated by a destructive capitalist system; indeed, as Raymond Williams noted, it is often the capitalists themselves who often put in a good word for their favorite exclusive spots (cf. Williams 1989, 220).

The Heimatschützer spearheaded a fundamental change in this system with their landscape conception of Heimat. On the one hand this involved a challenge to the purity of "Nature" as it had been commonly thought of: city parks were not "Nature," for instance, and distant lands were not necessarily Rousseauian paradises, either. On the other hand—and this is the important point from an environmentalist perspective—it also involved pointing out that real nature was all around, in every landscape: from the most sublime and untouched to the hardest-working and most quotidian. The lesson to be learned from this was that humans should learn to accept this kind of everyday nature and learn to cultivate it as an enrichment of life. In city parks it was clear that the mass of recreational needs demanded a nature pruned back somewhat from the nineteenth-century ideal of the wild; but in the working countryside it was conversely true that nature should be allowed to assert itself noticeably more than previously.

In summary, the Heimatschutz advocated a spectrum of uses in which people and nature would have locally varying emphases, but that overall would be marked by a spirit of recognition, integration, and cooperation. When we take into account its skepticism about visions of wilderness as well, it is clear that the landscape aesthetic of the Heimatschutz contained perhaps the first politically effective recognition of "socially constituted nature." This view is based on the knowledge that nature no longer exists without our express involvement: "The nature that is of practical significance for us today . . . is co-determined in its condition all over the globe by the existence of people, it is more or less an historical and social product" (Böhme 1989, 73). Confronted with a relatively small country in which virtually every square kilometer was either in production or threatened with a notoriously insensitive regimen of development, the German Heimatschützer grasped the truth of this statement long before any comparable group elsewhere in the industrial West. And as the Heimatschützer surveyed the bare and monotonous surfaces of twentieth-century life, it became apparent to them that humankind's complete domination demands that it be able to relax that domination. It must even consciously pursue a strategy of reducing pressure and finding ways to accept nature back into spaces and processes from which it had been all but eliminated.

Here the aesthetic orientation of the Heimatschutz contributed decisively to its intuitively correct understanding of the problem. The specifically aesthetic transmission of the Heimatschutz program has been reflected upon at length above; it suffices here to recall a few of the major findings. The first is that an aesthetic view of the environment was an organizational precondition given the specific schooling and background of many bourgeois citizens. Beyond that, though, an aesthetic discussion of landscape values can also be seen as a (not necessarily conscious) strategic adaptation to severe hegemonic pressure; in this language it became possible to overcome some of the axiomatic demands for industrial rationality and disposability, and instead to assert community rights to a well-cared-for environment. But going even further, it is also apparent that the Heimatschutz's aesthetic approach to the landscape was well suited to a high degree of concrete ecological perception and reflection.

It was in fact only a short step or two from a visual appreciation of the countryside to consideration of the "household of nature"—the material interchanges that make up the scenic hedgerows, forests, and rivers (Schlesischer Bund 1911). This last fact marks the Heimatschutz aesthetic as a significant advance, for as Peter Finke has argued, all subject-centered versions of landscape appreciation have participated in environmental destruction by discounting the importance of ecological "realia": "Our landscapes suffer not only from the wounds that we inflict on them. Long before that they are threatened by disdain for their realia, by ignorance of the fullness of life that fills them and builds them up" (Finke 1986, 273).

But while Finke is no doubt right about the significance of ecological facts, it still seems appropriate to emphasize the importance of all the non-factual aspects of environmentalism. Even today, but particularly in the historical setting of Wilhelmine Germany, the array of available facts will always have holes. In our daily lives our behavior is also not always linked to a rational analysis of facts, but rather to a layer of custom and cultural response. If environmentalism is therefore to succeed, facts will always have to be backed up by values, and these in turn will have to be embodied in certain practices.

This, I think, was a second and truly significant feature of the Heimatschutz aesthetic. In a time in which ecological facts were fewer and farther between than we can sometimes imagine today, the landscape aesthetic not only guided people to the study of ecological relationships, but also gave that study a real, culturally tangible meaning. It may have done so by merely "elaborating that which already is," but its effect was to give environmental concern that "rational spontaneity" that Gramsci emphasized as an important element of a counterhegemonic culture. In the next chapter we shall examine some further embodiments of this new culture of environmentalism.

CHAPTER 4

Toward an Environmentalist Culture of Home

Much of the preceding two chapters has been devoted to showing that the aesthetically interested Heimatschützer became truly concrete in their concerns. In a certain sense this chapter will proceed in the opposite direction: for while the Heimatschutz project can be identified with a kind of functional, hands-on ecological planning, it is important to see that this was not pursued simply as a matter of imposing a biotechnical fix. Instead, from the very beginning the idea of Heimatschutz was rooted in a broader kind of environmental education in which functional demands were balanced by efforts to work out patterns of social acceptance, conviction, and practice. These eminently cultural efforts can be divided into two main types. One type would correspond to a more traditional, artwork-centered idea of culture. Under this rubric we will look primarily at novels that take up some of the Heimatschutz concern for the environment in their fictional worlds. Along with these reflections, though, we will also consider a second type of culture: that of active daily practice, or *Lebenskunst.* Here we will be concerned above all with the practice of tourism, which the Heimatschutz recognized as a vital site of interaction between modern humans and their environment. In examining these aspects we will see that the Heimatschutz was crucially located in attempts to form a new environmentalist culture.

The interpretation to be offered here will require the reader to abandon a good portion of the conventional wisdom that has accrued on the subject of Heimat. Precisely in the realm of culture this is a word whose connotations have been almost universally negative. Often this can be traced to critics' generally modernist prejudices, of course: in their eyes any attempt to rebel against the industrial-urban system can be dismissed as "romantic escapism" (Achleitner 1989, 165). But there are more specific critiques that deserve somewhat more attention. These connect the idea of

Heimat in the arts more closely with an interest in nature, and thence with particular social or ideological agendas. Because of superficial similarities it is therefore important to make a distinction between *Heimatkunst*—"Heimat art"—and the cultural program of the Heimatschutz movement.

Although Heimat and art are concepts with many different overlaps and conjunctures, "Heimat art" has been a term connected above all to a circle of writers that gave new emphasis to the slogan of Heimat in the two decades around 1900. The new journal "*Die Heimat. Blätter für Literatur und Volkstum,*" published by Georg Heinrich Meyer beginning in 1900, provided a key focal point of discussion. Here theorists such as Friedrich Lienhard, Adolf Bartels, and Ernst Wachler claimed a new role for regionally distinctive cultures as a counterweight to the "predominance of Berlin" (Lienhard 1900). Heimat in their sense demanded above all a negative definition: it was to be a rebellion against the big city and industrialism generally, as well as against the style of naturalism, political parliamentarianism, and internationalism, to name just a few. What this meant in positive terms is less clear. Lienhard was the least biased of the theorists, arguing for a regionally concrete brand of national idealism that would attain to "Goethe's mature harmony of worldview" and breathe an ethereally optimistic "highland spirit" (Lienhard 1900, 2). According to Adolf Bartels, though, Heimat art pursued something far earthier. Taking his cues from Langbehn's widely read book *Rembrandt as Educator* (1890) as well as racialist thought, Bartels saw the key to all cultural development in the continued existence of a race of warlike German peasants. Under Bartels's influence Heimat stood for a program of nationalist blustering, explicit anti-Semitism, and a knee-jerk rejection of modern civilization.

Along with the theoretical views just mentioned, it was above all the novels of a few scores of loosely connected authors that have shaped our understanding of the cultural value of Heimat in turn-of-the-century Germany. Karlheinz Rossbacher has compiled a list of approximately 116 Heimat writers; Gustav Frenssen, Wilhelm Holzamer, Hermann Löns, Wilhelm von Polenz, Peter Rosegger, Heinrich Sohnrey, and Clara Viebig are perhaps the best known of the group (Rossbacher 1975, 65f.). Not surprisingly, their novels reflect many of the demands made at the theoretical level. Rejecting big-city settings and much of the thematics of modern civilization, the plots tend to unfold within the tightly circumscribed arena of a German farm family—a unifying characteristic that has allowed critics to speak of the "peasant novel" interchangeably with the "Heimat novel." Contemporary political dimensions are omitted, and the texts generally obscure an analysis of social mechanisms (Schwerte 1967, 133f.). In their stead, developments are generally driven by naturalized phenomena: mythical feuds, inborn character, age-old traditions, floods, and droughts.

These stories culminate in the successful propagation of yet another generation of proud, defiant German farmers, a familial event flavored with parallel significance for the continuing existence of a strong German nation. Through the plots of these novels the word *Heimat* thus acquired a very specific set of connotations centered on the construct of a national-chauvinist "eternal peasant" (cf. Zimmermann 1975, 54; Muthesius 1974, 124). A disturbing ideological direction latent in this genre is traced out in Hermann Löns's *Wehrwolf* (1910), a book whose peasant thematic is connected with an aggressively racist German nationalism.

But not everything that has come to be categorized as *Heimatkunst* corresponds to this picture of primitive, virulent chauvinism. A variety of styles and concerns existed here that defies most monolithic group characterizations. Even individual authors could vary in their tone and concerns from piece to piece. Hermann Löns, for instance, cannot be reduced to his *Wehrwolf* novel, but must also be appreciated as one of the most widely published *belles lettres* authors of the Heimatschutz and Naturschutz scene. It is significant that his shorter forms attracted attention in these circles, forms in which there was neither room nor any particular need for all of the extra ideological baggage that the *Wehrwolf* carried with it. Characteristically, it was above all his newspaper articles that were praised upon his death early in the war: his sharply worded attacks on "nature-pillagers of every kind," along with his impassioned critiques of a narrowly oriented "conwentzional" Naturschutz, were recognized as important contributions to the public debate ("Hermann Löns" 1915). The Austrian author Peter Rosegger presents a convincing parallel: as Charlotte Anderle has shown, it was in his journalistic writing that all the superfluous, moralistic "packaging fell away," enabling him to attack industrial pollution and even the inadequacies of the Naturschutz park idea with uncommon precision (Anderle 1983, 75f.). Latter-day critics would do well not to dismiss these writings by simple reason of association with the much-vilified idea of *Heimatkunst.*

One way in which we can come to a more evenhanded appraisal of the idea of Heimat is thus to acknowledge the diversity and the potentially progressive environmentalist aspects that have often remained hidden behind the category of "Heimat art." Jost Hermand (1991) has taken this tack in his reading of *Heimatkunst* works, arguing that there are important Green convictions contained in their allegiance to natural beauty and a national social ethic. But this strategy only relativizes the many negative qualities foregrounded by a host of other critics; it does not eliminate them. Hermand himself finds no other choice than to abandon the concept of Heimat to their accusations of "volkish mysticism, agrarian romanticism, pre-fascism" and the "false illusion of an idyllic world" (Hermand

1991, 123). If this were actually true, then it would be a devastating blow not only to our picture of the Heimatschutz movement, but also to environmentalism in general. Fortunately, though, there are good reasons to doubt this conventional wisdom.

The root of the problem lies in the fact that both sides in the debate seem willing to lump together all movements, art forms, and expressions that have the word "Heimat" in common. These critics often portray authors of the *Heimatkunst* movement in particular as members and comrades-in-arms of the Heimatschutz movement. Against this I would argue that there are very good reasons to distinguish between the two, both organizationally and ideologically.

The organizational reasons to distinguish between the two movements (discussed below for the case of the authors) can be given fairly solid demographic proof. Rossbacher conducted a very thorough sociological examination of the *Heimatkunst* phenomenon and found that authors and audience alike were generally located in a precarious position at the edge of the bourgeoisie. The prototypical *Heimatkünstler* was an elementary school teacher, and his readers occupied "much the same social position": their social prestige was low, their degree of education was not sufficient to gain them entrance into the higher echelons of society, they were often located in rural areas, and they were characterized by existential economic fears that they compensated with visions of strong independent peasants (cf. Rossbacher 1975, 74–75, 98). A comparison of this social profile with that presented in the data in chapter 2 reveals truly significant differences: the Heimatschützer were on the whole well educated, often secure in their civil service jobs, and very often urban; if the breakdown on social prestige is at all reliable, the Heimatschützer were noticeably better-off in this department as well.

We can underscore the divergences revealed in this sociological data by challenging the frequent assertion that *Heimatkunst* authors were avid participants in the Heimatschutz movement, and vice versa. Approximately fourteen moderately well-known authors were asked to sign the original "Aufruf zur Gründung eines Bundes Heimatschutz" in early 1904, but of these, only three—Friedrich Lienhard, Heinrich Sohnrey, and (perhaps) Peter Rosegger—could be characterized as belonging to the *Heimatkunst* movement.

After the founding this fairly low participation by the *Heimatkunst* movement in the Heimatschutz declined further still. Looking at the available membership lists and public appeals, a total of approximately ninety men and two women Heimatschützer either self-identified as "poet" (*Dichter*) or "writer" (*Schriftsteller*) or can be independently labeled as such. A significant number of the names here were those of older authors who were not at all part of the newer ideological grouping of *Heimatkunst:* these included Viktor Blüthgen (1844–1920), Wolfgang Kirchbach (1857–1906), Heinrich Seidel (d. 1906), Johannes Trojan (1837–1915), and Julius Wolff (b. 1834) from the Berlin/Brandenburg area, Martin Greif (1839–1911) and Wilhelm Jensen (1837–1911) from Munich, Heinrich Hansjakob (1837–1916) in Freiburg, Felix Dahn (b. 1834) from Breslau, and Theodor Pantenius (1843–1915) and August Trinius (1851–1919) from Saxony/Thuringia.[1] This group of older authors was complemented by a strong contingent of less prominent, younger writers born primarily in the 1860s. Most of these were journalists of one kind or another: representative figures would be Max Bittrich (b. 1867), a newspaper editor in Freiburg, Carl Meißner and Paul Schumann (d. 1927), both contributors to the *Kunstwart,* Hermann Löns (1866–1914), who worked for newspapers in Hannover, and Eugen Kalkschmidt (1874–1962), at this time an editor in Dresden. Wilhelm Bölsche (1861–1939) was a better-known author who corresponds somewhat to this last group as well. But except for Löns and Gustav Falke, a member of Hamburg's Heimatschutz association, none of these men can be considered to be a member of the *Heimatkunst* movement, or to have even a remote commitment to ideas centering on an "eternal peasant."[2]

One of the difficulties of defining the cultural dimension of the Heimatschutz comes from the fact that writers' and artists' commitment was often only temporary or superficial, limited to signing their names on a public appeal or paying a few marks per year in membership dues. Some of the best-known writers on the list belong in this category. Friedrich Lienhard is a good example. Often cited as prominent connection between *Heimatkunst* and Heimatschutz, his participation in the latter movement seems to have been limited to signing his name on the the original "Call for Founding." An analysis of his works reveals little concrete environmental concern and no mention, to my knowledge, of the Heimatschutz. His *Journeys in the Vogesen—A Book of the Times* can serve as an indication of his distance from the engaged environmental thinking of the Heimatschutz (Lienhard 1902). It is precisely distance that the narrator seeks, actually: as an elite poet it is beneath him to "get involved in social issues in the noise-filled haze down there" (16). A highlight of the journey is a night spent camping on a mountain top with an acquaintance. It is a clear night and Lienhard's narrator admires the "overwhelming scale" of the stars, before whose metaphysical backdrop the concrete earth fades away, creating a "profound desire for eternal clarity, for knowledge of the final harmony of all things" (118). The effect of all of this nocturnal reverence is shown to be negligible the next morning, however. Having woken up and stretched a little, the two men proceed down the mountain in something less than an awed and reverential frame of mind, shooting off a revolver to make the valleys "resound with a crack" (121). Lienhard's text is a good example for that environmental alienation that has often accompanied interpretations of nature as *sublime,* as opposed to concretely and emotionally *beautiful.*

Lienhard presents just one example of a writer whose work shows little reflection of Heimatschutz thinking, but he is not alone. Many of the writers who earned their keep as journalists or as producers of specific types of literature may have had little opportunity to weave their otherwise solid Heimatschutz conviction into their professional output. On the other hand it is also possible to discern some fairly clear references to Heimatschutz or environmental thematics in works by people who cannot necessarily be documented as members. In the following discussion of fictional "Heimatschutz" works this content-oriented definition of the works has provided the basis for selection in combination with an author's participation in the Heimatschutz milieu.

The first example of environmental thematics is a negative one. Max Haushofer was the author of a pamphlet on Naturschutz and was frequently cited in Bavaria's conservationist circles (cf. Haushofer 1906). He is representative of some scientifically schooled writers whose exposure to the broad sweep of natural history made them impatient with the narrow problematics of their present society. Their goal was instead to see the big picture, to find its ancient roots, and speculate on the earth's ultimate fate. In a move typical of this fascination, Haushofer wrote a series of short scenes and even an entire novel concentrating on the end of the world. Extrapolating from popular decadence theories, he asked, "Can times of an animalic regression overtake us? Times in which civilized peoples become savages again and finally sink from savages to half-monkeys?" (Haushofer 1908, 210). The answer he supplied in his collection *At the Limits of Existence* was a "yes"—given "completely changed external conditions of life," human life, too, would decline (1908, 210). Haushofer went on to apply this "ecological" approach in a number of short "images of the future." Interestingly, though, the stories all have a distinctly antienvironmentalist effect. The almost geological time span and the enormous magnitude of world-ending disasters that he describes are so large, so "sublime," in a way, that they dwarf any pitiful actions or attitudes over which humans might have power. Such is the theme of "World Autumn," a scenario in which a solitary wanderer ruminates on the inevitable, final, and universal changing of planetary "season." The story "Sunny Side" plunges us into consideration of geologic cycles and the impermanence of climate. Greenland and the Arctic, Haushofer observes, were once verdant; "today a frosty shroud lies over those long-gone fields!" The chief effect of this macroscopic view seems to be to desensitize the reader to here-and-now obligations; Haushofer himself noticed this effect, saying that "There is nothing that raises a person so far and large above his little sorrows and joys" than geological history (223). Significantly, in all of these scenarios it is never anything that humans do that precipitates disaster. The story "When Streams Run No Longer" presents a scenario in which the world simply runs out of water as the result of planetary cooling. "Worldwide Fire," on the other hand, gives us the hotter and much more dramatic prospect of a star or huge comet crashing into Earth.

It is interesting to note that the politically enervating effect of these stories is matched in recent times by a kind of ecology that emphasizes the long-term instability in nature. Reacting against an earlier generation's harmonious and balanced model of synchronic nature, some younger students of ecology tend to emphasize that nature displays radical changes if considered over very long periods of time. Donald Worster has noted that this kind of environmental relativism, while "valuable" intellectually, can also become a

> dangerous habit of thought, leaving people bewildered and uncertain to the point of paralysis, demolishing all the cherished myths of tradition but leaving nothing in their place. Now that it has become so thoroughly imbued with historical consciousness, ecology runs the risk of total relativism. A cutover land can be seen as good ecologically as a forested one. A landscape riddled with opencast coal mines, bleeding acid into streams, is as "natural" as any other. Only human subjectivity can decide which state of the earth is preferable to another. (Worster 1993c, 176)

As Worster rightly points out, the emphases of this particular branch of ecological research feed right into a kind of value-free transformationism and thus indirectly reinforce the dominant culture of environmental exploitation. Whether spatially or temporally, a too-distant perspective on things results in political quietism. Fortunately, the converse statement is also true: bringing the perspective closer to home and making issues more tangible in the present is often the first step toward effective political action. In the best of the following novels we will see that the Heimatschutz conception of Heimat embodied precisely that kind of concrete local focus that enables and encourages active environmentalism.

A local emphasis is certainly an important part of Hermann Burte's *Wiltfeber* (1912), subtitled the "History of a Heimat-Seeker."[3] Although Burte, whose real name was Struebe, was not a member of a Heimat group, in the overall thematic as well as a number of close observations we can see parallels to Heimatschutz concerns. The events of the story are set into motion when the protagonist Wiltfeber returns to his home village in southern Baden after a nine-year absence. Like the Heimatschützer, Wiltfeber carried with him the ideal of a landscape in which historical ambience and natural beauty would be maintained; the village of his memory was such a place, with its "lovely houses, beautiful running springs, blooming gardens, rustling linden trees, old, artistically mortared bridges." But upon his return Wiltfeber finds that this once-satisfying environment has been laid to waste: as he comes into the center of the village he recoils, seeing that the trees in the town square have been chopped down and replaced by reinforced concrete poles; the public fountains are in bad repair as well, and incompetent people have made grievous changes to the church and other ancient structures. Not long afterward Wiltfeber surveys what used to be a scenically wild flood plain. In language that was taken directly from the Heimatschutz, the narrator describes Wiltfeber's dismay at seeing that engineers have been at work regulating the river. The first change he notices is the "bare embankment":

> In his youth a wood had stood there, poplars, willows, elders with thick undergrowth, full of unusual plants, large moths, beautifully marked salamanders and snakes: it had been a kind of a jungle, a natural protector of fields of crops and hay against the raging of the swollen river. But the governmental hand had cut down the wood and put up stone embankments: sober, hard, and uninspiring, these geometric banks cut through the valley, always leaking somewhere, needing repairs, damaged; their sloped sides sported a sparse covering of green; no grass grows wherever the governmental hand strikes. (196)

Only a minor difference in tone indicates that this narrative will take a very different direction from that of the mainstream Heimatschutz: the rigid denunciation of the "governmental hand" suggests some of that antibureaucratic fervor that was building up within the New Right during the Wilhelmine era. But in its detailed observation there is that same close connection of aesthetic sensibility and ecological understanding that stood out as a trademark of Heimatschutz thought.

The narrative reaches a crescendo of environmental indignation in a scene in which Wiltfeber and a few loyal friends are exchanging stories in a riverside cave on a beautiful Sunday morning. They observe a wagon laden with a large tank crossing the nearby bridge; one of Wiltfeber's acquaintances identifies it as a load of "factory effluent" that is regularly dumped in the river. The description of this malicious act of pollution can be seen as a central passage in the book because it combines in exemplary fashion a clear aesthetic-ecological critique of the act with the narrator's increasingly misguided interpretation of the social and economic reasons for pollution. For here it is not the factory owner or an entrenched system of profit and exploitation that is seen as the cause of environmental degradation, but rather the implausible bogeyman of a resentful, surly lower class, as personified by the driver of the wagon:

> Cutting a broad swath for himself, the hireling strode self-importantly next to the horses, did not see the men in the cave, and brought the wagon to a stop in the middle of the bridge. And looked all around to make sure that nobody saw him; then he opened the faucets on the tanks, and a jet of golden solution shot into the clear water of the canal. And look! far out of the river jumped noble trout in their fight against death; their bodies shone in the sun like silver; there was a frightening jumping, bobbing, springing, coiling, slapping, writhing, and snapping up out of the water and a united falling back into the poisonous brew.
>
> But up on the bridge there stood the hireling and laughed and snapped with his whip at the leaping fish: "Go on and die, I'm not going to get sick eating you," he cried, "you are not there for guys like me!" The river, reeling from the potion and stench it had swallowed, threw the dead trout onto the shore, and they rolled up to the men's feet in a most sorrowful fashion. (267–68)

The book's tendency to "transfer all evil and negative properties from capitalism to the proletariat" is something that Peter Morris-Keitel has noted in his dissertation on environmental literature of the German youth movement (Morris-Keitel 1991, 138–39).

The slant becomes patently obvious in the story told subsequent to this pollution scene: one of the three men present in the cave is a now-propertyless farmer named Behringer who relates how, out of misguided sympathy with the egalitarian philosophy of Socialism, he agreed to run his farm on the principle of common property. What followed was a long decline marked by a virulently anti-Heimat disregard for natural and historical beauty. The chief villain in the farm's demise, a "Red" agitator named Sailer, incongruously mimics capitalist rationality with his demand to convert treasured household antiques into money for more equipment. Indeed, the new Socialist strategy seems to have been to turn the farm into a prototypically modern bit of agribusiness:

> Artificial fertilizer was thrown by the pile onto the fields, the crop growth shot up, but later it was worse than at the beginning; machines were acquired, real clockwork mechanisms that had to be fixed all the

> time because no one took care of them; rivers of cupric brew, true cloudbursts of greenish-blue suds were sprayed on the grapevines and eliminated the pests, it must be admitted, but also their enemies, creating along with the medication a new disease, and the vines weakened. (330–31)

The difference between this remarkably skewed and displaced analysis and that provided by the mainstream of the Heimatschutz movement is enormous. As we saw in chapter two, from the very beginning the Heimatschutz recognized systemic pressures of capitalism for what they were, and more specifically it targeted an entirely different class of wealthy "native vandals" as those most responsible for Germany's degraded environment. Burte's book elides this entire aspect and transfers the blame to the very class that had the least say in the process. In spite of the shared environmental concerns, *Wiltfeber* thus constitutes a monumental misappropriation of Heimatschutz ideas.

The book is also suffused with a level of racial argumentation that has nothing in common with the paradigms subscribed to by the wider movement. Here Germans' treatment of the environment is torn out of its cultural and economic context and traced back to a heavily racial scheme of "supermen" and "subhumans." A neglected cemetery becomes the sign of a "bastardized race," the village and Germany alike are in decline because the population has become a "third-string race of half-breeds," and the only source of salvation would be a return to the firm and autocratic leadership of a blond nobility (15, 50). Needless to say, this path toward an environmentally sound Heimat contrasts sharply with the solutions implied in most other Heimatschutz-style evocations of the theme, to which we now turn our attention.

Wilhelm Kotzde's 1917 novel *Frau Harke geht durchs Land*—an approximate translation would be "The Rake-Woman Goes through the Land"—takes a concrete Heimatschutz case and uses it as the basis of a romantically tinged novel of contemporary society.[4] At the center of the novel stands the regulation of the Havel river around 1909–12, an event that the author experienced firsthand as a resident of the river city of Rathenow. As someone who was interested in folklore, something of an amateur botanist, and a frequent companion of the local fishermen, Kotzde was a founding member of the Rathenow-based Havel Land Heimat Association in 1909; not surprisingly, he saw one of its main tasks to be an active program of "protecting the beauty of our Heimat," with a special emphasis on bird protection (cf. *Brandenburgia* 1909–10, 310f.). This background clearly contributed to his desire to write about the social and environmental costs of the massive regulation project, and as early as

1913 he published an article highly critical of it in the journal of the Brandenburg Heimatschutz association. This article and a follow-up article in which he rebutted a Prussian official's reply contain the sociopolitical essence of the later novel, with some passages being transferred almost verbatim (Kotzde 1913; Kotzde 1914). The two sides in the real-life debate reappear in the novel as well: the Prussian civil engineering technocracy is represented in the book by a technical adviser named Grothe; meanwhile it is a wise, middle-aged character named Hans Martin Wittenbusch who takes up the cudgels for the cause of the Heimat.

Wittenbusch's central role needs to be emphasized, since in the later stages of the novel members of the youth hiking movement, the *Wandervogel,* are introduced. This circumstance led at least one member of that group to label *Frau Harke* a "*Wandervogel* novel," a judgment that has been taken up in Morris-Keitel's recent scholarly examination as well (Gerlach 1917; Morris-Keitel 1991, 177f.) In general this would make far too much of the rather limited *Wandervogel* presence in the narrative, however: Gerlach herself conceded that the land was both "showplace—and hero" of the novel, an opinion in line with another contemporary reviewer who disputed the significance of the *Wandervogel* and preferred to call *Frau Harke* a "Heimat book" (Gerlach 1917; Kitzler 1918). With regard to the environmental thematics in particular, though, it would constitute an outright distortion to call this a *Wandervogel* book, for it is chiefly through Wittenbusch's reflective and informed perspective that we gain an insight into the environmental destruction that is wrought on the Havel river. The *Wandervogel,* on the other hand, are portrayed as a kind of outdoor hiking organization full of romantically self-centered (or nationalistically Bismarck-crazed) youth who must first be educated to appreciate and take an active role in protecting the environment. The novel thus provides a historically accurate representation of what Morris-Keitel himself found: it shows that environmental concern originated in the Heimatschutz and Naturschutz circles of the older generation and only slowly filtered through—one might say, in a distorted and insufficient fashion—to the youth movement.

We can appreciate this all the better if we return to Kotzde's original confrontation with river regulation, one that clearly took place under the sign of Heimat. In his articles for the Brandenburg Heimatschutz association's journal, Kotzde had been concerned to force at least an after-the-fact analysis of the regulation, which quite aside from its supposed benefits for shippers had incontrovertibly had an effect on many other aspects of life along the river: "It not only destroyed many beauties of nature and put an end to the unique animal life of the area, it also intervened deeply into economic life and deprived hundreds of families of their bread" (Kotzde

1913, 108). And even if this analysis came too late for the communities on the Havel, Kotzde insisted on looking at the mistakes that were made "with all due incisiveness, so that future regulations in other places do not steamroll over fishermen as they did here" (Kotzde 1914, 17). The arguments mirrored that of the *Bund Heimatschutz* in its fight against the Laufenburg dam, where it was proposed that any project had to be analyzed in terms of its total impact on a region and not just selectively or in terms of the most immediate financial return. As Kotzde emphasized, not even the project's backers could actually know whether the project would be beneficial in the long term: "The full benefit or cost that the regulation brought will only be determinable years in the future" (Kotzde 1914, 23).

Certain things were apparent already, though. The environmental costs of regulation, for instance, were clearly high:

> The unique beauty of the lower Havel region is for the most part destroyed. Where once the river's edge was lined with trees and shrubs and reeds, now there lies stone ballast. Everything was taken away, everything . . . People traveling down the Havel feel like they have walked into a field of debris. (Kotzde 1913, 113)

Kotzde also vehemently criticized the social and economic dislocations that regulation brought; from a politically conservative point of view he noted bitterly that fishermen who had once been "loyal to the king" were now voting almost without exception for the Social Democrats.

But there was more than a bit of the folklorist's and anthropologist's lament here, too: this was a uniquely rooted community, Kotzde wrote, with its own sayings and unique customs. Indeed it is from this store of figures of myth and superstition that he appears to have drawn the title figure of the "Rake-Woman," a Gaia- or Frigga-like goddess who is directly visible only to children. According to the one small girl who was lucky enough to glimpse her, the Rake-Woman appeared in "a long green dress, her hair was golden yellow like ripe wheat on the field, and her eyes, they shone just like the blue sky" (26). The environmental colors would be fully appropriate if, as a contemporary reviewer theorized, the Rake-Woman represents "the atmospheric magic that rests on the austere Brandenburg landscape" (Mielke 1917–18, 126).

Although this Heimat goddess slips into the narrative only for a moment or two at a time, and then only in the imagination of the characters, she nonetheless deserves the title role. For she represents precisely that quality that differentiates the novel from the articles that preceded it. They treated social issues and claims and counterclaims about the economic viability of regulation and in a certain sense only rhetorically touched on the value of an environmentally intact Heimat. The book, on the other hand, was accurately subtitled a "novel of a landscape." In living up to this name it had to do two things: first, it had to make a commitment to an exact, even biologically precise understanding of the land. This is guaranteed by the central figure of Wittenbusch, a university-educated Heimat enthusiast who maintains a large natural history collection and who also keeps in close touch with the folk wisdom of his fishermen friends. Characteristically, his adversary Grothe suffers from a limited and stereotypical understanding of the landscape. In Wittenbusch's words, Grothe sees only

> that the sky is blue and that white clouds sail by on it, and sometimes grey ones—then there's rain. When the white gulls scream and snap up a struggling fish, that looks quite lovely . . . but maybe you saw other things, too, for example, that the green pines stand out very nicely above the steep white banks and that the trunks glow red in the evening sun. About what a painter from Berlin sees when he comes out on his summer vacation. (36)

With this critique of superficial aesthetics, Wittenbusch shows how the ideal of Heimat demanded more from its proponents, a willingness to delve into more thorough, more rigorous, inevitably more scientific studies of home.

But at the same time the book impresses upon us that the landscape can never be equated with descriptions of the river and its environs, no matter how detailed. Instead it is the scenes in which humans express their deep-felt ties to the local environment—that "Heimat emotion" emphasized by one reviewer—that have remained the most powerful passages (Kitzler 1918).[5] Commenting on the first wave of regulation and bank reconstruction, for instance, one old fisherman says of the engineers:

> They chopped down all the willows. The whole time it was like looking into an open grave. I suppose we shouldn't hold it against them, they didn't grow up on the water and don't know all the things that they are destroying. Children will thrust their hands into a bird's nest where we treat it as if we were looking straight into the eye of God. (51; see also 160, 162)

To Kotzde's credit, his novel realizes that not every fisherman is so attuned to an environmentalist point of view and includes their differences of opinion with Wittenbusch over the question of predatory birds. Finding an osprey dead in his nets, a fisherman comments approvingly,

"That's one fish-robber less." Wittenbusch, who happens onto the scene, retorts that such birds are probably an insignificant factor in terms of the total fish population but are irreplaceable elements of the area's beauty. He then goes a little bit off the deep end with an overblown celebration of the osprey as a "precious inspiration of God"—a flight of rhetoric that is met with polite but stony silence on the part of the fishermen (144).

The great strength of the novel is thus that environmentalist sentiment, when it does appear, flows out of a complete and well-rounded picture of the fishing village's economic base, its social life, and its traditional and even lightly superstitious cultural background. Precisely in these latter subjective and human constructs the novel begins to communicate some of the "atmospheric magic" that lay in the landscape of Heimat. Romantically tinged as she is, the Rake-Woman, like Kotzde's novel itself, represents an all-important act of emotional investment in the beauty and vibrance of the environment.

The Naturalist writer Max Dreyer (1862–1946) was a member of the Heimatschutz association in Mecklenburg. His 1911 novel *Auf eigener Erde*—"On One's Own Soil"—is structured as a love story that starts with two children growing up on adjacent estates in Pomerania: Ursula von Eich is the only child of a moderately prosperous farmer, and her childhood friend is Anselm von Morveldt, youngest scion of the far older noble family that owns Rotenmoor. Both Ursula and Anselm have a powerful attachment to their respective pieces of land. When it becomes clear that Anselm's older brother will inherit the estate, Anselm complains that he now has "no Heimat"; no one understands this feeling of loss, the narration states, better than Ursula, "who herself is the heir to a rural estate, who holds her property with all her passion, who watches over everything with jealous tenderness: over every tree, every clump of earth, every rock, every blade of grass of her land" (17, 10).[6] In the course of the story the reader is confronted again and again with the contrast between Ursula's pragmatic but caring stewardship of the Eichhof estate and the progressive liquidation of Rotenmoor under the control of Anselm's disinterested and decadent older brother Jochem.

This opposition of a sober, more considerate, and implicitly more bourgeois regimen to the more wasteful and bizarre practices of the German nobility begins at the symbolic level of architecture. The Morveldts' buildings are a mishmash of styles going back to the twelfth century, each one the reflection of a new owner's caprice and desire to stamp his individuality on the estate. In a probable allusion to the architectural sins of the Second Empire's "Founders' Generation," this disharmonious aspect is linked particularly to an unhealthy influx of French money around

1690. This was a time when, as we are told, the young heir came home from service with the French, "equipped with a considerable amount of money and an even bigger fascination with Versailles." The young man thereupon "ruined the beautiful old park" and built a new wing on the mansion in a "bombastic Baroque style, with no consideration for what else stood there" (40–41). The contrast with its neighbor from the middle of the eighteenth century is striking:

> How different the Eichhof house is . . . Here there reigned, along with all spiritual freedom, a loyal sense of tradition; at the Morveldts, who are so much older than the Eichhofs, the connection was torn asunder again and again by a mercurial and stubborn temperament of envious pride. (42)

This contrast between a harmonious bourgeois spirit on the one hand and the nobility's self-centered conceits on the other takes on an environmental dimension in the respective owners' relationship to the land. The increasingly decadent Jochem von Morfeldt treats his estate as a source of quick income, cutting down an ancient grove of oaks and then selling off parcels of land in order to support his consumer and drug habits. As the narrator remarks disapprovingly: "What did he care about Rotenmoor? To him land was dust, and woods were lumber and ashes. For him there was only the soft music of his scintillating senses" (295). By contrast it is clear that the narrator of this "book of the earth," as it was labeled by one reviewer, attributes high value to Ursula's careful upkeep of buildings, her defense of her wheat fields against trampling sport hunters, and the way she accepts the naturally slow and unforced rhythms of the countryside (Meyer 1912, 86).

This attitude is expressed particularly well in a scene in which the father, who is something of an amateur scientist, has obtained a small vial containing radium and cannot stop admiring it and speculating with acquaintances about the possibilities it contains. But the idea that this new radioactive "original substance" could revolutionize life offends Ursula, who tells her father spontaneously:

> I don't like your substance. I just don't like that kind of an extract—that kind of artificial concentrate or condensate or whatever you call it. And when people go on about this sort of a substance, saying that it has all these properties and can do everything, then I just get mad about it . . . People should go out into the sunshine instead. And into the woods. The original substance is there, too, but alive and in freedom. And not just dust in a capsule. (69)

Ursula finds her opinion justified in the revelation, soon afterwards, that the radioactive "extract" can be lethal (70). The thematization of radium betrays some of Dreyer's Naturalist roots and a related interest in scientific theories. His best-known work before this had been the 1899 play "On Probation" (*Der Probekandidat*), which had defended the teaching of Darwinism in schools (Dreyer 1903, 1904). This particular respect for biological science, along with the strong Naturalist conviction that environments can have a decisive impact on human growth and development, forms an important backdrop for assessing Dreyer's 1911 novel.[7]

As Ursula's comments reveal, this later Heimat period is perhaps characterized by a certain desire to relativize earlier scientific commitments: not to give them up, but to reemphasize the wider human and natural settings in which they are employed. Ursula's dislike of the radium should thus be interpreted as a healthy skepticism vis-à-vis technical fixes and cure-alls of all sorts. Significantly, these doubts in principle do not prevent Ursula from taking advantage of technology. She is able to keep the farm going through several tough years after her father dies in part because of a patent he obtained on a process for converting sugar beet waste to fertilizer. But as hard as she works herself and as efficiently as she farms the Eichhof, she always returns to a patch of marshy land around a pond that is her personal site of "rootedness." After her firstborn dies and her first marriage is breaking up, it is at this pond that she finds her continued identity:

> It was at it had been and looked back at her as usual . . . But there were more flowers, especially irises.
>
> She leaned against a tree and closed her eyes and then embraced the trunk with sudden emotion.
>
> Here I am and here I belong, and whatever is out there is a foreign country. And again she closed her eyes, held by her tree. (324–25)

This passage is interesting because it seems to recognize an active, participatory relationship between humans and their local environment: both the pond and the tree end up "supporting" Ursula not just at the physical level, but also at the level of an emotional need.

Although these scenes in which Ursula communes with nature may strike the reader as sentimental, it should be remembered that they are scattered widely over the course of a narrative that shows Ursula to be a remarkably hard-nosed and independent woman. She spends long hours toiling over the farm's books, overseeing harvests, and worrying about liquidity. In contrast to a purely capitalist agribusiness, though, her sense of property carries with it a sense of environmental propriety—or as we read early in the story, her idea of property is one suffused with tenderness. Her almost dialogic relationship with the land is therefore not to be discounted as a bit of poetic fluff left over from the Romantic period. Instead it seems reasonable to see her as a model for the new environmental relations of Heimat, where it has become acceptable not only to talk about the economically functional value of the land, but also to give voice to the human and emotional stakes in its continued health and beauty.

A novel that takes many of the same themes into more socially concrete territory is Maximilian Böttcher's 1910 novel *Back to the Land* (*Heim zur Scholle*). Here the enlightened nobleman Dietrich von Eckardt eventually gives over much of his inherited estate in Brandenburg's Mark to his workers, who are able to turn the area into a garden-like landscape full of social harmony. Like Dreyer's *Auf eigener Erde, Heim zur Scholle* is structured as a combination love story and *Bildungsroman:* Dietrich must go off to the city, serve in the army and in the civil service, and also absorb the teachings of reformers before he can come back to the quiet but capable Waldtraut, who by this time is a socially engaged teacher in the small town near the estate. But what makes the novel of particular interest in our context is its advocacy for a new attitude toward the land. As a "Brandenburg Heimat enthusiast," the protagonist Dietrich makes the ties between humans and their local environment an integral part of his "green" reform plans (37).[8] Major sources of inspiration for him are the land reform and Garden City movements (503), accompanied by an early recognition of the "magic of the Heimat in the Mark" (10). Exaggerating only slightly, his best friend calls him a modern-day Antaeus (224).

As in Dreyer's novel, this new paradigm of environmental and social consideration is contrasted with an established system of exploitation. Berghausen, the estate that Dietrich stands to inherit, is endangered by his spendthrift and morally decrepit absentee father, a hot-tempered alcoholic, womanizer, and wife-beater. Not coincidentally, the father's attachment to the land is purely financial: at one point he suggests simply selling the estate and investing the proceeds in a more profitable scheme of buying Berlin apartment buildings. But the father is in fact only one of an array of figures who embody the socially conservative and environmentally unsound practices of the large landowners of Eastern Germany. As a critical figure in the novel observes, one of their key failings is their treatment of the land as "an object of speculation, a money-making venture" (108). One of the chief offenders in this category, a Major Bögendorf, subscribes quite logically to a landscape aesthetic typical not only of the nobility, but also of their increasingly mechanized and capitalistic agribusiness:

> Bögendorf's aversion to all rural smallholding came not only from his pride in his station, one which would have liked to see only the well-endowed nobility established on rural land—it also had something to do with his aesthetic taste. His eye loved the expansive fields that were often hundreds of acres in size. (114)

His views are seconded by a forester who espouses the value of rationalization; against the argument that a forest should be beautiful, he states baldly that in a modern forest there is "no room anymore for beauty" (55).

Dietrich's reform plans are aimed directly at overthrowing this large landowner system, in both its environmental and its social aspects. Strong and capable female figures stand by his side in this endeavor: Dietrich's mother is a sober and industrious counterweight to the father's irresponsibility, and his eventual fiancée Waldtraut proves her tenacity not only by running her ailing father's household, but by teaching and participating in community affairs as well—both she and Dietrich refuse, moreover, to let themselves be distracted by jibes about "modern emancipation witches" (500). The proximity of Waldtraut's domestic activities to merely conventional and sexist depictions of women might give one pause, but as with Ursula in the previous novel, Waldtraut demonstrates a self-aware and rational commitment to her multifaceted work that is not at all identical with superficial and exploitative stereotypes. The novel does, admittedly, attribute to her a pronounced nurturing quality. Walking through the woods together early in their acquaintanceship, Dietrich notices that "she stepped carefully around every sprouting little plant that had pushed itself out too far into the narrow path"; during the same pivotal walk he cannot help remarking to her, "How gently you treat everything green" (55, 57). Nevertheless this is not a permanent division of labor, for soon Dietrich has learned from her and begins to propound the rule "Let it grow if it is beautiful and does not hinder you" (161).

A characteristic feature of Boettcher's book, though, is the way in which it sets these laudable environmental convictions within a much wider and socially concrete commitment to reform. Dietrich expresses at one point the fervent hope that his planned model village will function like a "terrible inferno" to "reduce the dark, twisted alleys of reactionary feudal ownership to ash, and to force its masters to get down in line with their workers, shoulder to shoulder, to build a straight, airy, secure road into the future" (506).[9] Dietrich's mentor and uncle plays an important role in this sociopolitical "light and air" reform. An intellectual hermit of sorts, Dr. Friedrich Berndt was a professor of economics who hobnobbed with Marx and Lassalle; after being forced to step down he retired to the life of an independent scholar and farmer. His intensively farmed and conscientiously maintained fields of fruit in Brandenburg's Mark were no doubt meant to resemble those of the "Obstbaumkolonie Eden," one of Germany's first and most successful vegetarian communes, likewise located not far from Berlin. In the book something of the same spirit of socialism is gradually instituted in Berndt's switchover to employee profit sharing, all of which helps provide Dietrich with inspiration.

Another source of encouragement is Harald von Bögendorf, who upon inheriting the neighboring estate of Friedrichshof begins to institute a similar series of reforms intended to better the living conditions of agricultural laborers, including profit sharing (210f.). The contrast with Burte's novel is instructive: although in Boettcher's book the "socialized" scheme of Friedrichshof is labeled a "crazy way to run a business" in much the same way that Behringer's experiment is derided in Burte's book, Boettcher makes clear that the source of this negative characterization is a self-interested class of large landowners, who quickly ostracize the reformer Harald and charge him with harboring a lethal threat to "all agricultural firms in the German fatherland" (247). Harald's early death prevents his own experiment from being carried out, but it only strengthens the young Dietrich's resolve.

Dietrich's plans begin to take shape more firmly in two more confrontations with the limitations of existing German society. The first comes when he refuses to engage in a duel and is dismissed from the officer's career path he had embarked upon (405f.). Although a few of Dietrich's friends write of a new feeling of community with lower-class soldiers in battles in German Southwest Africa, Dietrich nonetheless comes to reject the militarist path to social unity: "Should one long for a war to take place in order to experience the wonder of this brotherliness here?" he asks himself—only to note that the "service of peace" offered "opportunity enough for awakening in the young sons of the people . . . a strong feeling of belonging together, a great national feeling" (415).

After renouncing the army Dietrich is spurred on toward active reform by a closer examination of the "narrow, pitiful holes" that were workers' apartments, and the emaciated, sickly proletarians themselves:

> He saw among them the ones with no energy left, the ones with pale, hollowed faces upon which death had already impressed its mark, the wretched ones who in the morning could not work up even enough energy to raise themselves up out of their miserable beds, and whom hunger and animal need, these terrible torturers of poverty, whipped up again and again, until under the whistling blows they died a groaning death. (425)

Although verging on the stereotypical, these close observations à la Käthe Kollwitz nevertheless confirm that Dietrich's rural reform project should be seen as a constructive answer to public health disadvantages in the capitalist-controlled cities, not simply as a knee-jerk form of big city–hating. The narrator tells us that it is in this context that Dietrich returns again and again to the legend of Antaeus; at his desk there also hangs a copy of Hans Thoma's famous "Sower," which seems to reinforce the lesson of the myth: that "lasting power" can only lie in closer contact with the soil (427).

It is this socially aware move "back to the land" that ultimately characterizes the reform community that Dietrich is able to found on the inherited estate. Although undeniably tinged with a certain amount of patriarchal benevolence, the closing scene of "spic-and-span" little houses echoes a societal vision that is consciously "green":

> Everywhere in the gardens men and women strolled around, the women mostly looking down at their beds of vegetables, the men for the most part looking up at the tops of their young fruit trees which had already produced a bold flower here and there. Would these young trunks have the energy to develop the flower into fruit, and bring the fruit to ripeness? The people would do their best to water them and care for them. And from many an arbor that was covered with webs of green-leaved wild grape vines there rang out the bright laughter of children at play. A picture of peace. (509)

The metaphorical parallels that Boettcher sets up here between the growth of the young fruit trees and the success of the community are perhaps a bit heavy-handed, but looked at in the context of the entire story, they are full of genuine significance. Unlike the large-landowner system with its featureless expanses of grain and its alienated labor, this is a community in which small-plot growers derive intense satisfaction from the abundant variety of plants and the trees under their personal care. Even the native vines seem to want to participate in the symbiotic relationship, freely winding their way up onto the arbors and giving throngs of children a splendid place to play. It is indeed a "picture of peace"—one that prophetically combines social justice with environmental harmony.

A similar integration occurs in Hermann Wette's humorous novel *Jost Knost—The Hercules of Latop* (1908). Land reform ideas are combined here with Heimatschutz environmental concerns in a portrayal of the labors of a farmer-citizen named Jost Knost who faces a modern Herculean task: that of ridding his small city of corruption, industrial pollution, and greedy land speculators. Knost's ideal is explicitly identified with the strong community-held resource base of the traditional German "*All-*

mende" system (345).[10] This admiration for a bygone era, combined with the fact that the hero is a farmer, may sound like elements of a *Heimatkunst* novel. But Jost Knost is no primeval peasant figure: he studies various natural sciences and economics at the university and sees the world while working for a rich uncle in San Francisco. Increasingly he combines the qualities of a larger-than-life noble savage with those of a sober social reformer who wants to devote his life to "works for the common good" (143).

The influence of the Heimatschutz can be observed in the fact that Jost's public-spirited acts of community service often have important environmental components. One of the first of these acts comes about when Jost has returned from the university to Latop for his Easter break; here he is informed of the corrupt city officials' plan to sell the municipally owned, marshy "salten field" to an industrial chemical concern owned by friends and relatives. Using his university training, Jost analyzes the hydrology of the area and discovers that any pollution on the site would quickly spread to other parts of the town (158–59). When this evidence fails to sway the townspeople and the vested interests, Jost convinces his father Heinrich to step in and buy the plot, donating all development rights to the community—in this manner "protecting the Heimat from any new attack" (161). In subsequent sections of the narrative the reader comes to see that his kind of far-sighted environmentalism apparently ran in the family.

When Jost goes to work for his uncle Wiprecht in California, he finds a man

> who did not with greedy exploitation ruthlessly suck dry the beneficial granary of this extravagantly productive land, as was usually done by most Americans, but who rather cared for and protected it with the conscientious thoroughness of the German farmer who also thinks to the future. (186)

This is a remarkably clear formulation of a pragmatic environmentalist attitude, and it is one that warrants seeing Wette's book as a kind of green *Bildungsroman.* In his "apprenticeship years" of working for his uncle, by now a world-class trader, Jost's horizons are considerably broadened. As a financial giant Wiprecht can even enforce a certain degree of environmental justice, as when he and Jost squash the disreputable firm of "Ajasses and Son." These "exploiters and profiteers," Wiprecht says, have committed *Raubbau,* "destructive exploitation, crimes against the forest of the most egregious kind. For miles and miles the most magnificent stand of trees was sacrificed to the moneybags of those two." Stepping into the role of a kind of Green free-market sheriff, Wiprecht orders his deputy

Jost to "Get those scoundrels!"—which Jost quickly does by means of stock purchases and manipulations (230).

At about this time, Jost's grandmother writes him a letter in which she expresses her hope that Jost will commit himself to serious reform even if he chooses to stay on with his uncle. It combines two appeals: the first is a Biblically flavored injunction to remain conscious of people's common "childhood in God," one that would require brotherhood in the social sphere along with a specifically environmental recognition that the earth was "given as a bond" for which humans would one day have to give an accounting. As if this were not enough, the letter went on to include a second level of environmentalist conviction with its excerpt from a contemporary newspaper article. In it an unnamed American conservationist argued for a new relationship to the land so that people "will henceforth not shamelessly exploit, plunder, and devastate this magnificent country's land and soil as pirates and industrial robbers" (211–12). His grandmother's letter apparently helps Jost see that Wiprecht's is still a capitalist environmentalism; and although he recognizes some positive sides to the United States, such as equal rights and self-aware women, it would remain for him the land of "unbounded egoism" (257).

Jost Knost therefore follows his heart home to Germany, at least in part because of the better prospects there for a more genuine sense of community. Of course his return is also predicated upon his growing fondness for his hometown sweetheart, Adda, who already in her childhood was an active protector of animals. After a four-year absence Jost finds a "terribly defiled Heimat": the beautiful town forest now features "stinking briquette-smoke chimneys" and nearby Jost spots "green-shimmering effluent streams of the chemical factory"; on top of this what remains of the forest has been ransacked, with "thousands of the most beautiful beech and oaks" having been sacrificed to meet financial needs, "without any consideration for a good regrowth through planting young trees." Jost also notices that the once-abundant animal life has vanished: deer and pheasants no longer inhabit the forest, and even the songbirds have been driven out due to the lack of underbrush (259f.). The city itself has also become a public-health and reform-movement case: houses stand empty or neglected, gardens and fences are ill-kept, "everywhere trash in the streets, in the gutters refuse! And the brook that used to be so cheerful and clear now a yellowish-green, cloudy and viscous mass full of stinking factory waste" (266). Even architecturally Latop has become a source of Heimatschutz concern, with the ostentatious new mayor's office a poor fit with the "Old German gable-roofed houses" that surround the town's central square (268). Combined with a town administration full of crooks and a deteriorating community fabric, these environmental problems constitute a true Augean stall that only this modern-day Hercules could clean.

One of Jost's first and most pressing tasks is that of taking over the family estate. Farmers all around are plagued by increasingly serious infestations of insects, and one summer stands out because of an enormous number of mice. Jost traces these problems to their source—a lack of a balanced habitat such as the town forest for birds and other predators—and in the best Heimatschutz manner proceeds toward shutting down polluters and carrying out a program of planting and reforestation "for the recuperation of the landscape" (295). The financial details of this program are of course something that belongs in a work of fiction: Jost wins a large sum from his uncle Wiprecht as the result of a bet and thus can single-handedly buy up and dispose of polluting industrial plants. The plan also only functions politically because the workers here were all imported "Polacks and Croats" with few ties to or rights in the German town.[11] And although Jost does go on to promote a railroad spur and to build a hydro-elecric and flood-control dam, the basic direction of his intervention is to simply get rid of all industry—thus avoiding tackling some of the thornier problems of an environmental build-down.

In spite of these drawbacks Wette's novel sketches out some of the broad commitments of a new, environmentalist Heimat sensibility with remarkable clarity. The social convictions that underly this conception are particularly notable. For at the end of the story Jost abjures his role as "Green capitalist" and donates all the land he owns to a municipal holding structure in perpetuity, thus following his conviction that all land should "again become the common good and property of the people" (346). As we saw already in Maximilian Boettcher's novel, the tendency here is toward a powerful linking of social and environmental agendas: home, according to these prognoses, would be a place where the ideals of conservation would be central to the economic structure of society.

Although we have concentrated on a literary transmission of Heimatschutz ideals above, it would be possible to submit the output of painters and illustrators who can be associated with the Heimatschutz to a similar analysis. As always, one would have to distinguish between different levels of participation. If a signature on the 1904 founding appeal can be believed, a number of well-known painters stood behind the Heimatschutz: these would include above all the Worpsweder Fritz Mackensen (1866–1953), Fritz Overbeck (1869–1909), and Heinrich Vogeler (1872–1942), as well as Hans Thoma (1839–1924). If we add in painters who became members of the *Bund Heimatschutz* in its early years the group can be enlarged to include Fritz v. Uhde (1848–1911) and Karl Vinnen (1863–1922). Among the moderately well-known painters in local or regional Heimatschutz organizations were the three *Elbier:* Fritz Beckert (b. 1877), Arthur Bendrat (b. 1872), and Ferdinand Dorsch (b. 1875), as well as Paul Ernst Hermann (b. 1870), all in Dresden. The available records show that the Heimatschutz movement also found support among graphic artists and illustrators: these included not only Otto Ubbelohde, the creator of the Heimatschutz logo, but also Wilhelm Hasemann, Hugo Petzold, Walther Strich-Chapell, and Hans Volkmann. It is likely that other painters and illustrators were involved in regional Heimatschutz activities or were influenced without ever becoming members. The range of material that could thus be examined is vast, and a thorough investigation of it would exceed the scope of this investigation.

From a cursory knowledge of the work done by Otto Ubbelohde and Hans Thoma—perhaps the two most active Heimatschützer in this group—it seems clear that "Heimatschutz painting," if there is such a thing, would resemble the landscape drawings discussed in chapter 3. Aside from an obvious thematic concentration on the countryside its main characteristic would be an uncomplicated, perhaps even conventional, but always direct technique, one aimed at drawing the viewer into an emotional rapport with the land.

The description fits Hans Thoma's generally bucolic paintings particularly well. To dismiss them as "kitsch" would be to miss the point: it is not what we see in them that is important, but rather what kind of significance they might have carried for contemporaries. It is striking that the all-around reformer Dietrich, in Maximilian Boettcher's novel, kept one of Thoma's most popular prints above his desk. And although this was only a fictional representation, there are indications that Dietrich's real equivalents in the Wilhelmine era drew similar inspiration. Part of this was related to the person of the painter, of course: Thoma enjoyed a very high reputation among Heimatschützer and Naturschützer due to his public engagement for conservation, having used his position in Baden's parliament, for instance, to voice demands for bird protection measures. A knowledge of such external dimensions might well explain why these people saw Thoma's images as icons of environmentalist sentiment. But other observers seemed to derive much the same meaning from Thoma's work. This is at least the conclusion that one must draw from a statement by another bourgeois reformer of the day, who, unconnected with the Heimatschutz, nevertheless praised Thoma's idyllic pictures for evoking a "noble and soothing feeling of community among humans, Heimat, and nature" ("Hans-Thoma-Feier" 1909, 301).

If this perception was, as seems likely, widely shared, then it is apparent that Heimatschutz in the pictorial arts achieved effects very similar to those of much of the literature discussed above. In both media artists played out their demand for Heimat in a fictional world, and in both, as we have seen, Heimat consisted precisely in an emotional and ethical involvement with the land.

It detracts not at all from these efforts to observe that a certain level of abstraction is built into their particular kind of cultural transmission: one that presumes the ability of the viewer to apply a set of feelings, attitudes, and even ethics in a completely different situation, one that is not hypothetical and "artistic," but rather quite real and intractably quotidian. And yet it is precisely in the common-sense and hegemonically structured daily world that change must take place. No movement for fundamental change can afford to ignore the gap that confronts people at this stage, and in this sense it can be said that one of the most significant areas of Heimatschutz cultural activity was that devoted to mediating individuals' real-time encounters with their own surrounding environment. The efforts here centered on tourism.

In the following section we will see that the Heimatschutz developed a *positive* conception of an environmental tourism. This may come as a surprise, for until now most critics have only perceived the Heimatschutz's admittedly far-sighted thematization of the *negative* consequences of modern tourism for the environment. Even these, though, were not just wild "rantings" (Dominick 1992, 71). Ernst Rudorff, for instance, demonstrated a clear understanding of the "exploitation and wearing-out system of tourism" in one of his earliest articles (Rudorff 1880, 264). The necessarily exploitative character of tourism was rooted, for him, in the dynamics of economic mechanisms. The laws of capitalism guaranteed that any location of exceptional beauty would gradually be exploited for its greatest profit potential through "the business-like operation of the enjoyment of nature" (264). More recent analysts might call it "the commodification of spectacle," but by any name its physical institutions—hotels, trains, roadways, and crowds—would then inevitably change the area; eventually they would result in the diminution or even disappearance of the nature that people had wanted to admire in the first place (Rojek 1993, 119). Very early, then, the Heimatschutz formulated that realization to which Hans Magnus Enzensberger would come back 90 years later: namely that tourism develops "a dialectic that in the end also turns against its beneficiaries" (Enzensberger 1973, 9).

Individual behavior was a key, constituting feature of the business dialectic: by streaming toward the prepared, paved, and popular attractions people not only encouraged businessmen to continue on a course of further touristic development, but they also learned and reinforced that characteristically modern, distanced way of seeing the environment. Many a later Heimatschützer seconded Rudorff's observations on this point with his or her own critique of tourist behavior: whether as litterers, rapacious wanna-be "botanists," or wantonly destructive louts, tourists often represented the very nadir of environmental awareness.

Far from cementing a pessimistic conclusion about the depraved character of modernity or the bad manners of the increasingly numerous proletarian class, though, this behavior's individual component opened up an entire area of positive cultural activity. Based on its pragmatic Enlightenment philosophy, the Heimatschutz calculated that the touristic passion could not be denied, but that it could well be refined in such a way as to "awaken a true joy in nature" and provide an "education in seeing" (Lux 1908–9, 637). But who would do the educating—or, put another way, what were the sources and reinforcers of current cultural attitudes? Paul Schultze-Naumburg identified travel guides as a pivotal element: they were, as he observed, a formidable "power, because travelers get to know the land through their eyes" (Schultze-Naumburg 1905, 65). And although in this passage Schultze-Naumburg went on to complain that the popular guide books had helped cause the "ruination" of Germany through their inability to value the most common kinds of beauty, it was apparent that the relation could be turned around. "A reform of travel guides" could actually be a first step toward a better kind of tourism, one that would eventually have a "furthering" effect even on conservation (Giannoni 1913, 157, 156). If we look more closely at what the Heimatschützer had to say about tourists, then, we will see the outlines of a positive cultural strategy of environmentalism.

The Heimatschutz position on this matter of tourism and travel deserves to be put in some historical and theoretical perspective. In Europe as a whole the decades around the turn of the century are generally recognized as a key era of transition into the structures of modern industrial life; among other things it was a time in which the character of travel was altered irrevocably. Increased social mobility found its geographical expression in a growing "democratization" of leisure travel; whole industries began to form around the profit potential of a steady stream of tourists. Technology spurred on this development with advances such as the personal automobile and highly efficient train, boat, and communication systems. The growing accessibility, superficiality, and banality of mass tourism led to a compensatory shift: the wealthy demonstrated their status by seeking out ever more distant destinations, and authors mimicked their change of focus by writing accounts of Japan, China, or India (Reif 1989). Along with their simple snob appeal, these foreign cultures were popular in large part because they seemed to offer that experience of exotic "otherness" that was in such short supply in the increasingly uniform, technologically driven industrial culture of Europe or North America. As we already saw in the preceding chapter, in terms of landscape and natural beauty the logical outcome of this frantic search for the "Other" of European civilization was a definition that valorized the wild, the untamed, and the highly romantic even at home.

The consciously foreign and exotic goals that were sought out by the touristic "avant-garde" of the fin de siècle help shed light on the overall function of modern tourism. Many recent theoretical treatments of tourism emphasize that this effort to leave the system had subtly affirmative and merely compensatory effects. This was crassly evident in the exotically costumed "dream worlds of mass consumption" developed by department stores and advertising firms (Williams 1982, 58ff.). But travel itself was implicated, too. The critic and author Hans Magnus Enzensberger provided a concise formulation in his much-cited "theory of tourism," which states that tourism is essentially a form of romantic compensation for the ugliness of bourgeois industrial society, a "flight from self-made reality" (Enzensberger 1962, 155). Tourism, from this neo-Marxist point of view, is actually just one example of the larger phenomena of recreation and leisure, which "create the illusion of freedom and self-determination" only in order to "adjust the working class to a subordinate position in society" (Rojek 1985, 107). There is much in this theory that makes sense. Fun-seeking on the beaches is routinely traced back even by participants to the simple urge to get away from everyday drudgery; most are resigned to the fact that nothing will have changed upon their return. The critics are thus probably correct in noting that much recreation serves to neutralize and channel the potentially powerful experience of lack into a self-perpetuating system of capitalist exploitation.

From the point of view of a less reductionist and more culturalist model of social action, though, it is not necessary to jump from the observation that travel *in some instances* is tightly linked with overt capitalist reproduction to the blanket assertion that *all* travel is necessarily system-affirmative and oriented toward increasing consumption. A Gramscian interpretation would agree with the Heimatschutz on the possibility (although not necessarily the inevitability) of a counterhegemonic cultural practice taking root in certain forms of tourism. Such an approach is moreover compatible with many of the insights contained in Dean MacCannell's theory of tourism. MacCannell has interpreted tourism—or "sight-seeing"—in much the same way that Baumgarten understood aesthetics: at its most fundamental level it is a "way of attempting to overcome the discontinuity of modernity, of incorporating its fragments into unified experience" (MacCannell 1989, 13). MacCannell's simultaneous conclusion, that tourism helps manufacture further differentiation and is thus "doomed to eventual failure," need not worry us here, for it is based on the continued dominance of those established forms of travel that the Heimatschutz attempted to modify. In the following we shall see that the Heimatschutz carried out a far-ranging critique of contemporary spectacle-oriented interest and attempted to institute a new kind of tourism distinguished by its environmental awareness.

From the very beginning the Heimatschutz consciously opposed the method of much of the tourism of the day. Its best-defined target, however, was easily that upper-middle-class brand of travel to foreign cities that emphasized canonized, clichéed attractions. In the German-speaking world it was a kind of tourism indissolubly associated with the name Baedeker, publisher of a highly profitable series of travel guides that included the innovative idea of designating "must-see" highlights with small stars. August Trinius, a prolific author of regional tour guides and an early Heimatschutz member, noted that this classical form of tourism not only cost a lot of money, but also consisted of nothing more than running around in "packs" from "*" to highly rated "*," and listening to the "mechanical explanations of professional explainers":

> Such travel is counted as an indispensable part of "*higher*" cultivation these days. One hurries, *exhausted,* from city to city, only in order to finally return home tired, with an empty purse and with a disorganized assemblage of the most variegated impressions in one's head. One flees back into one's familiar shelter with the words: never again! Only to join the herd again next year. (Trinius 1916, 315)

Heimatschützer objected to both the pace and the commercialized structure of this kind of journey. Speed was both an outcome and a cause of the narrow selection of *de rigueur* objects; the scattered hodgepodge of random information that remained at the end of the trip was itself part of the commercial apparatus, which could count on the customer to return for an equally insubstantial diversion the next year.

Against this built-in frivolity the Heimatschützer tended to see travel as "a very serious cultural undertaking" that rightly demands a certain amount of intellectual effort from the individual (Stauff 1909, 23). Foreign tourism in particular was not supposed to be all fun, but rather could involve hard work:

> For the real purpose of a trip . . . is that one receives a clearly articulated overall picture of the area that one is visiting. City and town, house and people, mountain and valley, typical clothes and customs, animal and plant world, economic life and culture: all these things belong together. But they are never available as a unit. They can only

> be absorbed singly; the task of bringing them together as a whole is given over to the individual traveller who proceeds intelligently. And only when he has intensively worked this out to the end will he have exhausted the potential of the trip. (Stauff 1909, 24)

The ultimate objective of Heimat-style tourism, according to this description, would be a thoroughgoing understanding of a complex whole of relations. These high expectations were bound to be disappointed in the case of a short foreign trip; certainly the majority of individuals could never be expected to pursue this kind of an exercise in *Bildung* with the prescribed tenacity.

Nevertheless the formulation of such an ideal was probably worthwhile as a demand directed at travel guides, whose authors could perhaps be expected to spend more time and effort achieving a reliable overall picture of a country or place. At the very end of the war the *Bund Heimatschutz* (now officially named the *Deutscher Bund Heimatschutz*) attempted to put together precisely this kind of a primer for travel guides to cities and even pointed to one of its own as a model (Wolf 1918). As its editor, Gustav Wolf, explained, the basic principles were clear: a highly functional section would be for orientation, but this would be supplemented by a series of aesthetically inspired digressions into history, natural history, customs, and so on, of the locale; further sources were to be listed in order to ease the reader's path into more involved studies. Repeatedly the goal of grasping the "whole" of the location was mentioned, along with its closest corresponding quality, the "typical"; by contrast Wolf warned against the common mistake of concentrating on the unusually "beautiful," or even the "bizarre" (Wolf 1918, 19). This, in fact, was the first rule of Heimatschutz travel: to reject the exotic—by which was meant less a foreign location than an isolated, fragmentary, monumentalizing mode of consideration.

As we saw already in chapter 3, the Heimatschützer were particularly sensitized to such issues through their experience with the 1902 and 1907 disfigurement laws. The central drawback of these laws was the same one they found in conventional tourism as a whole: both singled out certain limited areas for their "exceptional" qualities and then intensively exploited them, with negative consequences for both the tourist and the landscape. The corollary feature of highlight-oriented tourism was the no less negative fact that people did not take advantage of many other areas, and that these could all the more easily be singled out for industrial uglification. The disastrous 1913 court case in Spandau inspired an incisive commentary in this regard. The jury's willingness to deny the aesthetic value of the local countryside (thus depriving it of legal protection), Karl

Storck wrote, could be "put down to that most offensive star economy of the Baedeker" (Storck 1912–13). Storck's use of the word "economy" (*Wirtschaft*) is suggestive, for it hints at the general tendency of conventional tourism to assist in constructing organizations of value versus nonvalue: spatially as the exceptional scene, temporally as the height of the season, sociologically as "education" or "cultural capital." But as we have seen repeatedly, it is precisely this differentiation that could be exploited by businesses to the detriment of the environment.

Perhaps the only long-term solution to the harms caused by a value gradient is to try to get rid of the gradient altogether, or at least to reduce its magnitude. Such "dedifferentiation" was an inherent tendency of the Heimatschutz idea, which stressed that every location in Germany had its own charms and its own landscape deserving of protection.[12] But the location was just part of the innovation, for Heimat implied a new touristic method as well: an unhurried hike in the green surroundings of one's local Heimat. This method was distinguished not only by its pace and its reliance on "low tech," but also by the expectation that the "compleat hiker" would be able to gather and synthesize information about the environment. Wilhelm Bölsche noted that tourism was useless without this aspect of enrichment, which, when present, ennobled even the most humble neighborhood hike: "If someone cannot manage to come home full of enjoyment and edification from a simple Sunday walk through the sandy pine heath of the Mark," Bölsche argued, "then he will also not get anything out of having Stangen's Travel Bureau transport him around the globe for a small fortune" (cited in Kelly 1975, 114). This kind of homegrown wisdom was typical of the Heimat movement, which promoted a thorough understanding of one's hometown area—particularly the immediate natural environment—as the most important basis for personal and civic improvement.

It is important that we not confuse this touristic emphasis on the Heimat with chauvinism. Even in the middle of World War I, foreign tourism was not so much decried as put into perspective:

> The urge to go off into the distance is understandable, and nobody wants to work against it. But foreign lands should not be reduced to the status of a fashion. Today Norway and Spitzbergen, then a heath, next year northern Africa. Start to look around you in a limited area. Get to know your own German homeland throughly and well! And when you have become acquainted with the great green garden of Germany, then direct your senses abroad. For travel is always a good cultivator: it makes you freer, more independent, gets rid of prejudices, brings different nations closer together. But it also will show

> you more clearly how much beauty and how many riches you possess within your German borders. (Trinius 1915, 315)

The reason why the Heimatschützer concentrated on Germany in their program of tourism was thus not because they disliked other countries, but rather because they rightly saw exposure to the Heimat as the all-important key to local environmental responsibility. Lack of exposure on the other hand, would result in decisions such as that made in the Spandau case. The Heimatschutz association in Brandenburg commented that such decisions were a reflection of the jury members' lack of experience with the wide range of beautiful areas in Germany, as well as their own "nearest surroundings":

> If these people have never personally ventured out into the open, if they are only acquainted with the stereotypical vacation trips of times past, then it is not so surprising that they think of the lovely Mark as that proverbial "sandbox"—one that is at best suitable for making industrial profits out of forests they have never seen. For already during the on-site inspection trip for the press corps we heard from Spandau citizens: "Well, we've never been here before!" And that is the characteristic thing, the decisive thing. ("Reklame" 1913, 30)

Again and again Heimat advocates contrasted the ignorance and exploitation bred by such traditional understandings of tourism with that of a new Heimat tourism. Heinrich Pudor's 1910 article "Hiking in the Heimat" is representative. He could barely contain his disgust with the "mad rush" of tourists in the Alps, which for him had "something pathological" about it. Instead of this excessive effort to compensate for the boredom of working life, Pudor recommended a commitment to moderation and refinement within the sphere of day-to-day existence. The very idea of a vacation represented a kind of temporal exoticism that Pudor rejected: "The idea is not to devote ourselves for a few weeks out of the year to the most strenuous kind of mountain climbing, but rather to live reasonably the whole year through" (Pudor 1910–11). Integral to this program was a new kind of "Heimat care" in which the local walk or hike, a form of recreation, was simultaneously a means toward knowledge. Although Pudor considered a hike in the countryside to be his ideal, he also expressly recommended walks through one's native city as a good way to get to know its many different facets.

The complement of this emphasis on walking was a profound distrust of the fully streamlined mechanism of modern tourism. In one article Ernst Trojan wrote sarcastically of the modern fascination with machines, and showed how technology was already drastically altering the way people traveled. Here, modern tourism is depicted as a fundamental distortion in the way people relate to their own human body:

> For example: a man wants to go on a trip. He lets himself down to the ground floor by means of an elevator and gets into an automobile that takes him to the train station. Here the train receives him and takes him off into the distance without him having to get up from his seat even once. Arriving at his destination, the hotel automobile picks him up and takes him straight to the hotel elevator, which in turn moves him at high speed into his room. The next day he goes down with the elevator, into the automobile and to the station, from whence he is carried by the cog railway in the quickest possible manner up to the top of the highest possible peak. And so forth *ad infinitum.* Ask yourself seriously: Does such a modern man even need legs anymore? Of course not, for the machine is nearly ready to replace such superfluous and obstructive organs as the legs with much more intricate and practical apparatuses. In the future it is imaginable that the entire human will be considered superfluous, and technology will no doubt manage to install an artificial mechanical person in the stead of this mortal body. (E. W. Trojan 1910–11, 289)

The mechanized gadgets of mass tourism are extended here into a dystopian vision of the complete alienation of man from nature, whether as the surrounding environment or his own natural body. An important component of this critique is the insight into technology's ability to change or block perceptions. Ernst Rudorff had thematized the insidious effects of speed on environmental perception as early as 1897, when he noted: "The average traveler from the big city who races through the land in an express train hardly becomes aware of the wounds that field rationalizations and commons divisions inflict on the landscape" (Rudorff 1926, 94).

A primary function of the hike was thus to exorcise what E. W. Trojan called the "devil of speed" and make possible a more reflective relationship to the land (E. W. Trojan 1910, iv). In this sense another author expressed his disappointment with the bicycle, which had only helped people speed right by nature:

> For a while as the sport of bicycling was beginning optimists believed in a lasting reinvigoration of joy in nature. This hope has been betrayed . . . Given the way in which it is generally practiced it cannot replace hiking, as it certainly allows one to rove through in a fleeting fashion, but does not let one get to know the blooming landscape.

> The mass of bicycle riders still finds its greatest pleasure in gobbling up the kilometers. (Wagener 1913, 12)

Speed was, of course, not only related to the technological means, but also to the frame of mind. "Kilometer gobblers" did not have to have wheels. Confronted with youth hikers who were chiefly interested in racking up the kilometers, the Heimatschützer's first recommendation was that they slow down and learn to enjoy nature: "The *Wandervögel* ought not to simply march, but rather ought to also learn to enjoy the beauty of the landscape more than is frequently the case these days" (forestry professor Karl Eckstein, cited in "Naturdenkmalpflege" 1915, 67). Such warnings to the *Wandervögel* might seem pedantic and characteristic of an inflexible older generation. If the Heimatschützer overreacted, it was perhaps because they realized how rare the ability to see nature had become. They recited in horror the findings of a study done in the Berlin school system, which showed that 87 percent of students had never seen a birch tree, 75 percent had never seen a live rabbit, 64 percent had not seen a squirrel, and 82 percent had never heard a lark (Mammen 1909, 79). Children such as these would grow up without any real commitment to the natural world, it was feared; the program of Heimatschutz was thus inseparable from *Heimatkunde,* the close study of a local area. Here the Heimatschutz demanded—and increasingly received—support from the schools. *Heimatkunde* in its broadest form was a pedagogical method of introducing ideas at a manageable and tangible level: thus it might cover "all the humanitarian, social, communal, state, and economic structures of the local area" as preparation for an understanding of these things in the world at large (Strobel 1912, 543). But the more specific emphasis of *Heimatkunde* was clearly on the physical realm, on biological, botanical, and geological features. Here the idea of Heimat had slowly become equated with the ensemble of physical relations and had even begun to take on a strongly ecological character. This was due in part to the tradition of teachers such as Friedrich Junge, whose book *The Village Pond as a Biological Community* had made the study of interrelations a central concern of Heimat biological study as far back as 1885 (see Trommer 1990, in particular chapter 6). The new activism of the Heimatschutz, though, demanded that certain changes be made in science education. "Dry instruction" should be kept to a minimum, Heimatschützer argued; there should be

> above all in the earth sciences somewhat less systematic classification, fewer Latin names, and conversely more of the life history of animals and plants, perhaps fewer exotics and in their stead more natives. The trees, shrubs, and flowers in the woods and the fields should be good

> old friends of every child of twelve to fourteen years of age, as good as the animals which populate the woods and the fields. (Arnswaldt 1908, 119)

The necessity of finding a popular cultural strategy thus drove a similar demand for a more accessible form of education, one with less rote learning and more hands-on experience (cf. also Knaut 1993, 225f.).

Crucially, though, this program of education was inseparable from a particular mode of experience. Rather than remaining in their classrooms, teachers were now supposed to "go outside with their students, and on walks and gymnastics exercises open up young hearts for the beauty of the Heimat" (Arnswaldt 1908, 119). But teachers were just the tip of the iceberg, for in and out of school the open-air walk or hike took on a central role for the program of Heimatschutz. Contacts between hikers and Heimatschützer had always been congenial, and the Heimatschutz association in Württemberg probably spoke for all other regional associations in retrospectively acknowledging hikers' contributions. Referring to the *Schwäbischer Albverein,* Württemberg's regional hiking group, the Heimatschützer wrote in 1913, "We know that it was the one that first reawakened the appreciation of Heimat land and Heimat folk, thereby preparing the soil upon which our association has thrived" (*Schwäbisches Heimatbuch* 1913, 6; similar praise for a hiking organization in "Heimatschutz für das Sauerland" 1914). In Brandenburg hikers from the General League of Tourists of the Mark were, by 1919, being welcomed as "capable supporters of the Heimatschutz idea" (cf. *Heimatschutz in Brandenburg* 1919, 92; Schulze-Moering 1919). By this time it is apparent that hiking organizations had started to absorb some of the philosophy of the Heimatschutz, for which hiking was "not the ultimate goal, rather, the means to an end: it should inspire love of the earth, the soil, the land, and the Heimat" (Trojan 1910–11, 292).

To a great extent the Heimatschützer were themselves hikers. A notice from 1907 stated that local Heimatschutz groups from both Gera and Jena were in the habit of "often undertaking hikes that simultaneously serve a better knowledge of the Heimat in historical and scientific regard" (cf. *Mitteilungen des Bundes Heimatschutz* 1907, 81). Many other groups arranged similar outings, and their further lectures, articles, and discussions of landscape aesthetics all show how Heimatschutz was geared toward the direct, on-foot experience provided by the hike. In a very fundamental way it appears that we can thus say that the hike or walk constituted the perceptual and cultural basis of the Heimatschutz movement. This is plainly evident in works such as Clemens Wagener's *Nature and Heimat,* where an exposition of the various forms of travel is the very first topic, at the conclusion of which hiking is advocated as "the only path that leads to a knowledge of nature and a joy in nature, and it is in turn the precondition for the care of nature, for nature protection" (Wagener 1913, 15).

The local Heimatschutz association in Gotha tried to contribute to this development of knowledge with a pamphlet designed to introduce the hiker to the flora and fauna of the area (cf. *Heimatschutzfragen* 1913, 11). In Saxony the State Association managed to distribute one *million* copies of a one-page flyer with the title "Protect Nature" in just two years, many to hikers (cf. *Mitteilungen des Landesvereins Sächsischer Heimatschutz* 1913–14, 177). By their very structure many lectures and publications, too, targeted an audience that could learn to apply a more acute visual sense on walks and hikes. And these direct interchanges with the larger public were increasingly complemented by *Heimatkunde* in the schools, where students would begin to learn about "the significance of the forest and the waters for our climate" on their school outings. Given this particular emphasis on close and even scientific observation, the tendency of the Heimatschutz was thus toward a "discovery of the Heimat" in the sense described by the nature writer Raoul H. Francé: it was a tourism that could provide insight into the "balance of the biological community" and thus one that would simultaneously help reinforce our "duties to the environment" (Francé 1923, 53–54).

It is useful to think about these typical Heimatschutz hikes out into the countryside in relation to Dean MacCannell's theory of tourism. Much of what MacCannell has to say is oriented more toward the spectacle, and the traditional touristic "sight," of course; it is a system, he insists, which "no one escapes except by retreat into a stay-at-home, traditionalist stance" (MacCannell 1989, 43). One reason why the Heimatschützer were able to innovate was thus perhaps their emphasis on staying at home. But it was still a form of tourism, of exposing oneself to new corners of the environment or catching up with developments in already familiar ones. MacCannell's concept of a "work display" has continued relevance here. The notion refers to the fact that tourism is often guided by a curiosity about work and production: factory tours and visits to the interior workings of dams are just two examples of how businessmen and government agencies have learned to accommodate this widespread modern tendency. What these organized work displays reveal is, of course, never the whole truth; inevitably they become the province of public relations experts. Nevertheless the touristic urge to "unify economics and aesthetics" can be considered a significant force on the production process: "there is a restructuring of the work place in response to its display" (MacCannell 1989, 62).

What Heimatschutz hiking did was to apply an increasingly well-informed version of this touristic display pressure to whole new areas of production in the landscape. Karl Giannoni noted one example with his observation that the operators of "resort towns and holiday inns" often came to "support enlightened forestry practices and forest protection in their surrounding area" (Giannoni 1913, 156). This change in the form of agricultural production was only limited, to be sure, and it was also driven by opposing commercial interests; nevertheless it shows the direction in which the much broader and less profit-oriented pressure of home tourism would gradually push landowners all over Germany.

As the Heimatschutz seems to have realized, the key to maintaining this counterhegemonic pressure was to give the recreational and aesthetic components of tourism their full due alongside more conceptual or educational aims. Even the scientists in the movement supported this way of doing things, recognizing that "the enjoyment of nature is for many the door through which science enters into our relationship with nature" (Ratzel 1906, 62). Many Heimatschützer also continued to insist that an aesthetic contemplation of nature was the most powerful force for "chang[ing] the habits of life" (Giannoni 1919, 30). The hike was thus not seen as a mere science trip, nor primarily as exercise, but was in fact intended to be a balanced and holistic experience. Here it was the recognition of the beautiful and harmonious aspects of nature in particular that would eventually lead to an intuitive understanding of the need for environmental protection:

> People who were never taught as a child to look at the landscape with sensitive eyes, to feel their way into it and to understand its people, will find it hard to do when they are old. For that reason our youths' attention should be directed more and more to the harmonious quality, the aesthetics of the landscape. We must educate our children and ourselves, too, never to disturb this harmony of a landscape, nor to allow it to be disturbed. ("Reklame" 1913, 30)

The reference to "sensitive eyes" lays emphasis on an element of environmentalism that is often passed over in discussions of the topic. Particularly in recent times the available wealth of scientific data has tended to shift the entire mode of environmental discussion toward statements of ecological fact or prognoses based on them. A complementary phenomenon is the proliferation of ethical philosophies that, in varying degrees of rhetorical escalation, preach the requirements of ecological responsibility. Both are no doubt important constituents of the future world order, and they have the advantage that they can be printed neatly in a book today.

But it can be questioned whether they are truly effective—whether they will be sufficient, in other words, to tear us out of our well-trodden path of capitalist consumerism and our equally deeply ingrained mechanistic domination of nature. In fact the last twenty years would seem to indicate that they are not enough, by themselves, to effect any significant change.

A Gramscian interpretation of the situation would stress that systemic environmentalist change requires a strong cultural basis. For as good as the functional arguments for change are, and as noble as the ethical ideals to be embodied might sound, the decisive predispositioning and reinforcement can only come from an almost instinctive level of "common sense" and everyday perception. Here it is important to take expressions such as "sensitive eyes" seriously. As William Morris said more than one hundred years ago: unless the use of your eyes makes you suffer, nothing will be done.

In many ways the decisive deed of the Heimatschutz was to take this adage to heart and engage Germans in a new use of their eyes, one that would make them "sensitive" to the configuration of the environment. Aesthetic impressions were connected here to a certain level of emotional feeling: the Heimatschützer were fond of calling it "love of the Heimat," but its name is less important than the observation that it should not be dismissed as "romantic" or "sentimental" and instead should be encouraged as a necessary component of environmental culture. This is, at least, a conclusion that emerged from our examination of Heimat-oriented novels. In their more detailed character development, we managed to see more clearly what is often only implicit in the journalistic or analytical work of the Heimatschutz: the fact that "environmental heroes" are often motivated by an undeniable emotional attachment to the land.

Instead of reading about it, though, people could start to build up this kind of involvement for themselves through tourism, a practice that Heimatschützer embraced precisely because of its role in helping to create and sustain environmental commitment. Not just any tourism was meant here: the positive alternative of a local hike at slow speed and in a spirit of reflective enjoyment came on top of a far-reaching critique of the environmental alienation implicit in much exotic tourism. In this differentiated appropriation we got a sense of the optimistic, forward-looking spirit of the Heimatschutz, which clearly shared the opinion of Paul Shepard that the tourist "is apt to be a boob, but he may be the hope of mankind" (Shepard 1967, 156). In fact the emphasis put on a process of self-education indicates that the Heimatschutz saw this cultural practice in Gramscian terms as a "decisive site of a new Enlightenment" as well as the pathway toward the "development of a new common sense" (Haug 1988, 44).

This Heimatschutz strategy has something to say to us even today.

For most of us, our estrangement from the land is understood: "We are all tourists now" (Barrell 1972, 188). What the Heimatschutz approach to tourism lets us see, though, is that we have it in our power to re-functionalize this important cultural practice in an environmentally progressive direction. It is an attitude that has recently resurfaced in today's desire for "gentle tourism" (cf. Hasse and Schumacher 1990). One important difference, of course, is that the Heimatschutz version goes beyond vacation-oriented tourism and emphasizes walks or hiking as a regular local activity. Precisely this kind of cultural practice holds out the hope of developing those everyday reflexes, likes, and aversions that can change our environmental behavior. Given the continuing neglect and destruction of the environment in our modern society it therefore remains "a pressing demand of our time . . . to become at home again in the nature that surrounds us" (Schlesischer Bund 1911).

Conclusion

In chapter 2 the reader may well have noticed that we abandoned our systematic organizational history of the *Bund Heimatschutz* once we reached 1914. The decision can be justified partly by reference to the objective, which was to examine in detail the workings and effects of this movement with particular regard to its environmental program. By 1914 all of the key theoretical elements of this program were well in place and enough time had elapsed for the Heimatschutz to show how it could adapt to and use the obtaining political and social structures of Wilhelmine Germany. If, as I believe, the environmentalist strategies developed here were truly significant, then this result, in and of itself, would be sufficient legitimation for the form of the undertaking.

But the longer historical view is still relevant, and it remains a valid question to ask what became of the Heimatschutz and its ideals. Many of the regional organizations still exist today, joined together under the umbrella of the "German Heimat League"; indeed, one can even speak of a resurgence of Heimatschutz in the regions of former East Germany, where associations are being founded anew after decades of state and Party hostility. But in terms of their overall social significance these descendants of the Heimatschutz are no match for the original. Heimat organizations today are not counted among the leading forces for reform in any German state, and even the torch of environmentalism has long since passed to other groups in German society. The question is, why? When and how did the Heimatschutz lose the initiative that it had shown in carving out a sphere of influence for itself in the Wilhelmine years?

Surprisingly, perhaps, the year 1914 did not represent a defining break in the development of the Heimatschutz, and in fact our analysis can be extended in almost all important respects to cover the next four years of war. But it might be noted that two new war-related fields of activity opened up. The first was the effort to influence the kinds of graves and monuments that would commemorate soldiers and battles. In the wartime atmosphere of national pride everyone seems to have agreed that such fixtures would have to be proudly "German." The Heimatschützer managed to give this their own particular spin by arguing that "German" was something especially close to nature and the native landscape. The monuments they called for were thus often trees or simple boulders. In general they eschewed the urge to plaster the landscape with heroic edifices, and they opposed the popular idea of inscribing cliffs or large erratic blocks with the names and dates of battles. Hugo Conwentz's letter of 3 April 1916 to the *Magdeburger Zeitung* is a good example of the extreme reticence of Heimatschützer when it came to this kind of empty nationalist campaign. Why hurry around carving up natural features? he asked. In a pragmatic tone he noted that anybody who lived through the war would remember it well enough, and that there would still be plenty of time in which to erect monuments to inform succeeding generations, if that indeed seemed necessary.

A second mission of the war years was the comprehensive design and planning work devoted to reconstructing the parts of East Prussia that had suffered during battles with the Russians. Hartmut Frank (1992) has documented how the Heimatschutz reform milieu responded quickly and decisively to the massive rebuilding project, and how it used the opportunity to develop a modern, functional style of structure in harmony with traditional architectural forms. Both of these wartime issues were covered extensively in the journal *Heimatschutz,* and it can be speculated that they gave new life to the moribund national level of the *Bund Heimatschutz,* which had had to accept even stronger oversight by individual regional associations in a charter change of 1914 (Knaut 1993, 115f.). Since neither of the projects was tied to any of the existing regional groups, their strong national and patriotic overtones perhaps gave the new leadership a chance to reassert its role within what was now a profoundly regional Heimatschutz movement.

Alon Confino has analyzed the Heimat movement as a phenomenon that explains "why Germans from all classes came together in August 1914" (Confino 1993). His argument is that this unexpected solidarity was made possible by, among other things, a far-reaching agreement on the value of Heimat. Confino is dead wrong in his supposition that Heimat was an *empty* value, of course: as we saw in chapter 2, it instead represented genuine material and social solidarity. Given that we can grant the possible connection between Heimat and a strong defensive reaction. For to the extent that many Germans, taken in by governmental propaganda, felt that they were fighting off an unprovoked attack, a shared concern with the highly valued material domestic sphere of Heimat could well have united Germans from all walks of life in a military response. Certainly some Heimatschützer tried to get as much mileage as they could out of the situation by arguing that "love of the Heimat" was effectively a "helper in the war," and that their reforms and their overall "Heimatschutz attitude" had "a large hand in" Germans' "readiness to sacrifice life and property" (Giannoni 1914–15, 387–88).

Despite this kind of patriotism, though, the Heimatschutz as a whole never descended into a blind and hate-filled chauvinism. There were, of course, exceptions: the life reformer and antivivisectionist Paul Förster, for example, used the pages of the Brandenburg Heimatschutz journal to oppose the "German with healthy sensitivities" to inhabitants of Allied countries, "countries which will remain enemies forever" (Förster 1917, 42). One factor that helped limit this sort of feeling was the immensely strong respect that the Heimatschützer had always harbored for the reform programs of their "English cousins"; as late as 1917 the English were being called "our teachers" in Naturschutz as well (cf. Hagen 1914–15, 237; Salomon 1917).

As can be imagined, this positive view of the English and of other European nations was put to a severe test by the events and propaganda of World War I. A particularly sensitive nerve was hit early in the war when the Allies accused the "barbaric" Germans of willfully destroying the university in Leuven and callously damaging the cathedral in Reims. Unable to corroborate the charges, and under the influence of their government's counterprotestations, German Heimatschützer took these largely justified accusations as an insult to their nation's character. Tales of unnecessary German destruction simply could not be true, they reasoned. They proudly pointed to an article in an independent "Norwegian newspaper" that had countered Allied claims of material barbarity with the observation "that there is a Heimatschutz association in almost every region of Germany" (Kruse 1914, 218). Otto Ubbelohde quickly supplied the corresponding pictorial refutation (fig. 13): it consisted of a typically harmonious, well-cared-for Heimat vista equipped with the sarcastic caption "Barbarians' Land." In effect the Heimatschützer were using themselves as character witnesses for the entire German nation and, by extension, the army, too. There can be little doubt that this amounted to an uncritical stance, especially given the Heimatschützer's keen awareness, before the war, that their views were not by any means shared by all segments of the population, nor by all branches of the government. Still, the vigorousness of their response shows that the Allied accusation had hit the Heimatschützer perhaps harder than any other group in German society, not so much in its reference to a *nation* of barbarians, but rather in its implication that Germans might allow a *barbarian treatment of land* and historical and cultural assets.

„Barbarenland"

Fig. 13. Heimatschutz response to charges of "barbarian" destruction in World War I. (From *Heimatschutz* 1915, 18.)

The response thus confirms that the idea of Heimatschutz had not changed very much at all for the Heimatschützer: it was still a considerate stewardship of the land, and it remained a battle that, in the long term, could only be fought on the home front, against one's own neighbors and countrymen, even against oneself. The war's role in all of this was merely contrastive. Since it would forcibly expose soldiers to the physical infrastructure of unfamiliar areas, the war, it was hoped, might well open people's eyes to the concerns of the Heimatschutz. An article written by a soldier in the East portrayed the war experience as one that would give a greater sense of urgency to landscape protection in particular:

> On all fronts many German soldiers, like the author, will have observed the land and the people, nature and culture. Comparisons

> with their Heimat will have quite automatically sprung to mind. And the result thereof?—With pride they will have recalled the high state of culture in German valleys, one not reached anywhere else on the face of the earth. For many a German comrade precisely this foreign experience will have opened up a new understanding for the ideal efforts in the fatherland and especially in his Heimat. When the time comes he will return home with a completely different understanding of the significance of such tasks. (Otto 1916, 86)

While these lines are not without a certain superiority complex, their rhetorical structure clearly shows that Heimatschutz was profoundly oriented to coming home and to building up a commitment to the local environment. A similar article assured Germans on the home front that they had the sacred patriotic "duty" to "maintain the character" of their Heimat; they would be judged by the returning soldiers, it was argued, and "not one of the returnees would understand if even one single tree had been cut down for trivial reasons" ("Schonet" 1915; see also Löhner 1916).

Throughout the war Heimatschützer and Naturschützer alike thus pursued many of the same programs as before, albeit with far less manpower, money, and materials. A new threat in the area of landscape preservation was posed by the imperative to produce more food within Germany's borders. Crash cultivation programs exposed moors and other supposedly "unproductive" areas of the landscape to a politically and organizationally powerful threat. Heimatschützer such as F. W. Bredt pointed out that the "reservations with respect to overly impetuous land improvements" that the Heimatschutz had voiced before the war were still valid. The only thing that had changed, he wrote, was that now landowners and parts of the agricultural bureaucracy were using special wartime powers in order to steamroll their opponents and put the land into production. Against this misguided and shortsighted version of the common weal Bredt set the "outstandingly patriotic sensibility" that lay behind the Heimatschutz ideas of long-term environmental stability (Bredt 1916, 98).

It is perhaps too much of a presumption to ask that critics stop and take another look at this kind of national rhetoric. When we find a bourgeois group emphasizing its patriotism in the middle of a world war, this will inevitably be seen in the context of those arguments that link the bourgeoisie with precisely the kind of blind nationalism and Prussian militarism that caused World War I. But this only makes it more important that we learn to distinguish between different kinds of nationalism. One of the most important distinctions to be made is the very pragmatic one between a domestic, internally referenced standard and a more dangerous kind of external and oppositional reference, one that constantly marshals national unity against morally inferior foes. The Heimatschutz clearly belongs to the former category of nationalism. Its language was the language of self-critique and self-improvement; its concerns were with "native vandals," not foreign invaders. Here national unity was first and foremost that ethical framework needed for programs of social welfare and redistribution. Even in the depths of World War I it was this definition of the nation as a *sharing community* that inspired the Heimatschutz; a long article on "The Consciousness of Community as Starting Point, Content, and Goal of Heimatschutz Efforts," for instance, only reemphasized the importance of limiting the pursuit of "naked personal utility and business profits" and of reining in the related "freedom to ruin our beautiful Heimat." Against this kind of laissez-faire individualism the typical Heimatschützer argued for a "concept of community that also requires sacrifices"—and since the practical application of such ideas is always the most important part, it should be noted that the article went on to locate them in municipal architectural and planning matters (Rieß 1914–15, 452, 453).[1] As we have seen, countless other articles applied the same community ethic explicitly to the maintenance of the environment.

The aftermath of World War I resulted in an upheaval in German political and social life that all but crushed the Heimatschutz's bourgeois-reformist vision of how to achieve social and environmental harmony. As unjust and as undemocratic as the class system of the Wilhelmine era had been, it had nevertheless supported a common bourgeois milieu that gave the Heimatschutz ties to the government and to some leading opinion makers. Making full use of the cultural allegiances of the middle class, it had been able to exploit the obtaining illiberal administrative structures for environmental aims that indeed ran counter to the dominant "liberal" values of property and production at any cost. In the Weimar period a new mass democratic culture diluted the precarious influence that the Heimatschutz had built up. In some areas, of course, groups recorded an explosive growth that took them well beyond the status of elite bourgeois clubs: Saxony's State Association, for example, was successful in making the jump from 3,500 members to approximately 40,000 in the late 1920s (cf. Deutscher Bund Heimatschutz 1954, 32). But for Germany as a whole it seems that the new mass-political exigencies of the Weimar era did not allow the same kind of close, behind-the-scenes cooperation that had been the hallmark of earlier Heimatschutz relations.

This difficulty rooted in the formal political system was only reinforced by powerful hegemonic forces now working directly against the ideas of conservation and Heimat. For specific historical reasons the Weimar era was one of unique economic pressures on Germans, stemming first from the payment schedule of the Treaty of Versailles, and then from the pendulum-like swings of a brief industrial prosperity and Depression. We should also not forget that the twenties were a decisive phase of globalization and rationalization in the business world generally, one that saw hegemonic pressures to exploit the environment reasserting themselves with vigor. These pressures were only exacerbated by an intense cult of modern technology that united Communists, Social Democrats, liberals, and even large segments of the right wing.

All of these stresses took their toll on a movement that had been dedicated to critiquing unnecessary development or rationalization. Never truly popular to begin with, large segments of the Heimatschutz movement began to retrench and accommodate. A new "pro-economic direction" was openly encouraged by Werner Lindner, general manager of the *Bund Heimatschutz* (Linse 1986, 29). Lindner's engineering and architectural background led him away from many of the organization's earlier environmental concerns toward a questionable attempt to achieve "symbiosis" with industry. At the same time the organization was undercut by a new division of labor: just as city planners began to monopolize issues dealing with urban growth, so, too, state agencies and scientific bodies increasingly "professionalized" environmental concern, turning it into a patchwork of administrative fiefdoms. Private organizations helped accelerate the fragmentation; even before the war observers had noticed a trend toward more and more environmental groups, each one clamoring for a piece of the public's strained attention (Schäff 1910). During the Weimar years we can assume that the trend continued, and that it helped draw people's attention away from the characteristically holistic impetus of Heimat.

The increasingly polarized political atmosphere of the Weimar Republic had a similar effect of dividing and dispersing environmental reformers who had previously gathered under one banner. Progressive, social welfare–oriented planners such as Fritz Schumacher were attracted to the ambitious building programs of the Social Democrats; meanwhile, though, the unfavorable outcome of the war and the tough conditions of the Treaty of Versailles did their part to turn previously moderate nationalists into bitter chauvinists. Already in 1919 Georg Bonne was analyzing the cause of the 1918 Revolution as a combination of capitalist neglect and "the incitement efforts of foreign-race seducers of the people" (cf. Bonne 1919).

A strong nationalist and revanchist faction within the Heimatschutz thus slowly began to appropriate "Heimat" for use as an exclusively right-wing slogan (Applegate 1992, 69). By the mid- to late 1920s a static and inflexible position became apparent in such things as the celebrated "flat roof vs. pitched roof" debate, an architectural flap in which prominent Heimatschützer such as Emil Högg and Paul Schultze-Naumburg denounced the modernist flat roof as a sign of godless internationalism and racial degeneration (Pommer 1983; Lane 1985). In these same years a number of Germany's leading Naturschützer turned increasingly toward theories of race and imperialist social policies (Gröning and Wolschke-Bulmahn 1985). By the late 1920s or 1930 biologistic, racist, and anti-Semitic interpretations began to appear more and more frequently also in Heimatschutz discourse (Ehalt 1985).[2]

Thus by about 1930 *Heimat* and *Heimatschutz* could easily mean something quite different from the same words said in 1910. Although in many cases devoted to the same local conservation and preservation work as twenty years earlier, Heimatschützer tended now to see themselves as opponents of the Weimar government, and unfortunately they may have been driven further to the right by the lack of real progress on concrete Heimatschutz and Naturschutz legislation (Dominick 1992, 85). Their frustration with Weimar created a vacuum that the National Socialists were able to fill rather easily; Hitler's regime subsequently absorbed them without any particular resistance after assuming power in 1933 (Speitkamp 1988).

Nevertheless it would be too easy to fall into a slickly unidirectional narrative that says that "almost all" Heimatschützer helped "lift Fascism into the saddle," or that Paul Schultze-Naumburg's development into a virulent anti-Semite is representative of everyone in the movement (Richard 1984, 300; Achleitner 1989, 166). Instead it seems that external circumstances and personal convictions could often determine very different outcomes for individuals who had shared roughly the same set of "Heimatschutz" convictions.

Fritz Koch's personal career provides an instructive look into the web of forces and potentials that helped shape the postwar Heimatschutz. He had come only reluctantly to his original position in the *Bund Heimatschutz,* unhappy with the lack of security and the fact that it interfered with obtaining the necessary qualifications for a planned career in the civil service (Knaut 1993, 105f.). His marriage to Elfriede Eisenberg was dominated by the pair's selfless devotion to the Heimatschutz cause: for many years their small apartment in the Thuringian city of Meiningen served as national office and storeroom for the *Bund Heimatschutz,* and later the Foundation for Heimatschutz was run in a similar fashion out of their own personal quarters. Koch's development after this point is documented in a personal file kept in the Thuringian Main State Archive in Weimar (cf. THStA Weimar Personalakte Fritz Koch, 2 vols.). After serving in the infantry in World War I Koch returned to Meiningen and campaigned for the left-liberal (German) Democratic Party. He then filled temporary positions in various central Thuringian bureaucracies before he eventually ended up working in Weimar for the state of Thuringia as a Heimatschutz official. Despite his new agency's shoestring budget and precarious position, he fought tirelessly to gain acceptance for Heimatschutz thinking. Indeed, he argued in 1922 that the "great social mission" of the Heimatschutz had to be taken even more seriously in the "impoverished times" of the postwar era; now more than ever, he wrote, it was necessary to protect the Heimat landscape as "the common property of all" and "something more than just an object of exploitation" (Koch ca.1922).

But although Koch managed to obtain the desirable title of Governmental Councillor, he never was given a tenured civil service position. When the Nazis took over the Thuringian Ministry of Culture—one of their earliest positions of power in Germany—Koch was terminated. From this time on his marriage to a Jewish woman was repeatedly used against him; even after Elfriede Koch died in 1935, a National Socialist "Student Leader" in Weimar used her to denounce the Heimatschützer (letter of 6 November 1936). Only through his old contacts with Schultze-Naumburg, by now a highly influential National Socialist, was Koch able to find a position with the State Technical School for Building Art and ride out a series of threatened firings. A letter that Koch wrote in November of 1933, asking for a temporary reprieve, shows him desperately trying to portray his past Heimatschutz work in National Socialist terms. He highlighted a 1919 call for the "subordination of the individual to the whole," for example, and underscored his opposition to Weimar's Bauhaus in the "years of cultural Bolshevism"; nevertheless these lines with their pugnacious conformism do not ring as true as those in which Koch complained of the lack of appreciation he had reaped for an "entire life . . . devoted to the struggle for German culture and art without consideration for personal advantage." He had "always supported our German cause," he repeated, and cited again from an earlier article in which landscape beauty was portrayed as a "national treasure" (letter of 16 November 1933 to Staatsminister Wächtler). We can assume that Koch's Nazi employers, however, were less than overwhelmed with his rather old-fashioned brand of national solidarity. Although they kept him on in a variety of jobs, he was always one step away from falling irrevocably into their bad graces, and it is perhaps symbolically significant that they apparently refused his late application for membership in the Nazi Party (questionnaire of 29 October 1945).

It is in fact doubtful whether Nazism shared anything with the Heimatschutz that was more fundamental than a certain amount of anticapitalist sentiment. Racist imperialism and a mechanized, uniformized, permanently mobilized mass culture of fascism were virtually antithetical to the original Heimatschutz philosophy, one that revered the historical individuality of buildings, the patina of folk culture, and the personal edification provided by a highly aesthetic view of nature (cf. Applegate 1992, 74; Koshar 1991). Those Heimatschützer and Naturschützer who were nevertheless taken in by Nazi promises would ultimately be severely disappointed. There were some notable early decrees on Naturschutz, of course, that looked good on paper, and the relatively sensitive landscaping done on the new *Autobahnen* can be considered a victory for Heimatschutz environmental aims.[3] But in the end, the National Socialist regime's central goals were war and imperial subjugation. In the pursuit of those objectives it threw all environmental caution to the winds: through ruthless exploitation in preparation for war, and in the war itself, it thoroughly betrayed the phlegmatic, conserving piety of Heimat. Unfortunately, this is one discrepancy that Nazi propaganda was wildly successful in covering up; in postwar Germany, ideas of Heimat and nature were thoroughly tainted by their association with Hitler's regime, and, as we have seen, many critics today still routinely assume that the Heimatschutz was somehow an insidious proto-fascist movement. It is more than a little ironic that we must struggle against the Nazi legacy in order to reclaim those same ideas of Heimat and local environmental responsibility that the Nazis themselves so cavalierly ignored.

Summary

One of the chief criticisms made in the previous literature on the Heimatschutz has been the finding that this movement was too "aesthetic" and thus not truly engaged in real societal problems. Against this I have attempted to show that we should understand the movement's aesthetic concerns as a necessary feature of the cultural struggle against a dominant, capitalist-mechanist brand of rationality. We found important precedents for this in some of the bourgeois reform movements of the nineteenth century, notably in the English groups that formed under the influence of John Ruskin and William Morris in the 1880s and 1890s. Their social convictions reappear in the Heimatschutz, which promoted access for every German to a wholesome and aesthetically rewarding environment.

In terms of its class dimensions this was "positive environmentalism" at its most positive. The Heimatschutz was not interested in repressing the lower class, but rather implicitly stood with it by pointing to the real culprits in society: the "native vandals," the landowners and capitalists who directed the exploitation of Germany's countryside. Its critique was never truly revolutionary, of course, in the sense that it always stopped short of envisioning a completely different society. From start to finish the Heimatschutz remained optimistic that some way could be found to restrain capitalism and harmonize the needs of the environment with a "socially acceptable prosperity" (*sozialer Wohlstand*) (Gurlitt 1910–11, 15).

But the reformist nature of its overall program should not deceive us into minimizing the Heimatschutz's sharp and perceptive critique of a capitalist use of the land. The role of aesthetics here was arguably even more crucial than in city planning issues because of the rather poor state of ecological understanding and the apparent lack of interested parties. In a pure capitalist society—to which Wilhelmine Germany represented a fairly good approximation, actually—a given company's treatment of its sugar beet fields or its forests is nobody's business but its own. It will inevitably attempt to maximize production and will probably rationalize its fields according to the conventional wisdom of science. In Wilhelmine Germany this would have meant removing any and all disturbing bushes, perhaps regulating streams, draining wet areas, and, where possible, applying chemical fertilizers to the monocultured crop. An ecologist today might point out that these actions have negative consequences, but we have to remind ourselves that this kind of scientific expertise was almost completely lacking in Wilhelmine Germany. Among the few people who dared or even cared to venture an opinion about the condition of the land was the occasional, aesthetically interested passerby. Chances are that this passerby matched the social profile of the Heimatschutz: a man, he would perhaps be a local professional or civil servant, he would probably not have a personal financial stake in agribusiness, and moreover, out of an aesthetic sensibility cultivated through years of schooling and touristing, he would realize that this landscape lacked something. This background would quite possibly give him an aesthetic designation for that something: charm, perhaps, or loveliness, or liveliness.

This private aesthetic diagnosis could be the end of it. The man could go on his way with a slight shake of his head and never give much thought to this field or this section of forest, just as he might resignedly accept the presence of an enormous billboard in a previously scenic spot. Important paradigms of thought conditioned this lack of response: apart from the ever-present justifications of profit and property, a convenient set of nineteenth-century compensatory strategies would help channel the implicit protest away, toward "green islands," wild-looking parks, special reserves, scattered tourist landscapes, or even future utopias. The all-important impetus of the idea of Heimat was to reject these deferments and narrow constraints and to demand action here and now, no matter where that "here" might be.

This was all of a sudden a public and political matter: it was Heimatschutz. Aesthetic discourse functioned as the ethical and communicative glue here in two different ways. The catalog of bourgeois aesthetic traditions, above all the Romantic heritage, provided a common field of perceptions and values that was central to the organization of the movement. But the Heimatschutz did not merely appeal to and confirm a preexisting aesthetic preference. Instead it used a general aesthetic sensibility in a new way as a cultural vehicle, even a journalistic device, to get people to think about the land. Through pictures and photographs and rhetoric the Heimatschützer pressed their fellow Germans to face the concrete physical condition of the environment and to confront the consequences of tailoring land to the requirements of capitalist-mechanist "rationality." Ecological analyses confirmed much of what the Heimatschutz demanded at the aesthetic level. But far from avoiding this level of investigation, the Heimatschützer welcomed it, and, in their project of Heimat tourism, tried to develop the emotional and cultural habits that would make regular investigation a self-perpetuating force in society.

A time-honored tradition in summarizing and evaluating an historical movement such as the Heimatschutz is the blunt question What did it achieve? Contrary to repeated critical assertions that the Heimatschutz's brand of environmentalism was characterized by "helplessness" or "ineffectiveness," we saw that in the space of its first ten years the Heimatschutz managed to bring about a number of improvements (Hermand 1991, 130; Andersen 1987, 150). In the area of the built environment, Prussia's 1907 Disfigurement Law and countless local construction guidelines protecting neighborhood aesthetics can be attributed directly to the influence of the Heimatschutz; less quantifiable but every bit as important was the Heimatschutz's indirect influence on individual citizens, its energizing and empowering effect in giving ordinary people the cultural and discursive tools to stand up for their respective visions of a congenial and socially egalitarian environment. In this regard the Heimatschutz deserves to be recognized as a key contributing factor to the powerful progressive traditions of German city planning and municipal socialism.

Any fair assessment of the positive work of the Heimatschutz must also proceed from the twin realizations that, first, the historical record on this movement is far from complete, and in particular we still possess comparatively little hard information on the activities of local and regional Heimatschützer; second, there is the related fact that the entire informal milieu of influence, negotiation, and consciousness-raising through which the Heimatschutz sought to and did operate has not lent itself to easy analysis or the neat listing of achievements. Only occasionally, as in the obituary of a forgotten building official in Weimar, do we get a glimpse of the fact that a single Heimatschützer was able to design fifty schoolhouses and was probably in a position to exert additional formative influence over a number of other developments in his area of jurisdiction. If we were to assume that only a small fraction of the 30,000 or so Heimatschützer who were organized by 1914 had this kind of power, we would have to say that the impact of the Heimatschutz was tremendous—and that before trying to estimate the overall collateral effect transmitted through millions of postcards, flyers, newspaper reports, lectures, slide shows and other means of communication.[4]

All of these arguments apply doubly to the Heimatschutz efforts to protect the landscape. We can point to a number of concrete laws and measures here, too, of course; general Prussian guidelines on civil engineering and agricultural improvement projects show the influence of Heimatschutz thinking, and in Saxony measures aimed at rediversifying the forests were just the most visible effects of a governmental apparatus steeped in the goals of the Heimatschutz movement. Municipal forest management was profoundly impacted by the Heimatschutz demand for more livable surroundings, and scattered decrees and guidelines from all over Germany give additional hard evidence of Heimatschutz effectiveness in protecting hedgerows, stream banks, and even water quality.

But by and large the impact here was of an indirect and unquantifiable nature. This is not the same thing as saying that the movement was somehow negligible or unimportant, however. What the Heimatschutz in fact did with its popular aesthetic sensibility was to open up a whole new area of discussion and debate in modern Germany. The key to its success was a remarkably well-rounded environmentalist conception of home, one that did not stop at a program of factual enlightenment, but rather extended to the construction of a new, counter-hegemonic culture: a way of seeing and feeling, a way of walking, a way of relating to one's environment. Although we cannot measure it in the usual manner, this was truly an achievement—one whose cultural and historical significance, I would suggest, is not lost even today.[5]

The Heimatschutz on "Americanism"

Rushing in where the angels of scholarship often fear to tread, I would like to use these findings to illuminate a somewhat different cultural context, namely, the history of environmentalism in the United States. As the Heimatschützer themselves realized, this was a country with a very different conservation history from that of Germany. The nineteenth century in particular was one long and relatively ruthless drive to "tame the wilderness," a euphemism for consuming the forests and bringing the land under cultivation. Especially with respect to the U.S. forests German observers were generally aghast at the exploitation (*Waldraubbau*) that went on. One

Heimatschutz author decried the "murder" of California's giant redwoods as an especially poignant example of the "vandalism of the American high-speed civilization"; another wrote that it was "impossible" for his fellow Germans to "imagine it in its entire depressing dimensions" (Pudor 1912, 412; Wilmowski 1910, 61).

The corollary and second phase of the battle with the wilderness was complete instrumentalization. Especially on the plains every square yard was put into production. In a story titled "A Visit to Farmers in Ontario," the Heimatschutz author Johannes Trojan expressed his disappointment with the new agricultural landscape, "only fields and meadows" without variation or interruption. Even though he praised the productiveness of the farmland, he was clearly distressed by the lack of the abundant bird life he was used to Germany: "There was little to hear of that sound which gives our Spring such great charm! A silence lay over the fields, the song of the lark was missing, which in our country can fill even a blind man with joy" (Trojan 1912, 288–89).

By the turn of the century "Americanism" was a bogeyman of cultural criticism in Germany; it could mean just about any negative aspect of a modern, industrialized socety. But to the Heimatschutz it had a specifically environmental meaning. The American landscape was the landscape of unrestrained capitalism and ecological depletion; it was land controlled and exploited with "cool ruthlessness," as the Heimatschützer Tilo von Wilmowski wrote in 1910. He further found "the empty uniformity of the farmhouses in the Midwest which are lined up like beads on a string" oppressive, and greeted the sight of the first weather-beaten "fisherman's hut in Cherbourg with a sigh of relief" (Wilmowski 1910, 61).

As a Heimat enthusiast, Wilmowski could be forgiven for succumbing to a bad case of homesickness, but the simple truth of the matter is that he and the Heimatschützer were not that far off the mark in describing the vast majority of the U.S. landscape as an empty and exploited realm. Their conception of Heimat allowed them further to see that these environmental problems were directly related to the land's cultural or emotional significance for the people who lived on it or used it. They thus attempted to explain the American situation by reference to the extreme rapidity of the westward expansion and to the lack of established agricultural customs, two factors that would have reinforced settlers' drive to transform the wilderness. In fact this analysis is probably correct: in contrast to the situation in Germany, in the United States the capitalist expansion of the nineteenth century took place largely in an emotional vacuum: there were no ancient cities or legendary historical spots to be ruined (at least in the eyes of the white settlers), nor even any of the old agricultural traditions to be overcome. The entire process was thus more radical and more legitimate, in the public consciousness, than it ever was in Germany or the rest of Europe. This far-reaching hegemony in the realm of environmental values is no doubt the reason why no widespread movement that one could compare to the Heimatschutz arose.[6] Conversely, though, as Wilmowski wrote, it was precisely the frontier-capitalist mentality of the U.S. population that revealed most clearly the need for a program of Heimatschutz: "It is quite unnecessary to emphasize the necessity of Heimatschutz for a country in which nature, for the overwhelming majority, still constitutes the enemy which has to be subjugated and turned to useful ends" (1910, 61).

Historians have written much about the *Sonderweg,* that "special path" with bizarre and ultimately disastrous consequences that the German nation took over the course of the nineteenth and early twentieth centuries. But if we consider the widespread German and European movement toward environmental philosophies of home, it was the United States that went on its own disastrous *Sonderweg* toward the end of the nineteenth century. Before this time its environmental excesses could possibly be understood, although not excused: the open frontier fed a potent symbolism of unlimited growth and expansion that had never been so concrete and geographically tangible in the heavily populated countries of Europe. Before this time the nation could possibly be forgiven for ignoring the impassioned warnings of a George Perkins Marsh, who in his book *Man and Nature* (1864) spelled out the folly of resource exploitation in terms that even capitalists could understand. The official closing of the frontier in the census of 1890 meant that this cornucopian mentality had to go. Americans were faced for the first time with the necessity of conserving a stock of resources that was less than infinite. They were thus approximately even with the European situation.

At this point two things happened. One was a fascination with wilderness. The bleaker and less hospitable their own immediate surroundings became, the more people desired to visit and even protect wild nature. The national park system that followed in the footsteps of Yellowstone was, and still is, largely dedicated to the characteristically nineteenth-century ideal of a scenic, romantic wilderness. If before 1890 there were only a few influential supporters of this idea, after 1890 it began to gather more and more support (Nash 1977). With the disappearance of the frontier, wilderness areas were suddenly special; popular interest grew, and a whole mythology of the importance of wilderness to the American soul sprang up—furthered, on occasion, by touristing interests such as the railroads (Runte 1993). With that a popular environmental agenda was set; to this day it can be said that "the defense of the wilderness has been the focal point of American environmentalism" (Short 1991, 100).

What would the Heimatschutz think of this wilderness hang-up? With all due respect to the scientific importance of wilderness areas, and even with the fullest sympathy for its romantic and spiritual values, the Heimatschutz would and did say that a concentration on wilderness is not enough. In discussing the enthusiasm for nature preserves Ernst Rudorff, the founder of the Heimatschutz, put it quite bluntly:

> I would like to warn the advocates [of wilderness parks] in all earnestness not to harbor or encourage the impression that with this deed *everything* has been done which ought to be done for the "protection of nature." America itself presents the most disturbing example that this would not be the case. As beautiful as the great national park there may be in its own right, what good does it do the crushing majority of the population when, as is well known, all of the rest of the land is becoming horribly bleak, and the wild animals are being ruthlessly extinguished, every living thing destroyed and sacrificed to industry?! (Rudorff 1910, 82)

Rudorff was right to worry that the wilderness park idea would give people a false sense of environmental accomplishment. As a critical historian recently wrote, in the United States the wilderness idea has "made conservation convenient" and allowed us to "enjoy the benefits of industrial development and over-consumption, while salvaging our consciences by setting aside a few undeveloped remnants for nature" (Callicott 1991, 245). The wilderness hang-up, in other words, has turned out to be an environmental cop-out. It is curious that it has taken us so long to recognize what Heimatschützer knew already at the turn of the century: that it is the vast area of land that is not "special" or "scenically exceptional" that needs stewardship the most.

This brings us to the second feature of the situation in the United States. If the popular imagination was captured by the beauty of wilderness, a no less influential portion of the environmental spectrum devoted itself to scientific conservation from above. Once again we can identify 1890 as an approximate turning point. In October of that year Congress created a major forest reserve in the Sierra Nevadas; it was followed by the Forest Reserve Act of 1891, which gave the president special powers to proclaim forest reserves on public land. The early 1890s were not coincidentally a time in which the first vocal advocates of a kind of scientifically oriented production ethic came to the fore, men such as the forester Gifford Pinchot, the hydrologist W. J. McGee, and Charles Van Hise. The Forest Management Act of 1897 reflected the "new sobriety" of these professional resource managers with its explicit orientation toward production; monocultures were an important part of the foresting techniques employed, as was clear-cutting (Runte 1991, 43f.; Moneyhon 1980). What makes this development particularly interesting in a comparative perspective is the fact that they learned these strategies of rationalized production largely from Europe; in the case of forest management from Germany itself, whose forest management techniques were imported by key figures such as Bernhard Fernow and Gifford Pinchot (Twight 1990; Miller 1991). The utilitarian conservation that they learned to apply to trees characterized U.S. government policy throughout the twentieth century in a wide number of other areas as well, from flood control to fish and wildlife management.

There can be little doubt that the intervention of scientific conservationists of the Pinchot school was crucial for the environmental future of the United States. Had it not been for them and their uniquely market-oriented arguments, it is doubtful that federal lands could have been maintained against the pressure of market capitalism. They allowed the U.S. government to justify its control to powerful private interests that probably would have otherwise succeeded in engineering a disastrous, large-scale privatization. But the solution at which these public-lands managers arrived had clear negative effects as well. One was that the discussion tended to shift to federal lands exclusively, usually the large blocks of land in the West. What private interests did with their property remained for the most part bracketed out of the discussion and legislation.

A related problem was that Americans as a whole were more or less relieved of the burden of having to think for themselves on this important issue. Called in as top-level appointees, the scientific managers did not have to gain wide public support and did not try particularly hard to convert the population at large, which remained largely insulated from their values (Hays 1959, 4). Over time a narrow, constituency-oriented technocracy became entrenched: the U.S. Forest Service catered to logging companies, hunters and sportsmen received the optimum "game management" of their favorite species, farmers were taken under the wing of expert extension agents, and the average citizen took little notice of the whole process, trusting or hoping or perhaps not even caring that things were being looked after.

If we look at forests in particular, though, this "professionalization" was something that had taken place in Germany long before. Germany's absolutist rulers had turned to the intensive production of timber in monocultured fir or pine forests back in the 1700s, at first in order to alleviate shortages, but very quickly in order to maximize financial return. Given this specific precedent it is perhaps not so surprising that the political basis of scientific conservation in the United States was not democratic, and that it was essentially "imposed on the population at large" (Runte 1991, 51). It was an executive-branch "enlightened absolutism" that has continued to shape practical U.S. conservation efforts ever since.

American "calcification" in this arena deserves to be contrasted with Germany's relatively more innovative development. As in the United States, the period from about 1890 to 1910 was a key era in the environmental history of Germany, a time in which the growth of industry and urban zones brought on a very similar sense that the environment was both important and endangered. But in Germany a state-sponsored rationalized conservation was already the rule: over the course of the nineteenth century a system of bureaucrats, extension agents, and technical schools had long since established a system of intensive, scientific agriculture. The irony is that the United States copied this concept just as a growing number of Germans, through long exposure, were beginning to realize its severe limitations. In spite of the traditionally large role of the German states in such matters it was a realization, moreover, that largely percolated up from below, and that was founded in a wider, even popular, culture.

A comparison can be made here that is crude, but crude tools are tools all the same. Over nearly one hundred years the land in Germany and the United States has been dominated by the same ubiquitous forces of modernity. All the mistakes that have been made in the one country have probably been made in the other. If we find that the ordinary German landscape has weathered its century better—if we see areas where the land is not an "instrumental terrain" of agribusiness, but rather resembles something like a commonly held "home" that citizens "shall never weary of"—then this is at least partly a result, I would contend, of that complex value of Heimat that the Heimatschutz movement began to instill in Germans at the turn of the century.

The comparison suggests that we ignore the aesthetic and cultural dimensions of environmentalism at our own risk. The scientific high-production philosophy that took hold in the United States did not succeed in changing most people's shortsightedly utilitarian view of the land, which in any case was hardly touched by the regulations imposed on federal lands. From the Dust Bowl of the 1930s to the timber scandals of the 1980s and 1990s the "scientific" regulation of production has been embarrassed by one disaster after another. It is clear that a significant part of the problem is the lack of ethical-practical impact that is achieved by the scientific analysis. The land has remained a site of mechanistic manipulation or analytical investigation, devoid of any deeper emotional significance for most people in society. Wendell Berry has traced this schism in environmental thinking back to the level of language:

> No settled family or community has ever called its home place an "environment." None has ever called its feeling for its home place "biocentric" or "anthropocentric." None has ever thought of its connection to its home place as "ecological," deep or shallow. The concepts and insights of the ecologists are of great usefulness in our predicament, and we can hardly escape the need to speak of "ecology" and "ecosystems." But the terms themselves are culturally sterile. (Berry 1993, 35)

As will be remembered, though, not all representations of the environment are culturally sterile. Over the last century Americans have become increasingly involved with the idea of wilderness. The national park system has been expanded, and more and more money has begun to flow into the environmental movement. Popular interest actually exists—but it is largely directed away from private property, away from most people's hometowns and day-to-day existences.

What the United States may need now is to make up for the step it missed at the turn of the century: it needs its own version of Heimatschutz. It would entail, first of all, getting rid of the idea of untouched wilderness as our standard of environmental significance. As J. Baird Callicott has written, if this is finally "replaced . . . then we might begin to envision ways of creatively reintegrating man and nature" (Callicott 1991, 245). Callicott's suggested replacement is the idea of "land health," a concept that unfortunately has all of the starched-white appeal of a visit to the dentist. If environmentalism is to wrest more and more people away from the dominant culture of exploitation, it will need to balance out this somewhat clinical idea of health with a broad cultural program. "A suitable public aesthetic" would be a good place to start (Shepard 1967, 262; cf. also Wrede and Adams 1991; Rollins 1996). But it must also be a private aesthetic, an attitude and a way of seeing things that we can almost instinctively bring to bear over the course of our everyday life. In an age of capital that has rightly been described as "aggressively homeless," an important part of environmentalism will be to pursue this greener vision of home (Rogers 1994, 3).

Notes

Introduction

1. To his credit, Hermand has criticized these earlier statements as indicative of left-modernist bias against environmental concern (1991, footnote 4).

2. Bergmann's argument clearly dovetails with the similar views on the "life-reform movement" and "bourgeois escapism" espoused contemporaneously by Janus Frecot, Johann Friedrich Geist, and Diethart Kerbs (1972) in their book on the painter and graphic artist "Fidus" (Hugo Höppner). They speak of a consistent middle-class tradition of Nature as a "site of escape" and see all life-reform interest in terms of an "ersatz religion." Their fundamentally top-down and nonenvironmental view of politics is summarized in their disappointed observation that "fallow-lying political energy" was wasted in the fuss about nature.

Chapter 1

1. The environmentalist possiblities that are present in Baumgarten and other early aestheticians were unfortunately undercut in the increasingly conceptual-cognitive philosophies of Kant and Hegel, who describe a progression from a tolerant brand of dualism to an intolerant, highly anthropocentric brand of idealism. Hegel's standard of the self-comprehending *Geist* marks something of a low point, essentially telling us that we should ignore nature and landscapes, which, like gardens, can never be more than "mere surroundings" (*bloße Umgebung*) (cf. Hegel 1951, vol. 13, 352).

2. In terms of historical succession some historians divide the landscape garden into four phases: a "naturalistic" phase, then a "pre-romantic-sentimental" one, then a "classical," and finally a "romantic" phase (cf. Nehring 1979, 101f.). These phases are drawn largely from the longer and more clearly articulated history of landscape gardening in England, however, and do not necessarily describe what went on in Germany with any great precision. Especially in terms of humans' approach to nature it seems allowable to reduce the variety of different designs to the important opposition between the generally romantic appeal to untouched nature and the classical goal of a culture-nature symbiosis.

3. In this very important respect the classical landscape garden in Germany appears to differ from the mainstream of developments in England as described by Ann Bermingham (1986). In an analysis that generally points up the social status associated with the large and expensive landscape garden, she claims that the "landscape garden's aesthetic effect depended on a completely nonfunctional, non-productive use of land." Nature, according to Bermingham, took on a desirable snob appeal precisely in inverse relation to its real availability in an era of extensive new, rectangularly laid-out enclosures: "As the real landscape began to look increasingly artificial, like a garden, the garden began to look increasingly natural, like the preenclosed landscape." Bermingham then explains the patently obvious fact that landscape gardens wanted to include the countryside as a clever bit of ideological finessing. "As enclosure proceeded, leaving the 'country without' increasingly less wild, professional gardeners refined the planning of prospects to avoid the raw-looking new enclosures. Brown's and Repton's gardens excluded such unpleasant objects from view while allowing real views of the countryside only if they were pleasing and of a piece with the garden." To thus "collapse the distinction between art and nature" is, in Bermingham's eyes, the height of under-handed manipulation, as it amounts to "recreating nature in the image of certain class practices and values" (13–15).

Part of Bermingham's critique seems to be that this was hypocritical: by their social actions, the landowners destroyed what they aesthetically sought to uphold. But part of it goes in another direction and implies that the aesthetic ideal itself was untenable and could only be upheld by biased "class practices." Methodologically there remains a tension between a kind of ecological definition of the natural state of the land and a traditional economistic view that sees enclosures, industry, and "progress" as the only "natural" course of events. Either way, it should be remem-bered, Bermingham's critique focuses on a schism between art-garden and ordi-nary countryside. As Barthelmeß (1988, 87) pointed out, it is precisely this dis-crepancy that Germany's classical landscape garden concept sought to overcome and not merely to disguise.

4. Here it is necessary to differentiate between the inclusive social reform tra-jectory of a Wörlitz on the one hand and the status-oriented and exclusive nature of English landscape gardens. John Rennie Short rightly emphasizes that the sup-posedly "'typical' English countryside, small regular fields and rolling acres of parkland, . . . were the result of a radical change in power relations" during a key phase of privatization and capital accumulation. Ample evidence of the antisocial nature of English gardens is also provided by the cases in which large landowners, through undue political influence, were able to destroy entire villages in order to lay out their grandiose parks. "It is ironic that the typical English countryside, the supercharged image of English environmental ideology, which can still conjure up notions of community, unchanging values and national sentiments, is in reality the imprint of a profit-based exercise that destroyed the English peasantry and replaced a moral economy of traditional rights and obligations with the cash nexus of commercial capitalism" (Short 1991, 70–71).

5. Terry Eagleton (1990, 13f.) comes to the opposite conclusion in his recent overview of the role of aesthetics. He sees the "strange new discourse of aesthetics"

begun by Baumgarten as the continuation of political domination by other means and specifically connects Baumgarten with the "colonization" of material and bodily sensation by reason. We agree on the existence of hegemonic domination in society, and I would even agree with Eagleton that much of that domination is rooted in the "colonization" and co-opting of sense experience. But Eagleton fails to acknowledge that the same process can be turned against the dominant brand of "reason," in which case it would follow that the aesthetic realm is a key site of social perception and contention.

6. Cited here and in the following from Schiller (1962, 309–412); my translations.

7. On the ecological significance of the "Philanthropinum" school in Schnepfenthal/Gotha and its chief educators Christian Gotthelf Salzmann, Rudolf Zacharias Becker, and Johann Matthäus Bechstein, see Gerhard Trommer (1990, 75ff.; also Niedermeier 1993). Trommer emphasizes that the concepts developed here were at the same high level of ecological awareness as the better known and roughly contemporary work of Gilbert White in his *Natural History of Selborne* (1789). Interestingly, Trommer further points out that the uniquely effective educational approaches of both centered on the holistic principle of a local Heimat: "[In both] there dominated a *heimatlich* kind of nature education and a descriptive kind of research into nature, along with a feeling at ease and a familiarity with the surrounding cultivated landscape" (97).

8. Vorherr began his public *engagement* for *Landesverschönerung* with a pamphlet in 1808 "On the Beautification of Germany" and is labeled by Däumel (1961, 44f.) the "central figure of land beautification." Däumel's glowing assessment of Vorherr is given some healthy relativization by Gert Gröning (1992), who emphasizes Vorherr's schematicism and his unwillingness to listen to other parties in the planning process.

9. Berndt does not see this bourgeois stance as an unalloyed evil. She debunks the "vulgar marxist" denunciation of public health in part by arguing that it is unhistorical: Social Democrats came to be in favor of such programs, and when they were in power gave them their full support.

10. Ladd notes that the phrase "light and air," which arose in the 1840s, quickly found its way into Prussian building ordinances and became a permanent feature of the institutionalized discipline of city planning, continuing to influence decisions in 1910 much as it had decades before (1990, 231).

11. A recent monograph links the English Romantic tradition of which Morris was an indisputable part to a revolutionary new ecological consciousness centered precisely on the nonchauvinistic value of "home" (cf. Bate 1991).

12. One important exception to the rule of bourgeois dominance in conservation and nature organizations can be found in the working-class Socialist *Clarion* movement, whose "Ramblers" fought to maintain or regain access to paths unilaterally closed off by landowners (cf. Hill 1980). This British movement would appear to be the exact equivalent of the the Socialist *Naturfreunde* in Germany and Austria, who were involved in very similar battles for access.

13. Jefferies has compared the Heimatschutz to Britain's National Trust and the American Scenic and Historic Preservation Society (founded 1895) and states that

they show a "very similar social character." This is only partly accurate. As I will show in chapter 2, the Heimatschutz social base was noticeably broader and included more than those "persons of distinction" who made up, for instance, the Scapa Society (cf. Evans 1926, 51f.). On top of this its landscape objectives gave its work a decidedly different and more ecologically relevant character.

14. Gould (1988, 16) complains that the bourgeois English environmental associations "posed no threat to the continuance of the existing social order or the direction that it was taking." He fails to ask why Morris, who assuredly realized the true nature of the organizations he joined, still consented to participate in and in some cases lead them. The answer is not to be found in dogmatic ontological definitions of what is and isn't "revolutionary," but rather in Morris's more nuanced assessment of the historical possibilities.

Chapter 2

1. Koshar quotes the Rheinland preservationist Paul Clemen as postulating an even wider version of "monument"—one that included "the entire cityscape, the maintaining of historical towns, and the image of the countryside." My first point with regard to this remark is that it shouldn't be understood as the product of a solely preservationist development. Koshar does not attempt to draw a sharp distinction between preservationist and Heimatschutz agendas, a task that is admittedly not easy due to their interwoven nature. Clemen was a member of both movements, and his speech was given at a combined biennial meeting of the two, the *Tag für Denkmalpflege und Heimatschutz.* This is actually a very good indication that the wider meaning of the word "monument"—and specifically, its extension to the landscape—was due largely to the Heimatschutz impulse. Preservationists were steeped in a tradition of caring only for architecture and art; I argue that it was the Heimatschutz that crucially mediated this aesthetic tradition and turned it toward consideration of the entire landscape. My second point is that not everyone would have agreed with Clemen in applying the word *monument* to landscapes; as we shall see in chapter 3, the Heimatschutz thought it was quite possible to attribute value to ordinary working landscapes without going to the extreme of static preservationism that characterized "monumental" ways of looking at nature.

2. Touristic exposure to the ordinary German countryside—made possible, incidentally, by the efficient and increasingly complete railroad network—was probably one reason why it was the middle class in particular that took up the cry of "Heimatschutz!" Factory workers were allowed neither the means nor even the vacation time for this experience; on their one day of rest per week it was often all they could do to go to a city park or take a hike in the nearest bit of surrounding countryside. These immediate sites for recuperation were accordingly the focus of much of the Social Democrats' environmental interest. Understandably enough, workers were deeply engaged in their fight for political representation and decent wages and working conditions; for this reason, too, they tended not to get as interested in issues of the countryside. On the other end of the social scale we can assume that the situation was reversed; the nobility as a whole still had access to large entailed estates, hunting parties, and a range of other privileges that no doubt

prevented it from taking an existential interest in the problems of development. In general it was only the middle class that was in a position to both enjoy the countryside and to take to heart how endangered it was.

3. This is not to deny that in some circles of Heimat interest and even in some Heimatschutz supporters there was a more racial use of the term. Robert Mielke and a number of the men in the Heimat movement in Lower Saxony were among the most prominent here, often speaking of "Stämme" and "Stammesbewußtsein." These words are rightly troubling, but it would be simply inaccurate to portray racist thinking as in any way essential to or representative of the prewar stage of the Heimatschutz. Gerhard Fehl's contrary assertion, that "whenever Heimatschützer spoke of 'deutsch,' for example of 'deutsches Volkstum' (Rudorff) . . . then there usually peeped out from behind the national meaning also the racist meaning of Langbehn, that is, with 'deutsch' is meant 'Aryan,' 'purebred,' of 'German blood,' and 'German race,'" should be treated with suspicion (Fehl 1995, 148; see also Koshar 1996).

4. In this study I focus on the Heimatschutz proper and not on the Heimatbewegung, which has been the preferred category in both Werner Hartung's *Konservative Zivilisationskritik* and in Celia Applegate's book *A Nation of Provincials.* Part of the reason that Applegate preferred the other term is the fact that she dealt with the *Pfalz,* one of the very few areas of the German Empire that never proceeded to the stage of founding an explicitly reform-minded Heimatschutz group. My definition of the Heimatschutz movement is essentially substantive: it was comprised of groups involved in serious efforts to maintain an area's unique natural or historical ambience. Most contemporaries defined Heimatschutz in this straightforward fashion, equating it with a program to protect natural landscapes and built environments. This was the basis for identifying Heimatschutz movements in other countries that did not have the term, for example in the usage of Gaebel (1912, 1), who identified "nature and building construction" as the "two main branches of the Heimatschutz."

This substantive definition allows us to include organizations that pursued many of the same goals as the *Bund Heimatschutz,* some of which even included the word "Heimatschutz" in their title, and yet chose not to affiliate themselves in an explicit, structural fashion. The Bavarian Association for Folk Art and Folklore, founded in 1902, provides a good example of an organization that shared many goals with the *Bund Heimatschutz*—so much so that it chose to change its name in 1916 to the Bavarian State Association for Heimatschutz, Association for Folk Art and Folklore. In spite of this it was not until 1928 that the independent-minded Bavarians finally agreed to an explicit affiliation. The Heimatbund Niedersachsen was another organization that eschewed direct ties to the national *Bund Heimatschutz* and yet pursued largely parallel aims.

Any reasonable definition of the movement would have to include these substantively similar organizations. Conversely, there were also groups that took on the increasingly popular slogan of "Heimatschutz" either in their title or their charter and that nonetheless showed no signs of the activist commitment to reform that was essential to the movement. One example of such a "false" Heimatschutz organization is the *Verein für Heimatschutz und Heimatpflege im Kreise Schlüchtern.*

This Schlüchterner Heimatbund was composed almost entirely of pastors and teachers and—judging by the contents of its journal—devoted itself solely to a particularly quietistic form of Christianizing Heimat kitsch. But even with regard to less odious segments of the associational spectrum, the substantive distinction between Heimatbewegung and Heimat*schutz*bewegung is important. The Heimatbewegung is a far larger category of organizations with (typically) far less concern for physical surroundings; in essence it refers to the whole associational network out of which the Heimatschutz originally developed and specialized, stretching from historical and patriotic interests to folklorists, hikers, beautification societies, and butterfly collectors. All of these organizations, not to mention a wide range of politicians and hucksters, were prone to using the word Heimat in one way or another. Even though Heimatschutz representatives employed some of the same nebulous rhetoric about inspiring "love of the Heimat," their distinguishing characteristic is that they usually meant something by it. Their love for their regional environment did not exhaust itself in phraseology, but rather found concrete expression in measures to protect or reform the world around them.

5. Heimat associations often remained important local footholds for the Heimatschutz even after it had been established. The kind of institutional constellation that developed between generic Heimat interest and the newer Heimatschutz interest can be illustrated for the case of Brandenburg. One of the seminal organizations here was the *Brandenburgia,* a kind of catch-all Heimat organization founded in 1896 that united historical and local natural science interests. It supported the development of the Heimatschutz movement from very early on, and a number of the *Brandenburgia*'s members became leaders of the *Landesverein Brandenburg* when it was formed in 1907. Although local branches sprang up in cities like Berlin, Neu-Brandenburg, and Ruppin, the Heimatschutz organization in Brandenburg was like most of the large, influential Heimatschutz organizations in that it was essentially a regional association. In its structure it therefore resembled the older *Brandenburgia* and indeed made use of that group's contacts in different municipalities across the province. The *Verein für die Geschichte Potsdams, Gesellschaft zur Pflege der Heimatskunde* apparently took on the role of Potsdam's local Heimatschutz organization; in Eberswalde and in Cottbus it was each city's *Verein für Heimatkunde* (cf. *Heimatschutz in Brandenburg* 1909–10, 18f.). Other Heimat groups such as the *Havelländischer Heimatverein,* founded 1909 in Rathenow, and the *Verein für Heimatkunde und Heimatpflege,* founded in 1913 in Fürstenwalde, provided further local affiliations for the regional group, but did not expect nor even necessarily encourage individual members to join the *Landesverein Brandenburg.*

6. The precise name of the architectural organization that Schultze-Naumburg chaired soon began to elude even some of the people involved and has since become the subject of a number of puzzled footnotes. Knaut (1991) notes that Schultze-Naumburg, in his *Lebenserinnerungen* manuscript, identified it as the (1) "Deutscher Ausschuß für gesundes Bauen in Stadt und Land"—the title used by Gerhard Kratzsch and since repeated in the two English-language articles on the Heimatschutz by Christian Otto and Matthew Jefferies. Knaut prefers the alternate title (2) "Ausschuß zur Pflege der heimatlichen Bauweise in Sachsen und

Thüringen" because it is used by Robert Mielke in a 1929 article and was also entered as the name of a participating organization in the Heimatschutz. To make matters worse, though, it appears that a number of other variants exist: Mielke (1914) mentions Schultze-Naumburg in connection with a (3) "Sächsisch-thüringischer Verband zur Erhaltung der volkstümlichen Bau- und Kunstweise" in an earlier article; in 1904, though, he mentioned a slightly different version: the (4) "Ausschuß zur Erhaltung heimischer Bauweise in Sachsen und Thüringen" (cf. Mielke 1904–5); meanwhile, though, Ernst Rudorff was sure that Schultze-Naumburg had spoken in 1903 at a meeting of the (5) "Ausschuß für ländliche und bürgerliche Bauweise in Sachsen und Thüringen" (cf. letter of 23 October 1903, *Staatsbibiliothek Preußischer Kulturbesitz,* Nachlaß Mielke, 80). The confusion is not helped by the fact that the precursor to Saxony's Heimatschutz organization was the similarly titled (6) "Ausschuß zur Pflege heimatlicher Natur, Kunst und Bauweise" (cf. *Mitteilungen des Landesvereins Sächsischer Heimatschutz* 1908, 1), which was known earlier as (7) the "Ausschuß zur Pflege heimatlicher Kunst und Bauweise in Sachsen und Thüringen," according to an "Entwurf eines Aufrufes" from 16 November 1903 (cf. Sächsisches Staatsarchiv Dresden, MdI 17519).

7. The minutes of the Isis provide good evidence of the close networking that helped propagate Heimatschutz ideas throughout other associations. A May 1904 day-trip to Meißen organized by members of the Dresden Isis shows that scientific and hiking groups cooperated closely, since they invited not only members of Meißen's Isis, but also the Beautification Society "*Naturfreund*" and numerous mountaineering groups. At a common meeting it was once again Oscar Drude who took advantage of the opportunity in order to elaborate on "Purpose and Goals of the Heimatschutz League." Following up on this, a leader of the "*Naturfreund*" reported his group's efforts to protect a nearby mountaintop, and the chairman of the Meißen *Isis* declared his organization's desire to help with the goals of the Heimatschutz (cf. *Sitzungsberichte der Isis* 1904, 14).

8. Some of the states that closely followed the Prussian example were the Duchy of Saxony and the Principality of Lippe. Formulating its own law several years later, Lippe might well have profited from Prussia's experience and indeed asked the Prussians for advice. Already by 1910 some bureaucrats within the Prussian Culture Ministry were beginning to agree with many Heimatschützer in acknowledging the shortcomings of the 1907 law—particularly the problematic standard of "gross disfigurement"—and told the Ministry of Public Works to tell Lippe that the Saxon and Bavarian legislation without this phrase was better (15 August 1910). The Ministry of Public Works, however, disagreed, fearing a "impingement on significant economic interests" and a "strong negative opinion in broad segments of the population" if there were intervention in the case of "every disfiguring construction" (28 August 1910). In the end Lippe received the advice that the Prussian law was basically good and only a vague indication that more generous language could be of use "in a few cases" (12 October 1910) (cf. letters in PGStA I.HA Rep. 93B 1546).

9. The donors of Koch's salary were Imperial Ambassador Krupp von Bohlen und Halbach, a prominent supporter of the Rhenish Association for the Care of Monuments and Heimatschutz, and Freiherr von Wilmowski, Heimatschutz sup-

porter and government official from Merseburg (cf. *Heimatschutz in Brandenburg* 1909–10, 124).

10. Although we agree to a great extent on the description of what happened in this phase of the Heimatschutz, it should be obvious that I put a completely different emphasis on it than does Matthew Jefferies, who praises Lindner as the herald of a new sober pragmatism in the Heimatschutz that finally got rid of the "völkisch mysticism and anti-urban rhetoric of the turn-of-the-century prophets of doom" that supposedly characterized the early years (Jefferies 1995, 98).

11. The group that Christiansen examined, the *Verein zur Pflege der Natur- und Landeskunde in Schleswig-Holstein,* clearly belonged to the older, vaguer, and more passive Heimatbewegung. Its reporting on the national *Bund Heimatschutz* meeting in Lübeck in 1908 demonstrates the temporal and methodological gap, as the commentator confessed "a certain feeling of envy" with regard to the new competitor, since "what was possible for it in just a few years, we were unable to achieve in eighteen years" (Peters 1908; also Eckmann 1908). The split between Heimatbewegung and Heimatschutz became explicit with the founding of the Schleswig-Holstein State Association for Heimatschutz right in the older association's home city of Kiel that same year.

12. The exact meaning of this participation remains open to interpretation. Adherents of a strict class theory would probably argue that noble participation occurred because of that class's desire to use the Heimatschutz as a bulwark against all change. On the other hand, though, it seems possible to see the many nobles who serve in the Mecklenburg state administration largely as civil servants in an effort to mediate change and steer growth—much as did their more numerous bourgeois counterparts. This takes up one of the standard puzzles of Wilhelmine Germany, as to whether we see an "embourgeoisement" of the nobility, or the "feudalization" of the bourgeoisie.

13. Celia Applegate (1990, 103) has disputed descriptions of the Heimat phenomenon as a strictly bourgeois, class-oriented structure, saying it was "more genuinely inclusive than exclusive." Paul Krische threw some light on the social convictions of a Heimat organization with his strongly autobiographical novel *Hein Träumerlein. Aus dem jungen Leben eines Wahrheitssuchers.* The book is interesting because it presents the personal history of a young man whose formative pre-Heimat movement experiences were summers spent in a life-reform–inspired state of nudism and nature worship. After a period of self-centered idealism the university student learns to accept "the tremendous manifold character of valuable human culture" while maintaining an engaged "consciousness of the jarring dissonances of economic life." In this newly realistic phase the student hero rejects the class apartheid of many of his fellow *Bürger* and helps found a *Heimatbund;* here he associates with "artisans, shop girls, and similar 'little guys'" because he feels it is his "duty to give to [his] fellow world" (Krische 1910, 152f.).

Inspired by Freud, Kropotkin and apparently Wilhelm Reich, among others, in later years Krische went on to develop the anthropological and psychological aspects of Heimat feelings and repeatedly emphasized the importance of building real democratic communities in any such political-cultural effort (cf. Krische 1918; also Krische 1923). His development toward a nondogmatic left-Socialist position is not typical for Heimat enthusiasts of the turn of the century, but it does show the potential compatibility of Heimat ideals with some of the most progressive social thought to come out of the early twentieth century. Krische called for a "federative Socialism," a kind of political self-determination united with ecological tolerance, something evident in the demand for "Heimatschutz, the protection of animals, the dispersal of the great cities which are weighed down with living quarters unfit for humans, the possibility of a patch of earth for every person and thus light, air, health, a strong feeling of *Gemeinschaft* in humans and nature—these are the novel, practical emphases of this movement" (Krische 1923, 331).

14. This information may contribute toward determining the social background of environmentalism in Wilhelmine Germany, which to a certain degree overlaps with interest in various life-reform movements. Many critics have anecdotally stressed the role of educated Germans, the *Bildungsbürgertum.* More specific studies on naturopathic groups have shown a concentration toward the "new middle class" of lower-level white-collar workers; this result has seemed plausible because it mirrors research into present-day environmental concern. The issue has been muddied recently by the conclusion that support for environmental policies came from "almost every subcategory" of Wilhelmine society (Dominick 1992, 59). The results that I give for the Heimatschutz would seem to indicate a slightly higher social location than the lower-middle-class clientele of clerks and *Angestellte* that has been previously assumed.

15. Paul Schultze-Naumburg's lament of 1905 applied to many of his fellow Heimatschützer over the years: "We are all men who are kept very busy in our professions, who can only work directly for the Heimatschutz in our spare time, and who still cannot be compensated the least bit" (Schultze-Naumburg 1906, 4).

16. The Heimatschutz supported an aesthetic revaluation of architecture that, at least implicitly, stood for a redistribution of social prestige and a long-overdue appreciation for the needs and contributions of ordinary people. As Oskar Seyffert, a Heimatschutz leader in Saxony, put it, Heimat was precisely the opposite of the aristocratic sphere of life, something that he saw as "foreign territory" (*die Fremde*). "Our path leads us not into proud castles, but into workshops. The path takes us not through groves of laurel and sprays of acanthus, but rather through forest and field, where the birds sing, where the folksong cries and laughs, our path leads us not into foreign territory, but rather into our Heimat!" (Seyffert 1910–11, 10).

17. On Saxony see *Mitteilungen des Landesvereins Sächsischer Heimatschutz* 1914–15, 20; Jefferies 1995, 57. Most regional Heimatschutz organizations received comparatively little money from government sources. In 1914 the *Landesverein Brandenburg,* which was after all the Heimatschutz association closest geographically to the center of Prussian government, received only 300 marks—and that was out of the pockets of the city of Berlin. Prussia itself gave nothing, and the association complained that only one-fifth of Brandenburg's municipalities had even bothered to become members—this although they frequently made use of the *Landesverein* and the work "completely selflessly" contributed by a number of its members (Mielke 1914, 60).

18. The view that much of the forward-pointing developments in the Wilhelmine

era were made outside of the realm of explicit parliamentary politics is basic to this whole examination. It is for this reason that I have not assembled data on individual Heimatschützer's party affiliations. It can be guessed that many tended toward the conservative end of the bourgeois spectrum, but the widespread journalistic support for the movement suggests that educated and informed members of almost all factions could find common ground in the general idea of environmental reform, if not in all of its specific legal provisions. As is argued below, this common ground shifted slowly but surely to the left in the last years before the war.

19. Tendentially, then, I would lay an opposite emphasis on the movement than that given by Celia Applegate, who contends that "the original impulse, the informing and defining concept behind Heimat, was not antimodernism at all but belief in the importance of regional identity" (Applegate 1990, 106). To a certain extent this difference is rooted in our different objects of study (the specific Heimatschutz movement versus a much broader and longer-term Heimat interest), but it also goes to the heart of what we each see in Heimat. I see the "antimodernist" critique articulated by the Heimatschutz movement as its most valuable contribution, and any regional elements in a secondary role.

20. This Heimatschutz advocacy for zoning *and* aesthetics flatly contradicts the frequent schematic division of turn-of-the-century planning movements into three separate groups, one being the "city planners," a second the "housing reformers," and the last being adherents of the "art of city planning" or aesthetics. The Heimatschutz has most often been included in this last group, one dismissed by critics such as Gerhard Fehl and Juan Rodriquex-Lores for its lack of insight into technical matters; conversely they claim an "advanced" status for the "city planners" precisely because this group embraced the idea of zoning (Breuer 1985, 525; Fehl 1995, 30f.).

21. It may be that the favored usage (calls for a neighborhood of villas and quiet streets) reflected the Heimatschutz's own class/group bias, but it seems to me that it was meeting its democratic burden by fairly, clearly, and logically presenting its arguments to the public. The Heimatschutz always liked to say that it was pursuing the general good, and by laying its cards on the table it allowed the public plenty of opportunity to take exception to that kind of invocation.

22. The *Werdandibund* is an interesting and perhaps the first clear example of what has come to be called, in Jeffrey Herf's terminology, "reactionary modernism." The sides in this dispute were precisely the opposite of those that most simplistic cultural schemata would predict: the socially relatively liberal Heimatschutz defended more conservative materials and forms, whereas a self-consciously modernist, big-business style was promulgated by the increasingly far-right Werdandibund. The industrial support for the Werdandibund and especially its Werdandihalle was made no secret. As the Werdandis themselves proudly said, the quality of their work was proved by the fact "that there is by now hardly any leading industrial association which has not given significant assistance, support and recognition to the Werdandibund" (cf. "Werdandi-Unternehmungen" 1913).

23. Conferences of municipal officials were examples of one forum where Heimatschutz talks found a willing audience; an example is the thirty-second Brandenburg *Städtetag* in 1910, where three presentations were given by *Landesverein*

Brandenburg members Robert Mielke, Wilhelm Wetekamp, and a forest official named Wilski. At the Hessisch-Waldeckischer Städtetag in Arolsen in June 1914, a local Heimatschutz member, Dr. Gaebel, called for municipal Heimatschutz in the form of more and better parks, the preservation of trees and the varying of street width when laying out streets, and in general avoiding the temptation to sacrifice too much to the supposed needs of "traffic." At the end of his speech he suggested that Heimatschutz associations were an important part of the planning resources available to a small- or medium-sized city: "Big cities that have their own city planner will need us little; on the other hand, though, we can be of practical assistance to the smaller cities, which do not have architects who are trained for all building tasks and above all are not artistically trained . . . I invite all of you, particularly from smaller villages, to come see us if you need us" (Gaebel 1914, 13).

24. The Heimatschützer even devoted efforts to small things such as street names. They observed that much history and much local identity was tied up in the names given to various roads, and they came out strongly for retaining popular street names. In a time of accelerating urbanization they also promoted the practice of giving new streets the name of the field or area they were replacing. The street name would thereby become a kind of "historical document"—as one writer put it elegiacally, "When the automobile has long been tearing along the pavement, the street name will still be able to tell our grandchild what once was here" (Wehr 1915). Conversely there was vehement opposition to a number of newer trends in giving names: people who went around changing "Gasse" to the more elegant-sounding "Straße" were hopelessly out of touch with their roots; the same could be said for those who would want to honor a famous person by imposing the new name over an established old one. Daring to quibble with Prussian patriotism, one author asked why a perfectly good, descriptive, and popular name like "farmside road" (*auf der Bauernseite*) should be exchanged for the more elegant-sounding "Moltkestraße." Beyond the pale, too, was the American practice of identifying streets by mere numbers. None of these naming practices had anything to do with the "character or position of the street," and thus they reflected nothing other than the age "which we can thank for the ostentatious apartment house façades of the great cities, the age of the great land speculation" (Uhlig 1910–11, 276; see also Lühmann 1912; "Straßennamen" 1912).

25. Schmoller's lectures were heard by thousands in his Berlin years alone; in Munich crowds of 500 regularly attended Brentano's lectures, forcing the university administration eventually to build a bigger auditorium (Seifert 1990; Brentano 1931, 256f.).

26. Ulrich Linse (1988) has compiled an impressively detailed analysis of this particular environmental battle. A valuable perspective that he supplies is an overall ecological-economic profile of the Upper Rhine in the nineteenth century, one that shows the value and size of fish resources declining due to pollution and other factors. This helps explain the relatively cheap sell-out of the fishing industry. Probably the most important result of his study, though, is to confirm the power and public support that were behind the dam, something that went as far as pro-dam parades and the burning of a dam opponent in effigy. Although extreme, this reflects precisely the kind of hegemonic situation that I assume as a major charac-

teristic of the age. In this connection it is important to note that Linse's assessment of the role of the Heimatschutz changed: whereas earlier (1986) he had belittled the Laufenburg action as a kind of sad excuse for "real" eco-activism, the 1988 article expresses a good deal of respect for the perseverance of the Heimatschutz against very high odds.

27. Peter Morris-Keitel (1991) notes in his dissertation that Haupt's pamphlet was also published in a magazine of the *Wandervogel* in 1911. This sequence of events confirms Morris-Keitel's general observation that the *Wandervogel* movement derived virtually all of its active environmental commitment from the established organizations of its parents, namely the Heimatschutz and Naturschutz.

28. Boyer's conception of "positive environmentalism" is as an instrument in class warfare; environmental reform in his view was only "positive" in that its proponents did not sink to the level of repression and criminalization that marked other bourgeois efforts to impose their "moral order" on the unruly classes. As the discussion makes clear, I think that this would be an unfair characterization of the Heimatschutz movement for environmental reform.

29. Alon Confino (1993) has noted an essential tolerance among Germany's Socialists for the bourgeois idea of Heimat, a point with which I would agree. We disagree, however, on what this bourgeois-universal idea of Heimat meant. Echoing earlier critiques of Heimat as an integrationist "ideology," now termed "national memory," Confino sees Heimat as an empty rhetorical figure, "apolitical," "non-specific," "extracted from the here and now," containing "no real people"; in short, Heimat was a place where—in David Byrneian diction—"nothing ever happened." Only for the reason that it meant nothing, he contends, could all parties agree on it and share it. By contrast I think it is clear that Heimat took on very specific and concrete meaning as it was developed and applied in a program of environmental reform. If this appealed to both the bourgeoisie and workers—so much the better.

Chapter 3

1. Sieferle rightly points out the absence of any essentially different stance on industrial growth in the ranks of the Social Democrats, who became in a larger historical sense "collaborators of progress" along with their nominal class enemy. Dominick (1992, 62f.) challenges Sieferle's assessment by noting that workers praised the experience of nature, that reformist Social Democrats were open to supposedly "piecemeal" environmental reforms, and that some Socialist leaders showed knowledge of and interest in environmental matters in a few Reichstag speeches. Linse (1993, 139) goes further in arguing that "it was above all representatives of the workers' movement" who led the way toward a multidimensioned, more socially accountable view of, for example, the forest. But many of these same positive developments took place within bourgeois circles as well; as a group, then, it is reasonable to believe that Social Democratic workers were neither much more nor particularly less environmentally aware than the bourgeois segment of the society—which is Sieferle's point.

2. Here it is also obvious that I disagree methodologically with attempts to see landscape protection solely as a crude bit of class politics. John Rennie Short, for instance, has denounced the historic and landscape preservation programs of Britain's National Trust as an expression of class interests and class oppression; what is preserved here, he says, is a "landscape of power" that merely reproduces dominance through its preservation of cultural symbols (cf. Short 1991). The disturbing tendency of this line of interpretation is its inability to see the landscape as a real, objective ecological formation about which people justifiably can be concerned. Instead it reduces land to a political blackboard with different groups or classes fighting to write on it.

3. The proximity of Heimat to the idea of "homelands" that has begun to attain a certain currency among cultural geographers is suggestive (cf. Nostrand and Estaville 1993; Roark 1993). The cultural studies method employed in this study, though, would seem to have an explanatory advantage over the geographic methodology as it does not assume any "mystical homeland-bond" and is thus in a better position to show how Germans went about constructing or reviving some level of attachment to their land (Roark 1993, 8).

4. If we accept that the Heimatschutz was expressing the real demands and meeting the real needs of ordinary citizens with this effort to protect and maintain landscapes, then this cannot be affected by Jackson's dismissal of "elitist" aesthetic judgment, which he finds "too easy," going on to argue that landscape can only be understood "in terms of its inhabitants." On the contrary; the Heimatschützer seem to meet his criterion for a consciously realistic-democratic kind of landscape appreciation, since they "abandon[ed] the spectator stance and . . . ask[ed] [them]selves how any man would fare who had to live in it." Indeed the Heimatschutz would have agreed with Jackson's inside-out approach to the everyday landscape: "What we need is not an aerial perspective of the globe but an aerial perspective of our own backyard. It is no use telling us that the world is our home and that we should love it, until we have learned to love our own corner of it, until we have learned what that corner possesses in the way of beauty and potentiality for happiness" (cited in Meinig 1979, 224, 217).

5. This has relevance for the running German debate on an "ecological aesthetic" (cf. Böhme 1989; rejoinder from Wolschke-Bulmahn 1990, 242). My position comes down somewhere in the middle of the opposing camps: Wolschke-Bulmahn is correct in noting that "ecological" rules need to be treated with suspicion insofar as they are needlessly inflexible and prevent full and fair analysis of social issues; on the other hand, though, ecological considerations are undeniably also relevant and important to good decision making. What we need is therefore an environmentalist aesthetic, one that promotes above all else the general importance of our ecologically dimensioned surroundings.

6. The rather clear discrimination embodied in the 1902 law proves that, before the intervention of the Heimatschutz, it was not at all the case that "all types of landscape, including the ordinary and unspectacular, were now appreciated in their own right" (as asserted in Muthesius 1981, 38).

7. In purely practical terms it can be noted that at least one government district

responded to his call for fairer lake access: the *Regierungspräsident* in nearby Potsdam followed Keller's recommendation for planning in shoreline promenades with a directive dated 6 June 1911 (cf. Hess. StA Marburg 180 2603).

8. Alon Confino (1993) contends that Heimat was never anything more than a figment of political speech. In support of this argument he analyzes a smattering of Heimat imagery drawn from a number of different sources and finds a considerable sameness. This, he concludes, is proof positive that Heimat was not about real places, because (he implies) real places differ from one another. But Confino overlooks the fact that Heimat was about one's relationship to place, and not the place itself; more a way of seeing, and less an object of seeing. It is in this common perspective that we should seek the particular meaning of Heimat images.

9. With an understanding of Naturschutz that emphasizes its narrow focus on scientific matters I disagree with Raymond Dominick, who thinks it inadvisable to "attach too much precision to terminology in this early period," and who goes on to lump all conservationist efforts together. His cites a speech given by Conwentz in 1914 as evidence of the shared broad agenda, one that would include forest composition, landscape, billboards, stream regulation, and bird protection. Dominick fails to notice that none of these concerns was original to Conwentz and Naturschutz, but rather by 1914 had been absorbed from the broad program of the Heimatschutz (cf. Dominick 1992, 7).

The background of different scientists seems to have had significant effects on their interpretion of *Naturschutz.* Conwentz came from a project cataloging West Prussia's "remarkable trees"—individual trees of particularly odd shape, for instance, or trees of record-breaking girth. The geologist Hans Menzel pointed the way toward a broader program of preserving whole landscapes when he remarked on the scientific ineffectiveness of the natural monument approach, saying that individual "monuments" of nature were all too easily treated as "curiosities" (cf. Menzel 1906, 67).

10. Of course the popular accessibility of the Heimatschutz landscape—its beauty—was a two-edged sword that could actually go against the needs of a scientific program of Naturschutz. In the struggle to preserve moors, the people who wanted to drain and plant them argued that the standard of beauty was on their side; in this case the Heimatschutz was obliged to abandon its merely aesthetic standpoint, saying "It is not an issue of beauty in the generally accepted sense, but rather a question of features with significance for the natural science and Heimat quality of the area, whose aesthetic effects are only partially evident, and even then not to the feeling of everyone" (cf. Bredt 1916, 97).

11. F. Bloh (1912), a Heimatschützer and animal protectionist from Hamburg, demonstrated a more radical critique in his article "Against the Devastation of Nature." Drawing the founding phrase of bourgeois hegemony in France—*enrichissez-vous*—into the context of the environment, he opened sarcastically: "'Let us enrich ourselves and murder!' That is our motto . . . Into all areas of the globe we have pressed, and everywhere we have decimated those of our fellow creatures that had any commercial value, or even none at all . . . Mercilessly we rage against the scared and tortured beings . . . just fifty more years of this senseless capitalist exploitation of nature, and we humans will be left alone on the earth with

our pets and the rats and mice! What unimaginable short-sightedness, what unpardonable lack of conscience toward the numberless generations of the descendants who after us will walk on this raped and plundered earth!" Bloh concluded his article with an appeal to action that included a list of organizations concerned with effective Naturschutz—first on the list was his local *Verein Heimatschutz im Hamburger Staatsgebiet.*

12. Schoenichen praised Conwentz's extreme moderation as a political tactic: "In wise fashion he avoids any hint of insult to the economic sector." It is hard to share this evaluation: from today's standpoint it does not appear wise to kowtow to industrial interests. But Schoenichen is wrong at the descriptive level, too; this was not even a tactic on Conwentz's part. With his narrow conception of Naturschutz all he wanted or needed was for a very limited selection of features to be preserved. See similar criticism of Conwentz and his pupil Schoenichen in Pflug (1969, 258).

13. The *Bund Heimatschutz* apparently tried to discourage a species-protection organization named the "League for the Preservation of Natural Monuments in the Animal and Plant Kingdom" in its early days out of similar concern about competition. Konrad Guenther and Hermann Löns were two founding members of the new organization who were also active in the Heimatschutz; Hugo Conwentz joined soon after the founding. The social composition of this group was remarkably mixed: it seems to have been chiefly supported by a disparate conglomeration of low-prestige civil servants and white-collar workers on the one hand, and high-prestige nobility on the other. The uniform presence of the educated bourgeoisie that characterized the Heimatschutz was missing.

14. In Germany the first bird preserves were concentrated on the North Sea area. The Jordsand Association led the way in 1906 by leasing the small mud flat area called Jordsand; soon afterward it was able to purchase the island of Norderoog with the help of the Stuttgart-based League for Bird Protection (founded 1899). Meanwhile a third organization, the central German "German Association for the Protection of the Bird World," founded 1875, purchased the island of Memmert; not to be outdone, the local Ornithological Association in Hamburg bought up the hunting rights to an area on the northern tip of the large Isle of Sylt. In 1909 the Baltic Sea received its first bird preserve near the island of Poel (cf. Zimmermann 1910–11). Lina Haehnle, the activist chairwoman of Stuttgart's League for Bird Protection, was a member of the Württemberg Heimatschutz association's steering committee.

15. An article in the *Blätter für Naturschutz* brought an example of the nature preserve–exploitation dialectic to light by observing that a certain National-Liberal Reichstag representative named Eckert was a big proponent of land rationalization and property liberalization measures—but simultaneously sat on the board of the *Verein Naturschutzpark* (cf. "Sonderbar" 1913).

16. This quote definitively contradicts the impression conveyed by many planning and environmental historians that the Heimatschutz wanted only a "preservation of parts instead of protective measures for the whole" (Wey 1982, 130).

17. Gradmann went on to deny the charge that modernists still make when confronted with such ideas: that "restoration" of an environment is somehow anti-

thetical to historical progress. He emphasized that the goal was not to produce an "artificial wilderness" but rather to develop "a natural civilization (or one in accordance with nature)" (Gradmann 1910, 48).

18. A similar widening of the idea of "Naturschutz" in the direction of Heimatschutz was contained in a 1912 guideline issued by the government in the Thuringian district of Roda, which decreed that hedgerows and living fences were to be preserved if at all possible, as were the beech trees and juniper bushes that constituted important parts of the native edge habitat. The influence of the Heimatschutz was more directly palpable here, as the text of the guideline referred to the movement by name and extended the scope of "Naturschutz" to include historical features as well (cf. *Heimatschutzfragen* 1914 [1], 1).

19. It would thus be seriously overstating the case to assert that the agricultural establishment was suddenly transformed by Heimatschutz thinking; it was not. A fundamental problem was that the Heimatschutz was in no position to make claims against the agencies; rather, it was largely dependent on their good will and lower officials' desire to pursue some of the general directions that higher-ranking offices had opened up. After the war Fritz Koch noted the unsatisfactory results of this informal system in Thuringia, where local rationalization agencies had been "instructed by their superior Prussian offices to take Heimatschutz efforts into consideration, and they also emphasized repeatedly (to the press, among others) that they were taking these efforts into account. In practice, though, the results did not come anywhere close to satisfying the demands of the Heimatschutz." Koch drew the consequences from this prewar experience: "Even the best set of guidelines for the cultivation agencies would not be sufficient"—and demanded an explicit, structurally anchored, "continuous participation" in improvement projects (letter of 8 June 1925 to Minstry of Interior/Economy, THStA Weimar LADH 44/27).

20. Salisch's book went through a second edition in 1902 and a third in 1913. In its spirit of aesthetic reform and its far-reaching effects it belongs in the same league as Camillo Sitte's 1889 manifesto *City Planning According to Artistic Principles.* Andersen's assertion (1987, 146) that Salisch's *Forest Aesthetics* achieved "no impact on concrete discussions" is tenable only in the narrow sense that the book may have had little immediate impact on the professional conventional wisdom and on day-to-day forestry practices. If one looks at the broader group of forestry theorists and people interested in general questions of the landscape, though, the book and its highly aesthetic approach achieved a great resonance. Salisch defended his ideas at the Sixth Main Conference of the German Forest Association in Darmstadt 1905, for one; in Hesse, for another, his book was officially distributed by the Finance Ministry to all of its upper forestry offices (cf. "Pfleget die Wälder" 1904). Meanwhile, in Saxony, the discussion of forest aesthetics inspired by Salisch and the Heimatschutz was intense. Groß edited a volume called *Heimatschutz in the Forest* that gave extensive coverage of aesthetic issues at the forty-ninth meeting of the Saxon Forest Association in Marienberg in 1905; the forestry docent Franz Mammen also taught a course in the summer semester of 1906 on aesthetic, Heimatschutz, and Bodenreform questions applied

to the forest industry. Although they had their differences, Mammen referred to Salisch as the "old master" of forest aesthetics and recommended him to a wide circle of readers (cf. review in *Mitteilungen des Bundes Heimatschutz* 1907, 31f.; Mammen 1908).

21. It cannot be denied that the link between aesthetic landscape concern and intervention occasionally went in the direction of intervention. In the flat, lake-dotted provinces of Brandenburg and Mecklenburg a number of Heimatschützer were concerned about the silting up of their scenic, shallow lakes and recommended, as the title of one article put it, "A Battle for the Preservation of Our Lakes." Here it was a matter of more intervention, and not less, due to aesthetic considerations (cf. Kuhls-Goslich 1912).

22. All of the above provides evidence against Wolfram Pflug's assertion (1969, 265) that "close-to-nature waterway engineering" began in Germany only in 1936, with Alwin Seifert. Seifert was a member of the Bavarian Heimatschutz and absorbed virtually all of his aesthetic lessons from the older traditions of the movement.

23. In Dresden in 1913 the reception of Bonne's speech indicated that Heimatschützer shared the goal of clean water but were often paralyzed by the apparent lack of any feasible solutions to the wastewater problem. A bizarre but informative sidelight on the culture of the time is provided by the mayor of Hildesheim, who was obviously irritated by Bonne's call for a thoroughgoing plan of wastewater treatment and disposal. He repeated the long-since-discredited theory that rivers would simply clean themselves, and he reminded listeners that potassium mines were active all around his city, for which they were "tremendously important" in spite of their high-mineral effluents. Struckmann thus proposed building a huge canal right across Germany to gather wastes and dump them in the ocean. A second commentator argued against the plan, joking that such a system leading out into the North Atlantic would surely be taken as another hostile maneuver by England. Bonne then rebutted Struckmann's "Jules Verne" idea as extremely far-fetched (cf. *Stenographischer Bericht* 1913, 191f.).

24. Dominick later mysteriously downgraded the Heimatschutz to near obscurity, though: in the book-length version of his study he characterized the Heimatschutz as an "auxiliary to the Naturschutz crusade" somewhere on "the fringes of the conservation movement" (Dominick 1992, 57). I tend to agree with Dominick's first assessment; identifying Heimatschutz as an auxiliary to the Naturschutz in this early phase is rather like saying that a dog is an auxiliary to its tail.

25. Berlepsch himself was usually careful to mention the need for genuine habitat as part of the program of bird protection. But his pronouncements tended in the direction of an exaggerated self-effacement. Noting that many of the features of modern life had contributed to the decline of the bird population, Berlepsch was quick to assure his listeners in an 1896 presentation that he had no intention of asking for radical reform: "There is no way that we can want to reconstruct the earth on behalf of birds! No, certainly not that. But we can and we must create a replacement for what was robbed. The earth does not need to become again the way it

was, it must merely be changed around *somewhat* in a few places in a way that corresponds to the biological conditions of the individual bird species" (cf. Berlepsch 1896).

Chapter 4

1. Of these authors Johannes Trojan and August Trinius were the only ones who distinguished themselves with Heimatschutz-oriented literary production. Pflug (1969, 250) points out that Heinrich Seidel had created a pro-hedgerow character in an 1897 novel; Dominick (1992, 31) notes a poem directed against seal killing. Trinius is discussed below in the section on travel. Trojan, who for many years edited the humor magazine *Kladderadatsch,* was well known in Heimatschutz circles through his long acquaintance with Hugo Conwentz and Franz Goerke, and also for his image-laden book titled *Our German Forests* (cf. Storck 1915–16; Trojan 1913, 164f.; *Mitteilungen des Landesvereins Sächsischer Heimatschutz* 1911/1912, 285). After a visit to some Danish islands with a group of boorish German tourists Trojan characteristically denounced "national boasting" and the "typical pride in German romanticism" that his countrymen had inflicted on their hosts (Trojan 1913, 109–10).

2. Kalkschmidt's volume of essays on the modern metropolis, *Großstadtgedanken* (1906), is indicative of the Heimatschutz—as opposed to *Heimatkunst*—point of view expressed by these authors. A review of the book emphasized its pragmatic reform attitude toward the metropolis, and listed proposals such as "intelligent building laws, increasing the number of parks, more functional street facilities, a different arrangement with respect to museums, etc" (cf. *Sauerländischer Gebirgsbote* 1907, 35). Carl Meißner contributed not only to the *Kunstwart,* but also to some of the popular rural almanacs of the time; an article "Protection for the Heimat" by Meißner appeared already in the *Königsberger Goethe-Bund Kalender für das Jahr 1905.* Paul Schumann was an extremely active member of the Saxon State Association's section for Naturschutz. Bölsche's direct involvement with the Heimatschutz was by contrast only tangential; after signing the original "Aufruf" Bölsche appears not to have engaged himself in Heimatschutz activities to any significant extent. Nevertheless his environmental concerns overlapped more than this relatively meager cooperation would lead one to believe.

3. Cited here from Hermann Burte [= Hermann Struebe], *Wiltfeber—der ewige Deutsche. Die Geschichte eines Heimatsuchers* [1912] (Berlin: Deutsche Buch-Gemeinschaft, ca. 1927).

4. Quotations in the following from the edition *Frau Harke geht durchs Land. Der Roman einer Landschaft* (Leipzig: Erich Matthes, 1919. 2d ed.).

5. The same cannot be said for the nationalist sentiments that suddenly begin to play a large role as the plot works its way toward 1914 and the outbreak of the war, since they do not seem to be any sort of an integral component of this kind of Heimat feeling. Significantly, it is the hiking youth of the region, the *Wandervogel,* who become the most important vehicle for nationalism. As portrayed by Kotzde, they are in the apparent habit of going on pilgrimages from one Bismarck monu-

ment to the other all across Germany; in August of 1914 it is they who are portrayed as erupting into nationalist enthusiasm for the war. Their reaction can be contrasted to that of one of the young fisherman, whose first thought is of the local environment: "Now the Havel will be left alone," he says: "All of the ones who ripped out her heart with steam shovels and excavators are going off to war." In reply one of the most restless and nationalist members of the youth movement, a young man named Hans Kleuß, makes the ridiculous assertion that the Havel was ruined because of Germany's enemies, who wouldn't allow the country enough *Lebensraum*—thus turning a domestic environmental problem into an offensive demand for more land on Germany's borders: "Why did they [the German engineers] destroy our last piece of Heimat? Because we did not have enough earth left to feed the growing population. Because so unbelievably many lacked that mother earth out of which happiness could have bloomed for them. There is much land on our borders that is waiting for German hands. Let us fight for it!" (302). Here it is apparent that the *Wandervogel* is being used by Kotzde as the conduit for his own nationalist and imperialist sentiments, for as early as 1915 he expressed the same desire for "pure German newland in the East" (Kotzde 1915, 12f.). In Kotzde we also have one of the earliest examples of a Heimatschützer who has converted to militant racialism: already in an article devoted to "Youth Care and Heimat" in early 1914 he argued that part of making German villages into a true "life community" would be racial cleansing, for instance of "Poles with Mongolian blood" (cf. Kotzde 1913–14).

6. Cited here and in the following from Max Dreyer, *Auf eigener Erde* (Berlin and Vienna: Ullstein, 1911).

7. "Environmentalism" was latent in the positivist thinking of Naturalists, and it is arguably an element that led a number of former Naturalists to a later stage of Heimat art. An invaluable introduction to the subject is Mahal (1975); Möbius (1980) has recently given the "commonplaces" of positivist influence greater depth and exactitude in his study of Arno Holz.

8. Cited here and in the following from Maximilian Böttcher, *Heim zur Scholle* (Leipzig: Grethlein, 1910).

9. There is a delicate interplay in the novel between this kind of progressive social conviction and problematic ethnic thinking. This is particularly evident where feudal landowners are taken to task for their willingness to hire "Polish and Galizian half-animals." Nevertheless this racist characterization is linked largely to the large landowners, the enemies of Dietrich's enlightened plans, who are shown preparing a petition to allow the import of "Chinese coolies . . . who, because of their cultural low standing and their resultant low expectations, would probably be the most suitable for working the large fields of German large landowners. Negroes could also be considered; it appeared questionable, though, if they could be successfully brought through the cold north German winter, and if transport and burial costs would not make the savings in wages and accommodation illusory" (433).

10. Cited here and in the following from Hermann Wette, *Jost Knost—Der Herkules von Latop* (Leipzig: Fr. Wilhelm Grunow, 1908).

11. Wette's antipathies toward foreign workers—who were increasingly com-

mon in the industrial Rhine-Ruhr area at this time—arose from an unwholesome brew of political and ethnic/social prejudices. In this text Wette's main objections were political: he saw the underpaid and discontented foreign workers as the most fervent supporters of a Socialist revolution. The connection between an exploitative class of capitalists and a revolutionary class of workers was for Wette virtually axiomatic, so much so that he more than once extended the term "red robbers" to unscrupulous capitalists (who came, conveniently, from outside Germany). The negative portrayals of shiftless, lazy, and alcoholic Poles and Croatians constitute one of the book's least savory and most superfluous aspects. It is one repeated in Wette's play *Peter Schlemihl—Modernes Teufelsmärchen* (1910), where the reformer Peter Schlemihl is ultimately killed by foreign workers. The play ends with a far less egalitarian political impetus than the novel, as it portrays a German nobleman stepping into the role as leader of a German-Christian workers' movement.

12. Dedifferentiation is seen by some critics as the primary feature of a postmodern culture. If so we are in the curious position of having to identify the proudly *anti*modern Heimatschutz as one of the world's first *post*modernist movements (cf. Rojek 1993, 3f., 184). A distinguishing feature would of course be the consistent orientation of the Heimatschutz tourism toward a real and knowable material world, and not toward the consumption of signs and images, which has been recognized as a hallmark of postmodern tourism (cf. Urry 1990, 83f.).

Conclusion

1. An example of the domestic-social orientation of the Heimatschutz is the fact that in 1915 there appeared articles actually worrying about the possibility of a German victory and a "New Founders' Generation." As the author of one of these articles pointed out, the victory in 1871 had produced nothing but trouble: "a time of national and social embitterment whose consequences we are still battling today" (Fresenius 1915, 219; see also "Der Sieg der Heimat" 1916). The *Schwäbisches Heimatbuch* of 1915 (11), on the other hand, looked forward to Germany's victory as the beginning of a new era of altruism without the "hunt for the dollar" and "petty party politics" of the present; nor would Germany's undoubted military superiority "lead to us acquiring more and more countries and subjugating new peoples"; instead the generous general staff would "found a new era of world rule that would make people free in truth." It is easy to see that the naive reformer who wrote these words would have been disappointed no matter what the outcome of the war, since he was using this period in history as a blank screen upon which to project wishes. Nor is it clear whether the apparent generosity toward Germany's enemies would survive a defeat and a Treaty of Versailles.

2. Andersen (1987, 152) traces the rise of "volkish theories" to the war years, something that is definitely not true of the movement as a whole. Werner Hartung (1991, 299) has similarly claimed that the swing toward the far right was complete already by the end of World War I. This may well reflect the Lower Saxony movement's politically extreme position. Knaut (1993, 211) has identified Robert Mielke with a more extreme, radical-volkish faction within the prewar Heimat-

schutz that, however, was present at the national level before 1918 only in trace amounts; Knaut identifies Baden and Lower Saxony as points of concentration. Perhaps the most radical appropriation of the slogan of Heimatschutz during the prewar years came from a Stuttgart garden architect who was, as far as I have been able to tell, not connected with the actual movement. His article titled "Heimatschutz" was in fact a tirade—complete with occasional bold-printed words—on the subject of "thinking and living in a *German* fashion," on the "volkish uniqueness" of various German tribes, and on *"race feeling"* (cf. Rasch 1913).

The extreme racist right's aggressive co-optation of the slogans of Heimat and Heimatschutz in the postwar period is demostrated in painful clarity in works such as the Austrian Hermann Kämpfer's 1930 work *Heimatschutz and Judaism,* a hate-mongering pamphlet that stated unequivocally that Heimatschutz was "locked in a life-or-death struggle with the culture-enemy of Jewish Marxism" and that it was directed "against the further Judification of our Heimat and against the further excrefication of our racial genetic material." This virulent anti-Semitic brand of "Heimatschutz" was completely unknown in the era before the war.

In this regard it is important to emphasize that the earlier Heimatschutz did not exclude Jewish members. Knaut (1993, 70) writes that two signers of the 1904 "Call for Founding" were Jewish, although he also points out that Ernst Rudorff voiced his aversion to including the Jewish publisher Meyer-Cohn. Nevertheless it does not appear tenable to extrapolate from this to a "certain implicit anti-Semitism" in the movement as a whole, an assertion that is additionally so vague that it would be hard to disprove (234). What we can say for certain is that there were at least some Jewish members. The banker Robert von Mendelssohn, for instance, was characterized in his obituary as a man who had "followed the founding of the *Bund Heimatschutz* with great interest" and who had "remained true" to the Brandenburg Heimatschutz association until his death (cf. *Heimatschutz in Brandenburg* 1917/1918, 155); the Hessian League for Heimatschutz counted a rabbi among its members.

3. This statement should be taken to mean that it is possible to distinguish between some of the ecologically informed environmental measures undertaken between 1933 and 1945, on the one hand, and the perfidious and murderous Nazi regime on the other. There can be no doubt that almost all aspects of life in those twelve years were in varying degrees tainted by political blindness, venal expedience, and collaboration with a racist-fascist worldview. Environmentalists and conservationists, just like any other segment of the German population, need to be criticized for their complicity, which in some cases was extreme and catered directly to societal objectives of racism and imperialism. Nonetheless this process of critique should not go to the simplistic extreme of assuming that everything said and done in the area of conservation was automatically, inherently, and fully imbued with the repugnant goals of the National Socialist regime. For an initial consideration of the issues raised here see Dominick (1992, 106f.), and Gröning and Wolschke-Bulmahn (1985); on Nazi forest administration, Rubner (1985). On the German *Autobahn* in particular see Rollins (1995).

4. The quantification of this collateral effect would be difficult to carry out and probably pointless, since distribution can never reliably prove influence. On the

other hand a wide distribution is probably the best precondition for influence, and in this sense it can be noted that the journal of the *Bund Heimatschutz* had a circulation in the tens of thousands; it will also be recalled that the Heimatschutz organizations were responsible for a huge output of flyers and picture postcards, many of which carried their explicit message; in addition Norbert Borrmann has pointed out that Paul Schultze-Naumburg's pamphlet *Die Entstellung unseres Landes* had already sold 45,000 copies by 1911, and that even early on, in 1907, it was estimated that Germany's periodicals were carrying Heimatschutz-related articles at the rate of 300 per month (cf. Borrmann 1989, 64).

5. Why is it that the Green Party, for instance, gathered such strength in Germany long before it was able to do so in other European or Western, industrialized countries? The political science explanation, which concentrates on post-1968 social and political developments, can supply part of the answer here, but I think it can at least be speculated that some fairly deep-seated cultural-aesthetic tendencies may also be behind the phenomenon; I would even go so far as to suggest that these derive much of their impetus from the turn of the century and the Heimatschutz. This argument would still hold precisely against the correct observation that today's Greens typically know next to nothing about the Heimatschutz: for it is not historical or consciously intellectual knowledge we are talking about here, but rather a vaguer kind of cultural dispositioning that could well have survived the postwar era's explicit Heimat witch-hunt.

6. Wilmowski mentioned that he ran across only two Heimatschutz-like groups in the United States: one was the American Civic Association, led by J. Horace McFarland; the other was the American Scenic and Historic Preservation Society, founded by Andrew H. Green and led in 1908 by George F. Kunz. The City Beautiful movement, which was centrally based in the aforementioned ACA, is occasionally seen as corresponding to the Heimatschutz; it was a reform movement with a strong aesthetic component, to be sure, but it was resolutely urban and is often described as having had a pronounced upper-class and elitist bias; quite unlike the Heimatschutz concern with urban preservation, it advocated large civic buildings and broad open spaces (cf. Short 1991, 121; giving a more positive interpretation is Wilson 1989). Wilmowski himself seems to have discounted it, saying that "there can be no question of there being anything like an extensive Heimatschutz movement here" (1910, 61).

References

Archival Sources

Akten des LADH	Landesamt für Denkmalpflege und Heimatschutz
Hess. StA Darmstadt	Hessisches Staatsarchiv Darmstadt. G15 Friedberg
HStA Marburg	Hauptstaatsarchiv Marburg. 150, 180
HStA Stuttgart	Hauptstaatsarchiv Stuttgart. E151
PGStA	Preußisches Geheimes Staatsarchiv, Berlin I HA Rep. 84a, 87B, 87F, 93B
SBPK	Staatsbibliothek Preußischer Kulturbesitz, Berlin Handschriftensammlung, Nachlaß Mielke
StA Coburg	Staatsarchiv Coburg. Min. D. 18, 20
StA Dresden	Staatsarchiv Dresden. Min. d. Innern 17513, 17519, 17524, 17526, 17531
StA Hamburg	Staatsarchiv Hamburg. A 507/13
Stadtarchiv Dresden	Akte "Landesverein Sächsischer Heimatschutz"
Stadtarchiv Eisenach	Akten zur "Ortsgruppe Heimatschutz"
Stadtarchiv Gotha	Akte "Heimatschutz"
Stadtarchiv Solingen	S2663, Ve 005–001, Ve 005–004
THStA Weimar	Thüringisches Hauptstaatsarchiv, Weimar
TStA Meiningen	Thüringisches Staatsarchiv Meiningen. Ministerium des Inneren Akte Nr. 2021
UB Tübingen	Universitätsbibliothek Tübingen. Nachlaß Fuchs

Achleitner, Friedrich. 1989. Gibt es einen mitteleuropäischen Heimatstil?—oder: Entwurf einer peripheren Architekturlandschaft. *Österreichische Zeitschrift für Kunst und Denkmalpflege* 63:165–69.

Anderle, Charlotte. 1983. *Der andere Peter Rosegger. Polemik, Zeitkritik und Vision im Spiegel des 'Heimgarten', 1876–1918.* Vienna: Österreichischer Agrarverlag.

Andersen, Arne. 1987. Heimatschutz: die bürgerliche Naturschutzbewegung. In *Besiegte Natur. Geschichte der Umwelt im 19. und 20. Jahrhundert,* ed. F.-J. Brüggemeier. Munich: C. H. Beck.

———. 1990. Umweltgeschichte und Fortschrittsparadigma: Plädoyer für eine Erweiterung der Geschichtswissenschaft. In *Umweltgeschichte: Das Beispiel Hamburg,* ed. A. Andersen. Hamburg: Ergebnisse.

Andersen, Arne, and Franz-Josef Brüggemeier. 1987. Gas, Rauch und Saurer Regen. In *Besiegte Natur. Geschichte der Umwelt im 19. und 20. Jahrhundert,* ed. F.-J. Brüggemeier and T. Rommelspacher. Munich: C. H. Beck.

Andresen, Hans-Günther. 1992. Zur Entstehung und Eigenart der schleswig-holsteinichen Heimatschutzarchitektur. In *Hans Roß, 1873–1922. Zur Heimatschutz- und Reformarchitektur in Schleswig-Holstein,* ed. H. Barg. Hamburg: Junius.

Applegate, Celia. 1990. *A Nation of Provincials. The German Idea of Heimat.* Berkeley: University of California Press.

———. 1991. Localism and the German Bourgeoisie: The "Heimat" Movement in the Rhenish Palatinate before 1914. In *The German Bourgeoisie. Essays on the Social History of the German Middle Class from the Late Eighteenth to the Early Twentieth Century,* ed. D. Blackbourn and R. J. Evans. London, New York: Routledge.

———. 1992. The Question of Heimat in the Weimar Republic. *New Formations* (17): 64–74.

Arnswaldt, G. von. 1908. Naturdenkmäler für die kommenden Generationen. *Zeitschrift des Heimatbundes Mecklenburg* 2:62.

———. 1908. Über Naturdenkmalpflege. *Mecklenburg. Zeitschrift des Heimatbundes Mecklenburg* 3:113—24.

Asmus, Gesine. 1984. Wohnungselend 1901–1920. In *Exerzierfeld der Moderne. Industriekultur in Berlin im 19. Jahrhundert,* ed. J. Boberg. Munich: C. H. Beck.

Attfield, Robin. 1983. Christian Attitudes to Nature. *Journal of the History of Ideas* 44:369–86.

"Aus der Tätigkeit." 1910. Aus der Tätigkeit des Bundes: Die geplante Errichtung einer Lokomotiv-Zentrale am Rosensteinpark bei Stuttgart. *Mitteilungen des Bundes für Heimatschutz in Württemberg und Hohenzollern* 2 (1/2): 15–16.

Baer. 1909–10. Heimatschutz und Verkehr. *Mitteilungen des Landesvereins Sächsischer Heimatschutz* 1:153–54.

Barg, Hartwig. 1992. *Hans Roß, 1873–1922. Zur Heimatschutz- und Reformarchitektur in Schleswig-Holstein.* Ed. H. Barg. Hamburg: Junius.

Barrell, John. 1972. *The Idea of Landscape and the Sense of Place, 1730–1840. An Approach to the Poetry of John Clare.* Cambridge: Cambridge University Press.

Barth, M. A. 1831. *Über Landesverschönerung als Gegenstand der Staatsvorsorge.* Augsburg.

Barthelmeß, Alfred. 1981. *Vögel—Lebendige Umwelt.* Freiburg: Karl Alber.

———. 1988. *Landschaft—Lebensraum des Menschen. Probleme von Landschaftsschutz und Landschaftspflege geschichtlich dargestellt und dokumentiert.* Freiburg: Karl Alber.

Bate, Jonathan. 1991. *Romantic Ecology. Wordsworth and the Environmental Tradition.* London: Routledge.

Bauer, Friedrich. 1906. Gartenbau und Landschaft. *Gartenkunst* 8:109–13.

Bergmann, Klaus. 1970. *Agrarromantik und Großstadtfeindschaft.* Meisenheim: Verlag Anton Hain.

Berlepsch, Hans von. 1896. Die Vogelschutzfrage, soweit dieselbe durch Schaffung geeigneter Nistgelegenheiten zu lösen ist. *Natur* 45:163–68.

Bermingham, Ann. 1986. *Landscape and Ideology. The English Rustic Tradition, 1740–1860.* Berkeley: University of California Press.

Berndt, Heide. 1987. Hygienebewegung des 19. Jahrhunderts als vergessenes Thema von Stadt- und Architektursoziologie. *Alte Stadt* 14:140–63.

Berry, Wendell. 1993. *Sex, Economy, Freedom and Community. Eight Essays.* New York: Pantheon Books.

Blackbourn, David. 1987. Introduction. In *Populists and Patricians. Essays in Modern German History.* London: Allen and Unwin.

———. 1991. The German Bourgeoisie: An Introduction. In *The German Bourgeoisie. Essays on the Social History of the German Middle Class from the Late Eighteenth to the Early Twentieth Century,* ed. D. Blackbourn and R. J. Evans. London, New York: Routledge.

Blackbourn, David, and Geoff Eley. 1984. *The Peculiarities of German History. Bourgeois Society and Politics in Nineteenth-Century Germany.* Oxford and New York: Oxford University Press.

Bloh, F. 1912. Gegen die Naturverwüstung. *Dokumente des Fortschritts* 6:606–9.

Bödecker. 1915. Ein Beitrag zur Moor- und Heidekultur. *Zeitschrift des Vereins Heimatbund Niedersachsen* 1:82–88 and 158–62.

Boerlin, [Gerhard]. 1912. Überblick über den Heimatschutz in der Schweiz. *Heimatschutz* 8:94–96.

———. 1931. Aus der Geschichte des Heimatschutzes in der Schweiz. In *25 Jahre Heimatschutz. Festschrift zum 25jährigen Bestehen der Schweizerischen Vereinigung für Heimatschutz.* Basel: Frobenius.

Böhme, Gernot. 1989. *Für eine ökologische Naturästhetik.* Frankfurt a. M.: Suhrkamp.

Bollerey, Franziska. 1978. Der neue Alltag in der grünen Stadt. Zur lebensreformerischen Ideologie und Praxis der Gartenstadtbewegung. In *Kunst und Alltag um 1900,* ed. E. Siepmann. Gießen: Anabas Verlag.

———. 1990. *Im Grünen wohnen—im Blauen planen. Ein Lesebuch zur Gartenstadt mit Beiträgen und Zeitdokumenten.* Ed. G. Fehl. Vol. 12, *Stadt—Planung—Geschichte.* Hamburg: Hans Christians.

Bölsche, Wilhelm. 1920. *Weltblick. Gedanken zu Natur und Kunst.* Dresden: Carl Reißner.

Bonne, Georg. 1912. Die zunehmende Verunreinigung der deutschen Gewässer—eine dringende Frage für den Heimatschutz. *Heimatschutz* 8:8–23.

———. 1913. Die Verunreinigung unserer deutschen Gewässer und ihre Verhütung. In *Stenographischer Bericht der Zweiten Gemeinsamen Tagung für Denkmalpflege und Heimatschutz, Dresden, 25. und 26. September 1913.* Berlin: Wilhelm Ernst und Sohn.

———. 1915. Was lehrt uns der Krieg über die Bedeutung des Heimatschutzes? *Heimatschutz* 10:168–71.

———. 1919. Die Ursachen der Revolution und der Heimatschutz. *Heimatblätter der roten Erde* 1 (2): 51–53.

Borrmann, Norbert. 1989. *Paul Schultze-Naumburg, 1869–1949. Maler—Publizist—Architekt. Vom Kulturreformer der Jahrhundertwende zum Kulturpolitiker im Dritten Reich.* Essen: Richard Bacht.

Böttcher, Maximilian. 1910. *Heim zur Scholle.* Leipzig: Grethlein und Co.

Boyer, Paul. 1978. *Urban Masses and Moral Order in America, 1820–1920.* Cambridge: Harvard University Press.

Bredt, F. W. 1908. Das preußische Gesetz gegen die Verunstaltung von Ortschaften und landschaftlich hervorragenden Gegenden vom 15. Juli 1907. *Heimatschutz* 4:16–25.

———. 1916. Bodenverbesserung, Naturschutz, Bodenaltertümer. *Mitteilungen des Rheinischen Vereins für Denkmalpflege und Heimatschutz* 10:94–98.

Breitling, Peter. 1981. The origins and development of a conservation philosophy in Austria. In *Planning for Conservation,* ed. R. Kain. New York: St. Martin's Press.

Brentano, Lujo. 1931. *Mein Leben im Kampf um die soziale Entwicklung Deutschlands.* Jena: Eugen Diederichs.

Breuer, Rüdiger. 1985. Der Niederschlag der Wohnungs- und Städtebau-Reform in der Gesetzgebung, insbesondere im sächsischen Allgemeinen Baugesetz von 1900. In *Von Licht, Luft und Ordnung in der Stadt der Gründerzeit. Städtebaureform 1865–1900,* ed. J. Rodriguez-Lores. Hamburg: Christians-Verlag.

Bruch, Rüdiger vom. 1985. Bürgerliche Sozialreform im deutschen Kaiserreich. In *Weder Kommunismus noch Kapitalismus. Bürgerliche Sozialreform in Deutschland vom Vormärz bis zur Ära Adenauer,* ed. R. v. Bruch. Munich: C. H. Beck.

———. 1989. Gesellschaftliche Funktionen und politische Rollen des Bildungsbürgertums im Wilhelminischen Reich. Zum Wandel von Milieu und politischer Kultur. In *Politischer Einfluß und gesellschaftliche Formation,* ed. J. Kocka. Stuttgart: Ernst Klett.

Brückler, Theodor. 1989. Zur Geschichte der österreichischen Heimatschutzbewegung. *Österreichische Zeitschrift für Kunst und Denkmalpflege* 63 (3/4): 145–56.

Brüggemeier, Franz-Josef. 1990. Auf Kosten der Natur. Zu einer Geschichte der Umwelt 1880–1930. In *Jahrhundertwende. Der Aufbruch in die Moderne, 1880–1930,* ed. A. Nitschke, D. J. K. Peukert, G. A. Ritter, and R. v. Bruch. Reinbek: Rowohlt.

Buchwald, Konrad, ed. 1968. *Grundlagen.* Vol. 1, *Handbuch für Landschaftspflege und Naturschutz.* Munich: Bayerischer Landwirtschaftsverlag.

Buddin, Fr. 1907. Zur Heckenwirtschaft im Ratzeburgischen. *Mecklenburg. Zeitschrift des Heimatbundes Mecklenburg* 2:114–19.

Bullock, Nicholas. 1985. *The movement for housing reform in Germany and France, 1840–1914.* Cambridge: Cambridge University Press.

Bund für Heimatschutz in Württemberg und Hohenzollern. 1910. Eingabe des Bundes in Sachen Blechquadern, Blechziegeln und Blechschindeln an das Königliche Württembergische Ministerium des Innern. *Mitteilungen des Bundes für Heimatschutz in Württemberg und Hohenzollern* 2:26–27.

"Burschenplatz." 1913. Der Burschenplatz im Rautal bei Jena für die Errichtung einer Kneipe preisgegeben. *Heimatschutzfragen* 2:11–12.

Burte, Hermann. 1927. *Wiltfeber—der ewige Deutsche. Die Geschichte eines Heimatsuchers.* Berlin: Deutsche Buch-Gemeinschaft.

Callicott, J. Baird. 1991. The Wilderness Idea Revisited: The Sustainable Development Alternative. *The Environmental Professional* 13:236–45.

Chickering, Roger. 1973. A Voice of Moderation in Imperial Germany: The "Verband für Internationale Verständigung" 1911–1914. *Journal of Contemporary History* 8 (Jan.): 147–64.

———. 1984. *We Men Who Feel Most German. A Cultural Study of the Pan-German League, 1886–1914.* Boston: George Allen and Unwin.

Christiansen, Alb. 1907. Pflanzen- und Tierleben im Knick. *Die Heimat* 17:66–72.

Christiansen, Jörn. 1980. *"Die Heimat." Analyse einer regionalen Zeitschrift und ihres Umfeldes.* Neumünster: Karl Wachholtz.

Collins, George R., ed. 1986. *Camillo Sitte: The Birth of Modern City Planning.* New York: Rizzoli.

Confino, Alon. 1993. The Nation as a Local Metaphor: Heimat, National Memory and the German Empire, 1871–1918. *History and Memory* 5 (1): 42–86.

Corti, Alexander de. 1910. Industrie, Ästhetik und Heimatschutz. *Bitumen. Fachzeitung für die Asphalt-, Teer-, Erdöl-, Paraffin- und verwandte Industrie* 8:270–74.

Craig, Gordon. 1978. *Germany 1866–1945.* Oxford: Clarendon Press.

Cronon, William. 1992. A Place for Stories: Nature, History and Narrative. *Journal of American History* 78:1347–76.

———. 1993. The Uses of Environmental History. *Environmental History Review* 17 (3): 1–22.

Dahrendorf, Ralf. 1967. *Society and Democracy in Germany.* Garden City: Doubleday.

Dammer, Udo. 1916. Zur Meliorationsfrage. *Blätter für Naturschutz und Heimatpflege,* 12–13.

Danahay, Martin A. 1991. Matter Out of Place: The Politics of Pollution in Ruskin and Turner. *Clio* 21:61–77.

Däumel, Gerd. 1961. *Über die Landesverschönerung.* Geisenheim/Rheingau: Heinrich Debus.

Deutscher Bund Heimatschutz. 1954. *50 Jahre Deutscher Heimatbund.* Neuß: Deutscher Heimatbund.

Dießner, Br. 1918. Die Verseuchung unserer Flüsse durch giftige Abwässer. *Blätter für Naturschutz und Heimatpflege* 4 (11–12): 1–4, 7–9.

Dominick, Raymond. 1986. Nascent Environmental Protection in the Second Empire. *German Studies Review* 9:258–91.

———. 1992. *The Environmental Movement in Germany: Prophets and Pioneers, 1871–1971.* Bloomington and Indianapolis: Indiana University Press.

Dreyer, Max. 1903. On Probation. In *Poet-lore.* Boston.

———. 1911. *Auf eigener Erde.* Berlin and Vienna: Ullstein and Co.

Drude, Oskar. 1909. Die Entstehungsgeschichte des heimatlichen Landschaftsbildes. In *Heimatschutz in Sachsen. Vorträge der Dresdner Volkshochschulkurse,* ed. R. Wuttke. Leipzig: B. G. Teubner.

Eagleton, Terry. 1990. *The Ideology of the Aesthetic.* Oxford and Cambridge: Basil Blackwell.

Eckmann. 1908. Gründung des Landesvereins des Bundes Heimatschutz. *Die Heimat.* 18:274–75.

Eckstedt, Vitzthum von. 1914–15. Antwort Sr. Exz. des Herrn Staatsminister Graf Vitzthum von Eckstädt auf die Interpellation der Herren Brodauf, Dr. Dietel und Genossen—die Handhabung des sogenannten Heimatschutzgesetzes betr.—in der II. Kammer der sächsischen Ständeversammlung. *Mitteilungen des Landesvereins Sächsischer Heimatschutz* 4:160–71.

Eder, Klaus. 1990. The Rise of Counter-culture Movements Against Modernity: Nature as a New Field of Class Struggle. *Theory, Culture and Society* 7:21–47.

Ehrlich, Hans. 1917. Neue Gesichtspunkte für den Tier- und Vogelschutz. *Blätter für Naturschutz und Heimatpflege* 3 (10): 13–15.

Eley, Geoff. 1984. The British Model and the German Road: Rethinking the Course of German History before 1914. In *The Peculiarities of German History,* ed. D. Blackbourn and G. Eley. Oxford and New York: Oxford University Press.

Engelhardt, Ewald. 1913–14. Schlußwort. *Zeitschrift des Vereins für Heimatkunde und Heimatschutz von Artern und Umgegend "Aratora"* 2.

Enzensberger, Hans Magnus. 1962. Eine Theorie des Tourismus. In *Einzelheiten.* Frankfurt a. M.: Suhrkamp.

———. 1973. Zur Kritik der politischen Ökologie. *Kursbuch* 33:1–42.

"Erhaltet die Knicks." 1912. *Schleswig-Holsteinische Natur- und Vogelschutzblätter* 1 (2): 3.

Erman, Heinrich. 1910. Heimatschutz und Bodenreform. *Heimatschutz in Brandenburg* 2:181–84.

"Erworbene Heimatliebe." 1911. *Berliner Abendpost,* 2 April.

Evans, Richard J. 1987. Wilhelm II's Germany and the Historians. In *Rethinking German History. Nineteenth-Century Germany and the Origins of the Third Reich.* London: Allen and Unwin.

Evans, Richardson. 1926. *An Account of the Scapa Society.* London: Constable and Company.

Fehl, Gerhard. 1995. *Kleinstadt, Steildach, Volksgemeinschaft. Zum "reaktionären Modernismus" in Bau- und Stadtbaukunst.* Braunschweig, Wiesbaden: Friedrich Vieweg.

Felber, Theodor. 1906. *Natur und Kunst im Walde.* Frauenfeld: Huber.

Felix, Ludwig. 1883. *Der Einfluß der Natur auf die Entwicklung des Eigenthums.* Leipzig: Duncker und Humblot.

Finke, Peter. 1986. Landschaftserfahrung und Landschaftserhaltung. Plädoyer für eine ökologische Landschaftsästhetik. In *Landschaft,* ed. M. Smuda. Frankfurt a. M.: Suhrkamp.

Fischer, A. 1914. Heimatkunst und Heimatschutz. *Blätter für administrative Praxis* 64:323–38.

Focke, W. O. 1905. Heimatschutz. *Jahresbericht des Naturwissenschaftlichen Vereins zu Bremen* 40:508–12.

Förster, P. 1917. Entdeckung der Heimat. *Heimatschutz in Brandenburg* 8:41–45.

Francé, Raoul H. 1909. *Bilder aus dem Leben des Waldes.* Stuttgart: Kosmos.

———. 1910. Die Entdeckung der Heimat. *Österreichische Rundschau* 23:191–94.

———. 1923. *Die Entdeckung der Heimat.* Stuttgart: Kosmos, Gesellschaft der Naturfreunde.

Frank, Hartmut. 1992. Heimatschutz und typologisches Entwerfen. Modernisierung und Tradition beim Wiederaufbau von Ostpreußen 1915–1927. In *Moderne Architektur in Deutschland 1900 bis 1950. Reform und Tradition,* ed. V. M. Lampugnani and R. Schneider. Stuttgart: Gerd Hattje.

Franke, Ursula. 1972. *Kunst als Erkenntnis—Die Rolle der Sinnlichkeit in der Ästhetik des Alexander Gottlieb Baumgarten.* Wiesbaden: Franz Steiner.

Frecot, Janus, et al. 1972. *Fidus, 1868–1948. Zur ästhetichen Praxis bürgerlicher Fluchtbewegungen.* Munich: Rogner und Bernhard.

Fricke, C. 1910. Die Wichtigkeit der Laubhölzer für die Erhaltung der Wälder. *Die Woche* 12:1373–76.

Friedrich, Ernst. 1904. Wesen und geographische Verbreitung der "Raubwirtschaft". *Dr. A. Petermanns Mitteilungen aus Justus Perthes' Geographischer Anstalt* 50:68–79, 92–95.

Fuchs, Carl Johannes. 1899. Der heutige Stand der Arbeiterwohnungsfrage in Deutschland. In *Zur Wohnungsfrage.* Leipzig: Duncker und Humblot.

———. 1904. *Zur Wohnungsfrage.* Leipzig: Duncker und Humblot.

———. 1905. *Heimatschutz und Volkswirtschaft.* Vol. 1, *Flugschriften des Bundes Heimatschutz.* Halle a. S.: Gebauer-Schwetschke.

———. 1906a. Die Ausnützung der Naturkräfte. In *Heimatschutz. Bericht über die Jahresversammlung des Bundes Heimatschutz in Goslar am 12.–14. Juni 1905.* Halle a. S.: Gebauer-Schwetschke.

———. 1906b. *Die Stromschnellen des Rheins bei Laufenburg und ihre Erhaltung.* Vol. 3, *Flugschriften des Bundes Heimatschutz.* Halle: Gebauer-Schwetschke.

———. 1908. Die Gartenstadt. In *Die Wohnungsfrage vor und nach dem Kriege. Aufsätze und Vorträge zur Wohnungsfrage—Neue Folge.* Munich and Leipzig: Duncker und Humblot.

———. 1909a. Heimatschutz in Württemberg. *Mitteilungen des Württembergischen Bundes für Heimatschutz* 1 (1): 1–3.

———. 1909b. Heimatschutz im Ausland und in Deutschland. *Mitteilungen des Württembergischen Bundes für Heimatschutz* 1 (2/3): 4–8.

———. 1911. Heimatschutz und Wohnungsfrage. *Heimatschutz* 7:139–49.

———. 1911–12. Gartenstadt und Heimatschutz. *Mitteilungen des Landesvereins Sächsischer Heimatschutz* 2:315–16.

———. 1912a. Eröffnungsrede auf dem Zweiten Internationalen Heimatschutz-Kongreß. *Mitteilungen des Bundes für Heimatschutz in Württemberg und Hohenzollern* 4 (3): 3–6.

———. 1912b. Die Ausnützung der Wasserkräfte. *Heimatschutz* 8:97–112.

———. 1917. *Die Wohnungsfrage vor und nach dem Kriege. Aufsätze und Vorträge zur Wohnungsfrage—Neue Folge.* Munich and Leipzig: Duncker und Humblot.

———. 1920. *Volkswirtschaftslehre.* 3d ed. Berlin and Leipzig: Göschen/Vereinigung wissenschaftlicher Verleger.

Fuhs, Burkhard. 1990. Natur und Klassengesellschaft in Wiesbaden. In *Stadtgesellschaft und Kindheit im Prozeß der Zivilisation. Konfigurationen städtischer Lebensweise zu Beginn des 20. Jahrhunderts,* ed. I. Behnken. Opladen: Leske und Budrich.

Fuller, Peter. 1985. John Ruskin: A Radical Conservative. In *Images of God. The Consolations of Lost Illusions.* London: Hogarth Press.

Gaebel. 1912. Der Stuttgarter internationale Heimatschutzkongreß. *Mitteilungen des Vereins für Naturdenkmal- und Heimatschutz in Kurhessen* 2:1–4.

———. 1914. *Pflege des Heimatschutzes der Stadtverwaltungen. Vortrag gehalten bei der Jahresversammlung des Hessisch-Waldeckischen Städtetages zu Arolsen am 5. und 6. Juni 1914.* Kassel: Weber und Weidemeyer.

"Gefährliche Ästhetik." 1909. *Bitumen. Fachzeitung für die Asphalt-, Teer-, Erdöl-, Paraffin- und verwandte Industrie* 7:337.

"Gegen ein Denkmal." 1911. Gegen ein Denkmal auf dem Langen Berg bei Gehren. *Heimatschutzfragen* (1): 96.

Geinitz, E. 1911. Beobachtung des Grundwassers. *Archiv des Vereins der Freunde der Naturgeschichte in Mecklenburg* 65:151–54.

Geisenheyner, L. 1916. Naturschutz in der Kriegszeit. Aus dem Nahegau. *Mitteilungen des Rheinsichen Vereins für Denkmalpflege und Heimatschutz* 10:87–93.

"Generalverordnung." 1911–12. Generalverordnung an die Oberforstmeistereien, die Direktion der Forsteinrichtungsanstalt, den Rektor der Forstakadamie Tharandt und die Forstrevierverwaltungen vom 20. Mai 1912. *Mitteilungen des Landesvereins Sächsischer Heimatschutz* 2:292–95.

Gerlach, Dankwart. 1917. Review of Wilhelm Kotzde, *Frau Harke geht durchs Land. Wandervogelführerzeitung* (2/3): 43–44.

Gerndt, Siegmar. 1981. *Idealisierte Natur. Die literarische Kontroverse um den Landschaftsgarten des 18. und frühen 19. Jahrhunderts in Deutschland.* Stuttgart: J. B. Metzler.

Gesellschaft der Freunde des Deutschen Heimatschutzes. 1930. *Der deutsche Heimatschutz. Ein Rückblick und Ausblick.* Munich.

Giannoni, Karl. 1913. Heimatschutz und Fremdenverkehr. *Kosmos* 10:153–57.

———. 1914–15. Deutschland und Österreich im Heimatschutz Hand in Hand. *Mitteilungen des Landesvereins Sächsischer Heimatschutz* 4:387–402.

———. 1919. *Naturschutz und Verkehr.* Ed. Staatliche Stelle für Naturdenkmalpflege. Vol. 21, *Naturdenkmäler. Vorträge und Aufsätze.* Berlin: Gebrüder Borntraeger.

Glaser, Hermann. 1984. *Die Kultur der wilhelminischen Zeit. Topographie einer Epoche.* Frankfurt a. M.: S. Fischer.

Glogau, Arthur. 1903. Landesverschönerung. *Gartenkunst* 5:96–98.

Gould, Peter C. 1988. *Early Green Politics. Back to Nature, Back to the Land, and Socialism in Britain, 1880–1900.* Sussex: Harvester Press.

Gradmann, Eugen. 1910. *Heimatschutz und Landschaftspflege.* Stuttgart: Strecker und Schroeder.

Gramsci, Antonio. 1985a. For a Cultural Association. In *Selections from Cultural Writings,* ed. D. Forgacs. Cambridge: Harvard University Press.

———. 1985b. Literary Criticism [Paul Nizan]. In *Selections from Cultural Writings,* ed. D. Forgacs. Cambridge: Harvard University Press.

———. 1985c. Questions of Culture. In *Selections from Cultural Writings,* ed. D. Forgacs. Cambridge: Harvard University Press.

Graner. 1912. Einige Fragen der Stuttgarter Waldumgebung. *Jahreshefte des Vereins für vaterländische Naturkunde in Württemberg* 68:71–78.

Gravelius, H. 1905. Zur Anthropogeographie des Wassers I. *Mitteilungen des Vereins für Erdkunde zu Dresden* (1): 28–38.

Gregor, Mary J. 1983. Baumgarten's *Aesthetica. Review of Metaphysics* 37:357–85.

Greiter, Remigius. 1914. Schutz der Hecke. *Bayerischer Heimatschutz* 12:24–25.

Greverus, Ina-Maria. 1972. *Der Territoriale Mensch. Ein literaturanthropologischer Versuch zum Heimatphänomen.* Frankfurt a. M.: Athenäum.

———. 1979a. Statt einer Einleitung: "Heimat"—politische Vokabel oder politische Aufgabe? In *Auf der Suche nach Heimat.* Munich: C. H. Beck.

———. 1979b. Denkmalräume oder Lebensräume? In *Auf der Suche nach Heimat.* Munich: C. H. Beck.

Grimm, J. 1908. Zum Vogelschutz. *Der Gartenfreund* 42:179–81.

Gröning, Gert. 1992. The idea of land embellishment. As exemplified in the *Monatsblatt für Verbesserung des Landbauwesens und für zweckmäßige Verschönerung des baierischen Landes* from 1821 to 1829. *Journal of Garden History* 12:164–82.

Gröning, Gert, and Wolschke-Bulmahn, Joachim. 1983. Naturschutz und Ökologie im Nationalsozialismus. *Alte Stadt* 10:1–17.

Großklaus, Götz. 1992? *Natur—Raum. Von der Utopie zur Simulation.* Munich: Iudicium.

Günther, Willy. 1913. Naturzerstörung und Naturschutz. *Himmel und Erde* 25:439–49, 489–503.

Gurlitt, Cornelius. 1909. Schutz des Stadtbildes. In *Heimatschutz in Sachsen. Vorträge der Dresdner Volkshochschulkurse,* ed. R. Wuttke. Leipzig: B. G. Teubner.

———. 1909–10. Heimatschutz. *Mitteilungen des Landesvereins Sächsischer Heimatschutz* 1:10–15.

———. 1911a. Ein Angriff auf die Heimatschutzbewegung. *Neudeutsche Bauzeitung* 7:176–78.

———. 1911b. Nochmals Heimatschutz und Pappdächer. *Heimatschutz* 7:110–14.

———. 1911–12. Ein Angriff auf die Heimatschutzbewegung. *Mitteilungen des Landesvereins Sächsischer Heimatschutz* 2:46–51.

"Gutachten." 1905–6. Gutachten der großherzoglichen badischen Oberdirektion des Wasser- und Straßenbaus über das Lochersche Projekt für Laufenburg. *Mitteilungen des Bundes Heimatschutz* 2:68–74.

Hagen, Kurt. 1914. Die Beziehungen des Heimatschutzes zu den Fragen des modernen Staatslebens. *Mitteilungen des Landesvereins Sächsischer Heimatschutz* 4:235–40.

Hager, Kurt. 1913. Baupolizei und Heimatschutz. *Mitteilungen des Landesvereins Sächsischer Heimatschutz* 3:206–8.

Hahn, R. 1905–6. Der Entwurf eines preußischen Gesetzes gegen die Verunstaltung der Straßen und Plätze in geschlossenen Ortschaften. *Mitteilungen des Bundes Heimatschutz* 2:177–86.

Hajós, Geza. 1985. Heimatschutz und Umweltschutz. In *Zwischen Natur und Kultur. Zur Kritik biologistischer Ansätze,* ed. H. C. Ehalt. Vienna: Hermann Böhlau.

———. 1989. Heimatstil—Heimatschutzstil. *Österreichische Zeitschrift für Kunst und Denkmalpflege* 63:156–59.

Hamann, Richard, and Jost Hermand. 1977. *Stilkunst um 1900.* Frankfurt a. M.: Fischer Taschenbuch Verlag.

Hanisch, Fritz. 1909. Welche Bedeutung haben Stadtwaldungen für Industrie- und Großstädte? *Gartenkunst* 11:183–93.

"Hans Thoma-Feier." 1909. *Gemeinnützige Blätter* 11:300–301.

Hartung, Werner. 1989. Denkmalpflege und Heimatschutz im wilhelminischen Deutschland 1900 bis 1913. *Österreichische Zeitschrift für Kunst und Denkmalpflege* 63 (3/4): 173–81.

———. 1991. *Konservative Zivilisationskritik und regionale Identität—Am Beispiel der niedersächsischen Heimatbewegung 1895 bis 1919.* Hannover: Hahnsche Buchhandlung.

Hasse, Jürgen, and Frauke Schumacher. 1990. *Sanfter Tourismus. Über ein konstruktives Verhältnis von Tourismus, Freizeit und Umweltschutz.* Bunderhee: Institut für Umweltforschung.

Haug, Wolfgang Fritz. 1988. Gramsci und die Politik des Kulturellen. *Das Argument* 30:32—48.

Haupt. 1911–12. Die Reinhaltung der Gewässer, eine Aufgabe des Heimatschutzes. *Mitteilungen des Landesvereins Sächsischer Heimatschutz* 2:455–59.

———. 1917. Der Einfluß der Kriegsindustrie auf die Flußverunreinigung. *Mitteilungen des Landesvereins Sächsischer Heimatschutz* 6:270–72.

Haupt, Richard. 1910. *Über die Erhaltung der Fußwege und das Recht der Nation am Walde—Ein Vortrag, am 28 Juli 1910 zu Westerland auf Sylt vor den Vertretern der Bäder und Sommerfrischen Schleswig-Hosteins gehalten:* Schleswiger Nachrichten.

Hauser, Susanne. 1992. "Reinlichkeit, Ordnung und Schönheit"—Zur Diskussion über Kanalisation im 19. Jahrhundert. *Alte Stadt* 19:292–312.

Haushofer, Max. 1906. *Der Schutz der Natur.* Vol. 1, *Veröffentlichungen des Landesausschusses für Naturpflege.* Munich: Kgl. Hofdruckbücherei Kastner und Callway.

———. 1908. *An des Daseins Grenzen. Geschichten und Phantasien.* 2d ed. Munich: C. H. Beck.

Hays, Samuel. 1959. *Conservation and the Gospel of Efficiency.* Cambridge: Harvard University Press.

———. 1987. *Beauty, Health and Permanence: Environmental Politics in the United States, 1955–1985.* Cambridge: Cambridge University Press.

Hedemann-Heespen, Paul von. 1910. Sollen wir Heimatschutz erzwingen? Zur Philosophie und Taktik des Heimatschutzes. *Heimatschutz* 6:86–91.

———, ed. 1914. *Sechs Jahre Arbeit des Schleswig-Holsteinischen Landesvereins für Heimatschutz 1908–1914.*

Heick, G. 1910. Der Naturschutzpark in den Parkanlagen. *Gartenkunst* 12:223–27.

———. 1912. Die deutsche Naturschutzparkbewegung. *Gartenkunst* 14:65–70.

———. 1914. Die Gartenkunst in deutschen Mittel- und Kleinstädten. *Gartenkunst* 16:261–68.

"Heimatschutz, Baufach." 1910. Heimatschutz, Baufach und Baubehörden. *Bitumen. Fachzeitung für die Asphalt-, Teer-, Erdöl-, Paraffin- und verwandte Industrie* 8:367–68.

"Heimatschutz für das Sauerland." 1914. *Siegerland* 2:132.

"Heimatschutz im Walde." 1905. *Mitteilungen des Bundes Heimatschutz* 2:24.

"Heimatschutz und Landwirtschaft." 1910. *Braunschweigischer Heimatschutz* 1 (2): 62–64.

"Heimatschutz und Wegebau." 1915. *Pommersche Heimat* 4 (2): 7–8.

Hellmich, Walter. 1953. *Natur- und Heimatschutz.* Stuttgart: Kosmos. Gesellschaft der Naturfreunde.

Hempel, Richard. 1914. *Heimatschönheit und Bodenkultur.* Hannover: n.p.

Henning, Charles L. 1913. Der Amerikaner und die Natur. *Himmel und Erde. Illustrierte naturwissenschaftliche Monatsschrift* 25:225–32.

Hennings, Curt. 1908–9. Heimatschutz und Naturdenkmalpflege. *Verhandlungen des naturwissenschaftlichen Vereins in Karlsruhe* 22:101–21.

Herf, Jeffrey. 1984. *Reactionary modernism. Technology, culture and politics in Weimar and the Third Reich.* Cambridge: Cambridge University Press.

Hermand, Jost. 1991. Nationalistische Phrase oder Ausdruck ökologischen Bewußtseins? Das *Heimatschutz*-Konzept um 1900. In *Im Wettlauf mit der Zeit. Anstöße zu einer ökologiebewußten Ästhetik.* Berlin: Edition Sigma Bohn.

"Hermann Löns." 1915. *Blätter für Naturschutz und Heimatpflege* 1 (1): 4–6.

"Hervorragend schöne." 1912–13. Hervorragend schöne Landschaften. *Der Türmer* 15:940.

Heuss, Theodor. 1905. Heimatschutz. *Die Hilfe* 11 (30): 10–11.

Hill, Howard. 1980. *Freedom to Roam. The Struggle for Access to Britain's Moors and Mountains.* Ashbourne, England: Moorland Publishing.

Hirschfeld, Christian Cayus Lorenz. 1990. *Theorie der Gartenkunst.* Stuttgart: Deutsche Verlags-Anstalt.

Hoemann, Reinhold. 1907. Wie wir unsere Heimat sehen! *Gartenkunst* 9:99.

Hüttenberger, Peter. 1992. Umweltschutz vor dem Ersten Weltkrieg: Ein sozialer und bürokratischer Konflikt. In *Staat und Wirtschaft an Rhein und Ruhr 1816–1991. 175 Jahre Regierungsbezirk Düsseldorf,* ed. H. Hoebink. Essen: Klartext.

"Industrie und Heimatschutz." 1914–15. *Mitteilungen des Landesvereins Sächsischer Heimatschutz* 4:5–7.

"In Sachen." 1908. In Sachen "Heimatschutzgesetz". *Bitumen. Fachzeitung für die Asphalt-, Teer-, Erdöl-, Paraffin- und verwandte Industrie* 6:305–6.

"Ist keiner da?" 1912. *Der Sauerländische Gebirgsbote* 20:252–53.

Jackson, John Brinkerhoff. 1984. *Discovering the Vernacular Landscape.* New Haven: Yale University Press.

Jacobi, Hans Bernhard. 1912. *Die Verdrängung der Laubwälder durch die Nadelwälder in Deutschland.* Tübingen: H. Laupp.

Jefferies, Matthew. 1992. Back to the Future? The *Heimatschutz* Movement in Wilhelmine Germany. *History* 41:411–20.

———. 1995. *Politics and Culture in Wilhelmine Germany. The Case of Industrial Architecture.* Oxford, Washington: Berg.

Jensen, Johannes V. 1914. Die Umwandlung primitiver Landschaften durch die Technik und die Entstehung einer neuen landschaftlichen Schönheit. *Dokumente des Fortschritts* 8:131–34.

John, Oskar. 1908. Bodenreformers Bücherei. *Bodenreform* 19:699–707.

Jünger, Ernst. 1982 [1932]. *Der Arbeiter: Herrschaft und Gestalt.* Stuttgart: Klett-Cotta.

Jünger, Oswald. 1913. Deutschland als Kultursteppe. *Blätter für Naturschutz* 4 (9): 1–5.

———. 1915. Weltkrieg, Moorkultur und Naturschutz. *Blätter für Naturschutz und Heimatpflege* 1 (5): 1–5.

Kache, Paul. 1914. Über die Gestaltung des heimatlichen Landschaftsbildes. *Mitteilungen der deutschen dendrologischen Gesellschaft* 159–64.

Kaeflein. 1902. Gartenbesitzer schützt die Vögel. *Der Gartenfreund —Organ des Verbandes Badischer Gartenbau-Vereine* 36:29.

"Kalender." 1914. Kalender im Sinne des Heimatschutzes. *Heimatschutz* 10:97–104.

Kämpfer, Hermann. 1930. *Heimatschutz und Judentum, Leuchkugeln.* Vienna: Bund für Volkserklärung.

Kampffmeyer, Hans. 1910. Ortsstatute. *Gartenstadt* 4 (7): 83–84.

Kant, Immanuel. 1970. An Answer to the Question: What is Enlightenment? In *Kant's Political Writings,* ed. H. Reiss. Cambridge: Cambridge University Press.

Kaudelka-Hanisch, Karin. 1991. The Titled Businessman: Prussian Commercial Councillors in the Rhineland and Westphalia during the Nineteenth Century. In *The German Bourgeoisie. Essays on the Social History of the German Middle Class from the Late Eighteenth to the Early Twentieth Century,* ed. D. Blackbourn and R. J. Evans. London and New York: Routledge.

Keller, Karl. 1908. Naturschönheiten als Privatmonopol. *Bodenreform* 19:596–98.

———. 1910. Die Bedeutung des preußischen Verunstaltungsgesetzes für den Schutz des Landschaftsbildes. *Heimatschutz in Brandenburg* 2:177–81.

Kelly, Alfred. 1975. *Between Poetry and Science. Wilhelm Bölsche as Popularizer of Science,* University of Wisconsin-Madison.

Kemmerich, Max. 1910. Naturschutzparke. *Sauerländischer Gebirgsbote* 18 (7): 152–53.

Kettler, Julius Iwan. 1906. Die Ausnutzung der Naturkräfte vom Standpunkte der

Kultur. In *Heimatschutz. Bericht über die Jahresversammlung des BH in Goslar am 12.–14. Juni 1905.* Halle a. S.: Gebauer-Schwetschke.

Kitzler, Georg-Eugen. 1918. Frau Harke geht durchs Land. *Die Mark. Zeitschrift des Allgemeinen Märkischen Touristen-Bundes* 151.

Knauer, Friedrich. 1912. Für und Wider in der Naturschutzfrage. *Der Aar* 2:809–17.

Knaut, Andreas. 1991. Ernst Rudorff und die Anfänge der deutschen Heimatbewegung. In *Antimodernismus und Reform. Zur Geschichte der deutschen Heimatbewegung,* ed. E. Klueting. Darmstadt: Wissenschaftliche Buchgesellschaft.

———. 1993. *Zurück zur Natur! Die Wurzeln der Ökologiebewegung.* Ed. Arbeitsgemeinschaft beruflicher und ehrenamtlicher Naturschutz e. V. Vol. 1. *Jahrbuch für Naturschutz und Landschaftspflege, Supplement.* Greven: Kilda-Verlag.

Koch, Fritz. 1907. Heimatschutz-Postkarten. *Heimatschutz* 3:45–50.

———. 1908. Geschäftsbericht. *Heimatschutz* 4:40–46.

———. 1909. Von der Heimatschutzerbeit und vom Organisieren. *Heimatschutz* 5:1–7.

———. 1911. Wie man in Thüringen Bäume ausästet. *Heimatschutzfragen* 1:93–95.

———. 1912a. Kurzer Überblick über den Stand der Heimatschutzbewegung. *Heimatschutz* 8:69–75.

———. 1912b. Regierungs- und Baurat Krielke in Weimar. *Heimatschutzfragen* (4): 48.

———. 1912c. Einiges über die Organisation des Heimatschutzes. *Heimatschutzfragen* (3): 5.

———. 1913a. Heimatschutzkarten. *Heimatschutz* 9:104–9.

———. 1913b. Hecken und Raine. *Mitteilungen des Schlesischen Bundes für Heimatschutz* 1 (3): 30.

———. 1914. Heimatschutz und internationale Verständigung. *Korrespondenz des Verbands für internationale Verständigung* 3:10–11.

———. ca. 1922. Die Pflege der Schönheit und Eigenart der Heimat als soziale Aufgabe gerade für unsere arme Zeit. Unknown newspaper, in THStA Weimar, Akte Koch I.

Kocka, Jürgen. 1989. Bildungsbürgertum—Gesellschaftliche Formation oder Historikerkonstrukt? In *Politischer Einfluß und gesellschaftliche Formation,* ed. J. Kocka. Stuttgart: Ernst Klett.

König. 1915. Laubwald, Nadelwald und Heimatschutz. *Schwäbisches Heimatbuch* 144–46.

Korb, Rudolf. 1913. Flußregulierungen und Talsperren! *Heimatschutz in Brandenburg* 5:102–8.

———. 1915. Der Verein Naturschutzpark. *Heimatschutz in Brandenburg* 7:48–58.

Koshar, Rudy. 1991. Altar, Stage and City: Historic Preservation and Urban Meaning in Nazi Germany. *History and Memory* 3 (1): 30–59.

———. 1993. Against the "Frightful Leveler": Historic Preservation and German Cities, 1890–1914. *Journal of Urban History* (May): 7–29.

———. 1996. The Antinomies of *Heimat:* Homeland, History, Nazism. In *Heimat, Nation, Fatherland. The German Sense of Belonging,* ed. J. Hermand and J. Steakley. New York: Peter Lang.

Kotzde, Wilhelm. 1913. Die Regulierung der unteren Havel und ihre Folgen. *Heimatschutz in Brandenburg* 5:108–13.

———. 1913–14. Jugendpflege und Heimat. *Deutscher Volkswart* 1:347–55.

———. 1914. Zur Havelregulierung, insbesondere zum Untergang der Havelfischerei. *Heimatschutz in Brandenburg* 6:14–25.

———. 1915. *Was sollen wir tun? Wünsche für Deutschland nach dem Kriege.* Leipzig: Dieterichische Verlagsbuchhandlung.

———. 1919. *Frau Harke geht durchs Land. Der Roman einer Landschaft.* 2d ed. Leipzig: Erich Matthes.

Kozesnik, Moritz. 1904. *Die Ästhetik im Walde, die Bedeutung der Waldpflege und die Folgen der Waldvernichtung.* Vienna: Wilhelm Frick.

Kratzsch, Gerhard. 1969. *Kunstwart und Dürerbund. Ein Beitrag zur Geschichte der Gebildeten im Zeitalter des Imperialismus.* Göttingen: Vandenhoeck und Ruprecht.

Krische, Paul. 1910. *Hein Träumerlein. Aus dem jungen Leben eines Wahrheitssuchers.* Berlin: Schuster und Loeffler.

———. 1918. *Heimat! Grundsätzliches zur Gemeinschaft von Scholle und Mensch.* Berlin: Gebrüder Paetel.

———. 1923. *Gemeinschaftskunde. Stoffdarbietung zur Einführung und zum Unterricht für Eltern, Lehrer und Jugendliche sozialistischer und freigeistiger Kreise.* 2d ed. Dresden: Verlagsanstalt für proletarische Freidenker.

Krüger, Dieter. 1983. *Nationalökonomen im wilhelminischen Deutschland.* Göttingen: Vandenhoeck und Ruprecht.

Kruse. 1914. Der Krieg und der Heimatschutz. *Siegerland* 2:217–19.

Kuhls-Goslich, Marie. 1912. Ein Kampf um die Erhaltung unserer Seen. *Heimatschutz in Brandenburg* 4:165–69.

Ladd, Brian. 1990. *Urban Planning and Civic Order in Germany, 1860–1914.* Cambridge: Harvard University Press.

Landesverein Reuß des Bundes Heimatschutz. 1913. *Jahresbericht 1912.* Gera: Geraer Verlagsanstalt.

"Landwirtschaft und Naturdenkmalpflege." 1907. *Berliner Korrespondenz,* 20 July.

Lane, Barbara Miller. 1985. *Architecture and Politics in Germany, 1918–1945.* Cambridge: Harvard University Press.

Lange, Konrad. 1911. Die Zukunft der Lustnauer Allee in Tübingen. *Mitteilungen des Bundes für Heimatschutz in Württemberg und Hohenzollern* 3 (1): 17–21.

Langewiesche, Dieter. 1990. German Liberalism in the Second Empire, 1871–1914. In *In Search of a Liberal Germany. Studies in the History of German Liberalism from 1789 to the Present,* ed. K. H. Jarausch and L. E. Jones. New York: Berg.

"Laufenburg." 1904–5. Laufenburg. *Mitteilungen des Bundes Heimatschutz* 1:69.

"Laufenburger." 1904–5. Die Laufenburger Stromschnellen. *Mitteilungen des Bundes Heimatschutz* 1:130.

Le Dinh, Diana. 1992. *Le Heimatschutz, une ligue pour la beauté. Esthétique et consciences culturelle au début du siècle en Suisse.* Université de Lausanne.

Lees, Andrew. 1975. Debates about the Big City in Germany, 1890–1914. *Societas* 5:31–47.

———. 1988. The Civic Pride of the German Middle Classes, 1890–1918. In *Another Germany: A Reconsideration of the Imperial Era,* ed. J. a. R. Dukes, Joachim. Boulder: Westview Press.

———. 1991. Berlin and Modern Urbanity in German Discourse, 1845–1945. *Journal of Urban History* 17:153–80.

Lehne, Andreas. 1989. Heimatstil—Zum Problem der Terminologie. *Österreichische Zeitschrift für Kunst und Denkmalpflege* 63:159–64.

Leibniz, Georg Wilhelm. 1985. *Theodicy.* La Salle, Ill.: Open Court.

———. 1989. On the Ultimate Origination of Things. In *Philosophical Essays,* ed. R. Ariew. Indianapolis and Cambridge: Hackett.

Leisi, Ernst. 1932. Ein Vierteljahrhundert Heimatschutz im Thurgau. In *25 Jahre Heimatschutz im Thurgau.* Amriswil: O. Zimmermann.

Lepsius, M. Rainer. 1968. Demokratie in Deutschland als historisch-soziologisches Problem. In *Spätkapitalismus oder Industriegesellschaft,* ed. T. W. Adorno. Stuttgart.

———. 1973. Parteiensystem und Sozialstruktur: Zum Problem der Demokratisierung der deutschen Gesellschaft. In *Die deutschen Parteien vor 1918,* ed. G. A. Ritter. Cologne.

"Leuchtende Stunden." 1911–12. *Mitteilungen des Landesvereins Sächsischer Heimatschutz* 2:284–86.

Lichtwark, Alfred. 1909. Die Zweckfrage bei öffentlichen Parkanlagen. *Gartenkunst* 12:75–79.

Lienhard, Friedrich. 1900. *Die Vorherrschaft Berlins. Litterarische Anregungen.* Leipzig, Berlin: G. H. Meyer Heimatverlag.

———. 1902. *Wasgaufahrten—Ein Zeitbuch.* 3d ed. Leipzig and Berlin: G. H. Meyer.

Linse, Ulrich. 1986. *Ökopax und Anarchie. Eine Geschichte der ökologischen Bewegungen in Deutschland.* Stuttgart: dtv.

———. 1988. "Der Raub des Rheingoldes": Das Wasserkraftwerk Laufenburg. In *Von der Bittschrift zur Platzbesetzung. Konflikte um technische Großprojekte: Laufenburg—Walchensee—Wyhl—Wackersdorf,* ed. R. Falter, W. Kretschmer, U. Linse, and D. Rucht. Berlin, Bonn: J. H. W. Dietz Nachfolger.

———. 1993. Das Proletariat—Komplize kapitalistischer Naturzerstörung? In *Mit den Bäumen sterben die Menschen. Zur Kulturgeschichte der Ökologie,* ed. J. Hermand. Cologne: Böhlau.

Lipp, Wolfgang. 1986. Heimatbewegung, Regionalismus. Pfade aus der Moderne? *Kölner Zeitschrift für Soziologie und Sozialpsychologie, Sonderheft* 27:330–55.

Lippischer Bund für Heimatschutz und Heimatpflege. 1909. *1. Jahresbericht des Lippischen Bundes für Heimatschutz und Heimatpflege.* Detmold.

———. 1910. *2. Jahresbericht des Lippischen Bundes für Heimatschutz und Heimatpflege.* Detmold.

Lloyd. 1989. John Ruskin—Storm Clouds of the Nineteenth Century and the Ozone Layer of the Twentieth Century. *The Ruskin Gazette* 1 (3): 25–27.

Löhner. 1916. Heimatschutz in Krieg und Frieden. *Bayerischer Heimatschutz* 14:123–24.

Löns, Hermann. 1911. Verkoppelung. *Braunschweigischer Heimatschutz* 2 (3): 101.

———. 1913. Naturschutz und Rassenschutz. *Blätter für Naturschutz* 4 (2): 1–4.

Lörcher, Ernst. 1915. Geschäftsbericht für das 6. Geschäftsjahr 1914. *Schwäbisches Heimatbuch:* 17–22.

Lüdtke, Hermann. 1905. Etwas über den Stadtwald. *Gartenkunst* 7:100–101.

Lühmann, H. 1912. Vom Flurnamen zum Straßennamen. *Braunschweigischer Heimatschutz* (Sonderheft): 15–19.

Lux, Joseph August. 1906–7a. Aussichtswarten. *Hohe Warte* 3:166–67.

———. 1906–7b. Baumpredigt. *Hohe Warte* 3:58–61.

———. 1907. Ansichten und Gedanken. *Gartenkunst* 9:77–78.

———. 1908–9. Naturempfinden und Touristik. *Berner Rundschau* 3:634–40.

———. 1908a. Die Gartenkunst und die Landschaftsgärtnerei. *Gartenkunst* 10:171–73.

Maaß, Harry. 1910. Knicks im Landschafts- und Städtebild. *Gartenkunst* 12:205–9.

MacCannell, Dean. 1989. *The Tourist. A New Theory of the Leisure Class.* New York: Schocken.

MacKail, J. W. 1899. *The Life of William Morris.* Vol. 2. London: Longmans, Green and Co.

Mahal, Günther. 1975. *Naturalismus.* Munich: Wilhelm Fink/UTB.

Mammen, Franz. 1908. Forstbotanische Merkbücher. *Heimatschutz* 4:62–70.

———. 1909. Heimatschutz im Walde. In *Heimatschutz in Sachsen. Vorträge der Dresdner Volkshochschulkurse,* ed. R. Wuttke. Leipzig: B .G. Teubner.

———. 1910–11. Heimatschutz im Walde und Waldschönheitspflege. *Mitteilungen des Landesvereins Sächsischer Heimatschutz* 1:33–50.

McIntosh, Robert P. 1985. *The Background of Ecology—Concept and Theory.* Cambridge: Cambridge University Press.

Meißner, Carl. 1908. Heimatschutz 1908. *Bitumen. Fachzeitung für die Asphalt-, Teer-, Erdöl-, Paraffin- und verwandte Industrie* 6:312, 316–17.

Menzel, Hans. 1906. Das Bild der Landschaft, seine Entstehung und seine Erhaltung. In *Heimatschutz. Bericht über die Jahresversammlung des BH in Goslar am 12.–14. Juni 1905.* Halle a. S.: Gebauer-Schwetschke.

———. 1910–11. Vom Heimatschutz. *Heimat und Welt* 1:11–13, 77–80.

Merchant, Carolyn. 1980. *The Death of Nature: Women, Ecology and the Scientific Revolution.* New York: Harper and Row.

Meyer, Oswald. 1912. Max Dreyer. *Die Grenzboten* (41): 86–92.

Meyer-Abich, Klaus Michael. 1990. *Aufstand für die Natur. Von der Umwelt zur Mitwelt.* Munich and Vienna: Carl Hanser.

Mieck, I. 1981. Luftvereunreinigung und Immisionsschutz in Frankreich und Preußen. *Technikgeschichte* 48:239–51.

Mielke, Robert. 1900. *Der einzelne und seine Kunst—Beiträge zu einer Ökonomie der Kunst.* Leipzig, Berlin: G. H. Meyer.

———. 1904. Friedrich Ratzel . Mitteilungen des Bundes Heimatschutz 1:75.

———. 1904–5a. Bericht über die konstituierende Versammlung. *Mitteilungen des Bundes Heimatschutz* 1:1.

———. 1904–5b. Unser Arbeitsplan. *Mitteilungen des Bundes Heimatschutz* 1:66–68.

———. 1906a. Zur Separation. *Mitteilungen des Bundes Heimatschutz* 2:129–35.

———. 1906b. Geschäftsbericht. In *Heimatschutz. Bericht über die Jahresversammlung des BH in Goslar am 12.–14. Juni 1905.* Halle: Gebauer-Schwetschke.

———. 1907. Aufgaben, Mittel und Wege des Heimatschutzes in der Provinz Brandenburg. Vortrag, gehalten auf der konstituierenden Versammlung der Landesgruppe des Bundes Heimatschutz. *Brandenburgia* 21:225–33.

———. 1908. Heimatschutz und Landesverschönerung. *Gartenkunst* 10:143–45, 156–60, 182–86.

———. 1912. Das flache Dach und der Werdandibund. *Heimatschutz in Brandenburg* 4:204–7.

———. 1913a. Jahresbericht. *Heimatschutz in Brandenburg* 5:54–55.

———. 1913b. Industrie und Heimatschutz. *Heimatschutz in Brandenburg* 5:74–82.

———. 1914a. Zehn Jahre Heimatschutz. *Heimatschutz in Brandenburg. Mitteilungen der Landesgruppe Brandenburg des Bundes Heimatschutz* 6:1–9.

———. 1914b. Aus der Arbeit. *Heimatschutz in Brandenburg* 6:58–63.

———. 1917–18. Frau Harke geht durchs Land. *Heimatschutz in Brandenburg* 8:126.

———. 1929. Meine Beziehungen zu Ernst Rudorff und die Gründung des Bundes Heimatschutz. *Brandenburgia* 38 (1/2): 1–16.

Milchert, Jürgen. 1987. 100 Jahre DGGL im Spiegel der Fachdiskussion—Auf dem Weg in die großdeutsche Landschaft, 1933–1945. *Garten und Landschaft* 97:29–31.

Möbius, Hanno. 1980. *Der Positivismus in der Literatur des Naturalismus. Wissenschaft, Kunst und soziale Frage bei Arno Holz.* Munich: Wilhelm Fink.

Moewes, Franz. 1926. Zur Geschichte der Naturdenkmalpflege. In *Wege zum Naturschutz,* ed. W. Schoenichen. Breslau: Ferdinand Hirt.

Moneyhon, Carl. 1980. The Environmental Crisis and American Politics, 1860–1920. In *Historical Ecology: Essays on Environmental and Social Change,* ed. L. Bilsky. Port Washington, N.Y.: Kennikat Press.

Morley, David. 1990. No Place Like *Heimat:* Images of Home(land) in European Culture. *New Formations* 12 (winter): 1–23.

Morris, William. 1896. Untitled comments. *A Beautiful World—The Occasional Journal of the Society for Checking the Abuses of Public Advertising* 3 (Dec.): 16–18.

———. 1914a. The Beauty of Life. Delivered Before the Birmingham Society of Arts and School of Design, Feb. 19, 1880. In *Collected Works,* ed. M. Morris. London: Longmans Green and Co.

———. 1914b. Art and the Beauty of the Earth. A Lecture Delivered at Burslem

Town Hall on Oct. 13, 1881. In *Collected Works,* ed. M. Morris. London: Longmans Green and Co.

———. 1947. The Prospects of Architecture in Civilization. In *William Morris on Art and Socialism,* ed. H. Jackson. London: John Lehmann.

———. 1969 [orig. 1882]. Art: A Serious Thing. In *The Unpublished Lectures of William Morris,* ed. E. D. Lemire. Detroit: Wayne State University Press.

Morris-Keitel, Peter. 1991. *Naturkonzepte in der Literatur der frühen Jugendbewegung,* University of Wisconsin-Madison.

Mugler, Edmund. 1911. Über Sinn und Bedeutung des Heimatschutzes. *Siegerland* 1 (1): 24–28.

———. 1912. Zur Heimatpflege. Theoretische Erwägungen und praktische Forderungen. *Siegerland* 1 (3/4): 73–83.

Muschamp, Herbert. 1993. Fear, Hope and the Changing of the Guard. *New York Times,* 14 November 1993, 37.

Müßiggang, Albert. 1968. *Die soziale Frage in der historischen Schule der deutschen Nationalökonomie.* Vol. 2, *Tübinger Wirtschaftswissenschaftliche Abhandlungen.* Tübingen: J. C. B. Mohr.

Muthesius, Stefan. 1974. *Das englische Vorbild. Eine Studie zu den deutschen Reformbewegungen in Architektur, Wohnbau und Kunstgewerbe im späteren 19. Jahrhundert.* Munich: Prestel-Verlag.

———. 1981. The Origins of the German Conservation Movement. In *Planning for Conservation,* ed. R. Kain. New York: St. Martin's Press.

Nash, Roderick. 1977. The Value of Wilderness. *Environmental Review* 3:14–25.

———. 1989. *The Rights of Nature. A History of Environmental Ethics.* Madison: University of Wisconsin Press.

"Naturdenkmalpflege." 1915. *Mecklenburg. Zeitschrift des Heimatbundes Mecklenburg* 10:66–67.

Naumann, Arno. 1909–10. Praktische Wege des Heimatschutzes. *Mitteilungen des Landesvereins Sächsischer Heimatschutz* 1:399–425.

Nehring, Dorothee. 1979. Stadtparkanlagen in der ersten Hälfte des 19. Jahrhunderts. Ein Beitrag zur Kulturgeschichte des Landschaftsgartens. Hannover: Patzer Verlag.

Niedermeier, Michael. 1993. Natur—Ökonomie—Sexualität. Philanthropen zwischen Paradies und Plantage 1770–1810. In *Mit den Bäumen sterben die Menschen. Zur Kulturgeschichte der Ökologie,* ed. J. Hermand. Cologne: Böhlau.

Nipperdey, Thomas. 1976. Verein als soziale Struktur in Deutschland im späten 18. und frühen 19. Jahrhundert. Eine Fallstudie zur Modernisierung I. In *Gesellschaft, Kultur, Theorie. Gesammelte Aufsätze zur neueren Geschichte.* Göttingen: Vandenhoeck und Ruprecht.

———. 1986. War die Wilhelminische Gesellschaft eine Untertanen-Gesellschaft? In *Nachdenken über die deutsche Geschichte.* Munich: C. H. Beck.

———. 1988. *Wie das Bürgertum die Moderne fand.* Berlin: W. J. Siedler.

———. 1990. *Arbeitswelt und Bürgergeist.* Vol. 1, *Deutsche Geschichte 1866–1918.* Munich: C. H. Beck.

"Norddeutsches Flachland." 1914. Das norddeutsche Flachland—keine hervorragende Gegend. *Heimatschutz in Brandenburg* 6:33–34.

Nordhausen, Richard. 1913. Die Begradigung der Natur. *Mitteilungen der deutschen dendrologischen Gesellschaft* 318–20.

———. 1914. Bretter und Drahtzäune vor der Welt. *Der Tag,* ca. 10 July.

Nostrand, Richard L., and Lawrence E. Estaville. 1993. Introduction: The Homeland Concept. *Journal of Cultural Geography* 13 (2): 1–4.

Orr, David W. 1992. *Ecological Literacy. Education and the Transition to a Postmodern World.* Albany: State University of New York Press.

O'Sullivan, Paddy. 1990. Struggle for the Vision Fair: Morris and Ecology. *Journal of the William Morris Society* 8 (4): 5–9.

Otto, Christian F. 1983. Modern Environment and Historical Continuity: The Heimatschutz Discourse in Germany. *Art Journal* (summer): 148–57.

Otto, Hugo. 1916. Betrachtungen über Natur- und Heimatschutz in Polen. *Mitteilungen des Rheinischen Vereins für Denkmalpflege und Heimatschutz* 10:80–86.

Peckelhoff, Fr. 1908. Schutz unserer heimischen Vogelwelt. *Bericht des Vereins für Heimatschutz [Lübeck]:* 12–15.

Peters. 1908. Bericht über die Tagung des Bundes Heimatschutz in Lübeck am 23. September 1908. *Die Heimat* 18:274–75.

Petsch, Joachim. 1984. Heimatkunst—Heimatschutz. Zur Geschichte der europäischen Heimatschutzbewegung bis 1945. *werk-archithese* 27–28:49–52.

Pflug, Wolfram. 1969. 200 Jahre Landespflege in Deutschland. Eine Übersicht. In *Stadt und Landschaft—Raum und Zeit. Festschrift für Erich Kühn zur Vollendung seines 65. Lebensjahres,* ed. A. Boettger. Cologne: Deutscher Verband für Wohnungswesen.

Pickardt, Ludwig. 1911. Heimatschutz und Reklame. *Berliner Tageblatt,* 26 April.

Plessner, Helmuth. 1959. *Die verspätete Nation.* Stuttgart: W. Kohlhammer.

Plochmann, Richard. 1985. Bemerkungen zur Waldkultur Mitteleuropas. *Zeitschrift für Politik* 32:195–207.

Pogacnik, Marco. 1991. The Heimatschutz Movement and the Monumentalization of the Landscape. In *The Architecture of Western Gardens. A Design History for the Renaissance to the Present Day,* ed. M. Mosser. Cambridge: MIT Press.

"Polizeilicher Schönheitssinn." 1911. *Berliner Tageblatt,* 8 September.

Pommer, Richard. 1983. The Flat Roof: A Modernist Controversy in Germany. *Art Journal* 43:158–69.

"Postkarte." 1910. Die Postkarte als Lehrmittel. *Gemeinnützige Blätter* 12:121–23.

Pudor, Heinrich. 1910–11. Wandern in der Heimat. *Heimat und Welt. Monatshefte der Vereinigung Heimat und Welt* 1:292–93.

———. 1912. Waldpolitik. *Die Gartenwelt* 16:411–13.

Raabe, Wilhelm. 1985. *Pfisters Mühle. Ein Sommerferienheft.* Frankfurt a. M.: Insel Taschenbuch.

"Raiffeisen." 1911. Raiffeisen für die Heimat. *Raiffeisen-Bote. Mitteilungen über Genossenschaftswesen und Landwirtschaftsbetrieb von dem Verbande ländlicher Genossenschaften Raiffeisenscher Organisation für Rheinpfalz, Großherzogtum Hessen und Baden e. V.* 12:302–3.

Ranlett, John. 1983. "Checking Nature's Desecration": Late-Victorian Environmental Organization. *Victorian Studies* 26 (winter): 197–222.

Rasch, Edgar. 1913. Heimatschutz. *Gartenkunst* 15:59–63.

Rasser, E. O. 1910–11. Die Abnahme des Waldreichtums der Erde und der deutsche Wald. *Heimat und Welt* 1:377–82.

Ratzel, Friedrich. 1906. *Über Naturschilderung.* Munich, Berlin: R. Oldenbourg.

Reepel, Martin. 1917. "Landschaftlich hervorragende Gegenden." *Pommersche Heimat* 6 (9): 33–34.

Reif, Wolfgang. 1989. Exotismus im Reisebericht des frühen 20. Jahrhunderts. In *Der Reisebericht. Die Entwicklung einer Gattung in der deutschen Literatur,* ed. P. J. Brenner. Frankfurt a. M.: Suhrkamp.

"Reklame." 1913. *Heimatschutz in Brandenburg* 5:29–30.

"Reklame in der Landschaft." 1915. *Vossische Zeitung,* 2 July.

Ribbeck, Konrad. 1910–11. Die Naturschutzparkbewegung. *Heimat und Welt* 1:124–28.

Richard, Winfried. 1984. *Vom Naturideal zum Kulturideal. Ideologie und Praxis der Gartenkunst im deutschen Kaiserreich,* 19, TU Berlin.

Richards, Bernard. 1978. Ruskin and Conservation. *Texas Quarterly* 21 (1): 65–83.

Riehl, Wilhelm Heinrich. 1861. *Land und Leute.* 5th ed. Vol. 1, *Die Naturgeschichte des Volkes als Grundlage einer deutschen Social-Politik.* Stuttgart: J. G. Cotta.

Rieß. 1911–12. Straßenbau und Heimatschutz. *Mitteilungen des Landesvereins Sächsischer Heimatschutz* 2:169–76.

———. 1913. Die Kunst des Photographen. *Mitteilungen des Landesvereins Sächsischer Heimatschutz* 3:19.

———. 1914–15. Das Gemeinschaftsbewußtsein als Ausgang, Inhalt und Ziel der Heimatschutzbestrebungen. *Mitteilungen des Landesvereins Sächsischer Heimatschutz* 4:451–63.

Ringbeck, Birgitta. 1991. Architektur und Städtebau. In *Antimodernismus und Reform. Zur Geschichte der deutschen Heimatbewegung,* ed. E. Klueting. Darmstadt: Wissenschaftliche Buchgesellschaft.

Ritter, Joachim. Landschaft. Zur Funktion des Ästhetischen in der modernen Gesellschaft. In *Subjektivität. Sechs Aufsätze.* Frankfurt a. M.: Suhrkamp.

Roark, Michael O. 1993. Homelands: A Conceptual Essay. *Journal of Cultural Geography* 13 (2): 5–11.

Robien, Paul. 1917. Der Unkrautanger—Ein Wort zum Vogelschutz. *Pommersche Heimat* 6 (5): 18–19.

Rogers, Raymond A. 1994. *Nature and the Crisis of Modernity. A Critique of Contemporary Discourse on Managing the Earth.* Montreal: Black Rose Books.

Rohland. 1912. Heimatschutz für unsere Flüsse! *Mitteilungen des Bundes für Heimatschutz in Württemberg und Hohenzollern* 4:1–3.

Rojek, Chris. 1985. *Capitalism and Leisure Theory.* London: Tavistock Publications.

———. 1993. *Ways of Escape. Modern Transformations in Leisure and Travel.* London: Macmillan Press.

Rollins, William. 1995. Whose Landscape? Technology, Fascism, and Environmentalism on the National Socialist *Autobahn. Annals of the Association of American Geographers* 85:494–520.

———. 1996. The Land Ethic as Aesthetic-Cultural Process: Aldo Leopold and the German Landscape of Home. In *Aldo Leopold: His Land Ethic and Influence in Germany and the U.S.*, ed. H. Geitz. Madison: Max Kade Institute.

Rommelspacher, Thomas. 1987. Das natürliche Recht auf Wasserverschmutzung. Geschichte des Wassers im 19. und 20. Jahrhundert. In *Besiegte Natur. Geschichte der Umwelt im 19. und 20. Jahrhundert,* ed. F.-J. Brüggemeier. Munich: C. H. Beck.

Ross, Hartmut. 1987. August Rode und das Dessau-Wörlitzer Reformwerk. In *Der englische Garten zu Wörlitz,* ed. S. S. and G. W.-O. Luisium. Berlin: VEB Verlag für Bauwesen.

Rossbacher, Karlheinz. 1975. *Heimatkunstbewegung und Heimatroman. Zu einer Literatursoziologie der Jahrhundertwende.* Stuttgart: Ernst Klett.

Rubner, Heinrich. 1985. *Deutsche Forstgeschichte, 1933–1945. Forstwirtschaft, Jagd, und Umwelt im NS-Staat.* St. Katharinen: Scripta Mercaturae.

Rudorff, Ernst. 1880. Das Verhältnis des modernen Lebens zur Natur. *Preußische Jahrbücher* 45:261–76.

———. 1904–5. Zur Talsperrenfrage. *Mitteilungen des Bundes Heimatschutz* 1:175–80.

———. 1910. Gutachten zur Frage der Naturschutzparke. *Heimatschutz* 6:81–83.

———. 1926 [orig. 1897]. *Heimatschutz.* Berlin: Hugo Bermühler.

———. 1954 [orig. 1904]. Aufruf zur Gründung eines Bundes Heimatschutz. In *50 Jahre Deutscher Heimatbund.* Neuß: Deutscher Heimatbund.

Runte, Alfred. 1991. *Public Lands, Public Heritage. The National Forest Idea.* Niwot, Colo.: Roberts Rinehart.

———. 1993. Railroads Value Wilderness. In *Major Problems in American Environmental History,* ed. C. Merchant. Lexington, Mass.: D. C. Heath.

Ruskin, John. 1906. The Storm-Cloud of the Nineteenth Century. In *Collected Works,* ed. Cook and Wedderburn. Vol. 34. London: George Allen.

———. 1985. *Unto this Last.* In *Unto this Last and Other Writings,* ed. C. Wilmer. London: Penguin Classics.

Salisch, Heinrich von. 1902. *Forstästhetik.* 2d ed. Berlin: Julius Springer.

Salomon, H. 1917. Der Naturschutz bei unseren Feinden. *Blätter für Naturschutz und Heimatpflege* 3 (11): 4–5.

Schaefer, Bernhard. 1909–12. Der Schutz des Waldes, besonders in Hessen. *Abhandlungen und Berichte des Vereins für Naturkunde zu Cassel* 53:178–93.

Schäfer, Wolfgang. 1985. *Die unvertraute Moderne.* Frankfurt a. M.: Fischer Taschenbuch.

Schäff, Ernst. 1910. Kritische Betrachtung zur Frage der Naturschutzparks in Deutschland. *Welt und Wissen. Hannoversche Blätter für Kunst, Literatur und Leben,* 10 August, 1.

Schier. 1904. Waldästhetik. *Gartenkunst* 6:223–25.

Schiller, Friedrich. 1962. Briefe über die ästhetische Erziehung des Menschen. In *Werke,* ed. B. v. Wiese. Weimar: Hermann Böhlaus Nachfolger.

———. 1969. *Über naïve und sentimentalische Dichtung.* Stuttgart: Reclam.

Schimank, Uwe. 1983. *Neoromantischer Protest im Spätkapitalismus.* Bielefeld: AJZ Verlag.

Schlesischer Bund für Heimatschutz. 1911. *Aufruf zum Beitritt in den Schlesischen Bund für Heimatschutz.* Breslau: Schlesischer Bund für Heimatschutz.

"Schleswig-Holsteinischer." 1909. Schleswig-Holsteinischer Landesverein für Heimatschutz. *Die Heimat* 19:252.

Schliepmann, Hans. 1917. Krieg und Reklame. *Heimatschutz in Brandenburg* 8:72–76.

Schloßmann, Georg. 1914. Pflicht des Landmannes. *Raiffeisen-Bote für Brandenburg, Pommern und beide Mecklenburg* 16:110.

Schmidt. 1913. Schweizer Gartenstädte. *Mitteilungen des Landesvereins Sächsischer Heimatschutz* 3:49–59.

Schmidt, L. F. Karl. 1919. In eigener Sache. *Mitteilungen des Landesvereins Sächsischer Heimatschutz* 8: endpaper.

Schmidt, Rudolf Karl. 1965. Zur Heimatideologie. Bemerkungen über die gesellschaftliche Relevanz des Heimatbegriffs in der deutschen Literatur (I). *Das Heft. Zeitschrift für Literatur und Kunst* 6:31–40.

Schneider, Camillo Karl. 1907. Landesverschönerung und Heimatschutz. In *Landschaftliche Gartengestaltung.* Leipzig: Carl Scholtze und W. Junghans.

Schneider, Ernst. 1913. Erziehung und Pflege des Schönheitswaldes—Mit Rücksicht auf die in der Nähe der Städte zu schaffenden und zu erhaltenden Stadtwälder. *Gartenkunst* 15:249–62.

Schoch. 1902. Heimische Schutzgebiete und Landesverschönerung. *Gartenkunst* 4:65–71.

Schoenichen, Walter. 1954. *Naturschutz—Heimatschutz. Ihre Begründung durch Ernst Rudorff, Hugo Conwentz und ihre Vorläufer.* Vol. 16, *Grosse Naturforscher.* Stuttgart: Wissenschaftliche Verlagsgesellschaft.

"Schonet." 1915. Schonet die alten Bäume! *Pommersche Heimat* 4 (2): 7.

"Schonet die Hecken." 1916. Schonet die Hecken und Sträucher zwischen den Feldern! *Blätter für Naturschutz und Heimatpflege* 2 (7): 28.

Schorler, Bernhard. 1909–10. Der Sahnwald bei Crimmitschau. *Mitteilungen des Landesvereins Sächsischer Heimatschutz* 1:298–303.

Schramm, Engelbert. 1991. Kommunaler Umweltschutz in Preußen (1900–1933): Verengung auf Vollzug durch wissenschaftliche Beratung? In *Stadt und Gesundheit. Zum Wandel von "Volksgesundheit" und kommunaler Gesundheitspolitik im 19. und frühen 20. Jahrhundert,* ed. J. Reulecke and A. G. z. Castell Rudenhausen. Stuttgart: Franz Steiner.

Schube, Theodor. 1911. Naturdenkmäler aus der Baumwelt der preußischen Oberlausitz. *Abhandlungen der Naturforschenden Gesellschaft zu Görlitz* 27:89–116.

———. 1914. Die letztjährigen Erfolge des Schlesischen Bundes für Heimatschutz in der Naturdenkmalpflege. Bericht, erstattet auf der am 17. Mai 1914 abgehaltenen Jahresversammlung: Schlesischer Bund für Heimatschutz.

Schultze-Naumburg, Paul. 1902. Das Ende unseres deutschen Landschaftsbildes. *Körper und Geist* 11:132–35.

———. *Die Entstellung unseres Landes.* Vol. 2, *Flugschriften des Bundes Heimatschutz.* Halle a. S.: Gebauer-Schwetschke.

———. Jahresbericht. In *Heimatschutz. Bericht über die Jahresversammlung des Bundes Heimatschutz in Goslar am 12.–14. Juni 1905.* Halle a. S.: Gebauer-Schwetschke.

———. 1906. Kraftanlagen in ihrer ästhetischen Bedeutung, insonderheit der Talsperren. In *Heimatschutz. Bericht über die Jahresversammlung des BH in Goslar am 12.–14. Juni 1905.* Halle a. S.: Gebauer-Schwetschke.

———. 1917. *Die Gestaltung der Landschaft durch den Menschen, Teil III.* Vol. 9, *Kulturarbeiten.* Munich: Callwey.

Schulze, Ernst. 1914. Die Zerstörung des Waldkleides der Erde. *Mitteilungen der deutschen dendrologischen Gesellschaft* 88–94.

Schulze-Moering, Georg. 1919. *Wanderbewegung und Heimatschutz.* Berlin: Alexius Kießling.

Schumann, Wolfgang. 1919. *Über den Dürerbund. Bemerkungen über Geschichte, Wesen und Aufgabe des Dürerbundes und seiner Unternehmungen.* Munich: Callwey.

Schüren, Reinhold. 1989. *Soziale Mobilität. Muster, Veränderungen und Bedingungen im 19. und 20. Jahrhundert.* St. Katharinen: Scripta Mercaturae.

"Schutz den Havelufern!" 1910. *Heimatschutz in Brandenburg* 2:219–20.

"Schutz der Heimat." 1913. Schutz der Heimat in Feld und Wald. *Ortsgruppe Heimatschutz für den Kreis Herzogtum Lauenburg und Umgebung—Jahresbericht* 14–22.

Schwabe, Friedrich. 1915. Schutz der Vogelwelt. *Zeitschrift des Heimatbundes Niedersachsen* 1:138–41.

Schwarz. 1903. Gartenbau und Vogelschutz. *Der Gartenfreund* 37:12–16.

Schwerte, Hans. 1967. Deutsche Literatur im Wilhelminischen Zeitalter. In *Das Wilhelminische Zeitalter,* ed. H. J. Schoeps. Stuttgart: Ernst Klett.

Seesselberg, Friedrich. 1910. Der Heimatschutz als Charakterangelegenheit. *Werdandi-Zeitschrift* 3 (3): 3–13.

———. 1911. Erklärung der Denkschrift des Bundes Deutscher Architekten. *Heimatschutz in Brandenburg* 3:60–61.

Seidel, Bruno. 1967. Die Wirtschaftsgesinnung des Wilhelminischen Zeitalters. In *Das Wilhelminische Zeitalter,* ed. H. J. Schoeps. Stuttgart: Ernst Klett.

Seifert, Eberhard. 1990. Kultur versus Natur? Anmerkungen zu einer Grundfrage der Wirtschaftsethik. In *Gustav Schmoller oggi: lo sviluppo delle scienze sociali in Germania e Italia / Gustav Schmoller heute: Die Entwicklung der Sozialwissenschaften in Deutschland und Italien,* ed. M. Bock, H. Homann and P. Schiera. Berlin: Duncker und Humblot.

Seyffert, Oskar. 1910–11. Volkskunde. *Mitteilungen des Landesvereins Sächsischer Heimatschutz* 1:8–10.

Shepard, Paul. 1967. *Man in the Landscape: A Historic View of the Esthetics of Nature.* New York: Knopf.

———. 1985. The Conflict of Ecology and Ideology. In *Modernization. The*

Humanist Response to its Promise and Problems, ed. R. L. Rubenstein. New York: Paragon House.

Shinar, Margalit. 1992. Feminist Criticism of Urban Theory and Design. *Journal of Urban and Cultural Studies* 2 (2): 29–39.

Short, John Rennie. 1991. *Imagined Country. Environment, Culture and Society.* London: Routledge.

Sieferle, Rolf Peter. 1984. *Fortschrittsfeinde? Opposition gegen Technik und Industrie von der Romantik bis zur Gegenwart.* Munich: C. H. Beck.

———. 1985. Heimatschutz und das Ende der romantischen Utopie. *Arch* (81): 38–42.

———. 1986. Entstehung und Zerstörung der Landschaft. In *Landschaft,* ed. M. Smuda. Frankfurt a. M.: Suhrkamp.

"Sieg der Heimat." 1916. Der Sieg der Heimat. *Bayerischer Heimatschutz* 14:183–88.

"Siegerland." 1913. Das Siegerland unter dem Verunstaltungsgesetz. *Siegerland* 2:49–52.

Sombart, Werner. 1904. *Gewerbewesen.* Vol. 2. Leipzig: G. J. Göschen.

———. 1920. *Der Bourgeois. Zur Geistesgeschichte des modernen Wirtschaftsmenschen.* Munich and Leipzig: Duncker und Humblot.

Spahr. 1913. Heimatschutz im ländlichen Bauwesen. *Mitteilungen des Vereins für Naturdenkmal- und Heimatschutz in Kurhessen* 3:1–4.

Spear, Jeffrey L. 1984. *Dreams of an English Eden: Ruskin and His Tradition in Social Criticism.* New York: Columbia University Press.

Speitkamp, Winfried. 1988. Denkmalpflege und Heimatschutz in Deutschland zwischen Kulturkritik und Nationalsozialismus. *Archiv für Kulturgeschichte* 70:149–93.

Stauff, Philipp. 1909. Touristensinn—Photographensinn. *Sauerländischer Gebirgsbote* 17 (2): 23–25.

Stauth, Georg, and Bryan Turner. 1988. Nostalgia, Postmodernism and the Critique of Mass Culture. *Theory, Culture and Society* 509–26.

Stein, Philipp. 1910. Sozialpolitik und Heimat. *Abhandlungen der Senckenbergischen naturforschenden Gesellschaft* 32:459–63.

Steindamm. 1909. Baupolizei und Ästhetik. *Die Hilfe* 15:299–300.

Stenographischer Bericht. 1911. *Stenographischer Bericht der Gemeinsamen Tagung für Denkmalpflege und Heimatschutz, Salzburg, 14. und 15. September 1911,* at Salzburg.

———. 1913. *Stenographischer Bericht der Zweiten Gemeinsamen Tagung für Denkmalpflege und Heimatschutz, Dresden, 25. und 26. September 1913,* at Dresden.

Storck, Karl. 1912–13. Hervorragend schöne Landschaften. *Der Türmer* 15 (1): 940.

———. 1915–16. Pantenius und Trojan. *Der Türmer* 18 (1): 396–98.

"Straßennamen und Heimatschutz." 1912. *Siegerland* 1:223.

Strobel, Joseph. 1912. Heimatkunde in der Volksschule. *Pharus. Katholische Monatsschrift für Orientierung in der gesamten Pädagogik* 3:542–50.

Tenbruck, Friedrich. 1983. Modernisierung—Vergesellschaftung—Gruppenbildung—Vereinswesen. *Kölner Zeitschrift für Soziologie und Sozialpsychologie,* 25 (*Sonderheft Gruppensoziologie*): 65–74.

Thompson, E. P. 1977. *William Morris—Romantic to Revolutionary.* 2d ed. London: Merlin Press.

Trinius, August. 1916. Recht für die deutsche Heimat. *Mitteilungen des Landesvereins Sächsischer Heimatschutz* 5:314–16.

Trip, Julius. 1908. Darlegung des Standpunktes, welcher für die Bewirtschaftung von Waldkomplexen in unmittelbarer Nähe der Städte maßgebend sein sollte. *Gartenkunst* 10:27–33.

Trojan, Ernst W. 1910. *Wanderkunst—Lebenskunst.* Munich: Gustav Lammers.

———. 1910–11. National-hygienische Grundlagen des Wanderns. *Heimat und Welt. Monatshefte der Vereinigung Heimat und Welt* 1:289–92.

Trojan, Johannes. 1912. *Erinnerungen.* Berlin: Verein der Bücherfreunde.

———. 1913. *Fahrten und Wanderungen.* Berlin: Verlag des Vereins der Bücherfreunde.

Trommer, Gerhard. 1990. *Natur im Kopf. Die Geschichte ökologisch bedeutsamer Naturvorstellungen in deutschen Bildungskonzepten.* Weinheim: Deutscher Studien Verlag.

Tutenberg, F. 1903. Landesverchönerung oder Landschaftsverschönerung? *Gartenkunst* 5:140–41.

Uhlig. 1910–11. "Bodenständige" Straßen- und Platznamen. *Mitteilungen des Landesvereins Sächsischer Heimatschutz* 1:275–76.

Urry, John. 1990. *The Tourist Gaze. Leisure and Travel in Contemporary Societies.* London: SAGE Publications.

"Verband österreichischer." 1912. Verband österreichischer Heimatschutzvereine. *Heimatschutz* 8:112.

Verein Deutscher Gartenkünstler. 1909. *Der Waldpark, seine Gestaltung und seine Erhaltung. Drei Vorträge.* Berlin: Paul Parey.

Verein deutscher Portland-Zement-Fabrikanten. 1912. *Protokoll der Verhandlungen des Vereins deutscher Portland-Zement-Fabrikanten e.V. am 29. Februar, 1. und 2. März 1912.* Berlin: Tonindustrie-Zeitungsverlag.

"Verunstaltung." 1907. Die Verunstaltung landschaftlich hervorragender Gegenden. *Frankfurter Zeitung,* 4 March 1907.

Viehöfer, Erich. 1988. *Der Verleger als Organisator. Eugen Diederichs und die bürgerlichen Reformbewegungen der Jahrhundertwende.* Ed. Historische Kommission des Börsenvereins des Deutschen Buchhandels. Vol. 30, *Archiv für Geschichte des Buchwesens.* Frankfurt a. M.: Buchhändler-Vereinigung.

"Vogelschutz im Sauerland." 1909. *Sauerländischer Gebirgsbote* 17 (12): 267.

Voit. 1822. Einige Worte über Landesverschönerung. *Monatsblatt für Bauwesen und Landesverschönerung* 37–38.

"Von der Berliner." 1912. Von der Berliner Vertreterversammlung des Bundes Heimatschutz. *Heimat-Blätter. Vierteljahresbeilage zur Fürstlich Reuß Geraer Zeitung* 4 (1): 2.

Vondung, Klaus. 1976. Zur Lage der Gebildeten in der wilhelminischen Zeit. In *Das wilhelminische Bildungsbürgertum. Zur Sozialgeschichte seiner Ideen,* ed. K. Vondung. Göttingen: Vandenhoeck und Ruprecht.

Wagener, Clemens. 1913. *Natur und Heimat. Eine praktische Einführung in die Natur- und Heimatpflege.* Mönchen-Gladbach: Volksverein-Verlag.

Walther. 1906. Einiges vom Walde und der Waldschönheitspflege. *Gartenkunst* 8:6–9.

Wangerin, Walter. 1917. Schutz der Moore. *Petermanns Mitteilungen* 63:218–19.

Wappes, Lorenz. 1950 [orig. 1887]. Über die ästhetische Bedeutung des Waldes. In *Festschrift zum 90. Geburtstage von Lorenz Wappes,* ed. T. Künkele. Munich: Bayerischer Landwirtschaftsverlag.

"Was geht." 1911. Was geht an der Havel vor? *Heimatschutz in Brandenburg* 3:84–85.

"Was sind." 1913. Was sind "hervorragend" schöne Landschaften? *Die Hilfe* 19:46.

"Wasserbauten und Heimatschutz." 1911–12. *Mitteilungen des Landesvereins Sächsischer Heimatschutz* 2:239–41.

Weber, Paul. 1910–11. Heimatschutz und römische Rechtsanschauung. *Mitteilungen des Landesvereins Sächsischer Heimatschutz* 1:226–27.

Wedding, W. 1910. Die Bedeutung großer, freier Plätze für das Volkswohl. *Heimatschutz in Brandenburg* 2:161–63.

"Weg mit dem." 1912. Weg mit dem "landschaftlich hervorragend"! *Sauerländischer Gebirgsbote* 20 (12): 271.

Wehler, Hans-Ulrich. 1985. *The German Empire, 1871–1918.* Trans. Kim Traynor. Leamington Spa: Berg.

Wehr, Georg. 1915. Die Erhaltung alter Straßennamen. *Pommersche Heimat* 4 (9): 36–37.

Weiß, Alexander. 1894. Die Schönheit der Städte. *Zeitschrift für Gartenbau und Gartenkunst* 12:225–26, 234–35.

"Werdandi-Unternehmungen." 1913. Die Werdandi-Unternehmungen. *Werdandi-Jahrbuch:* 130–32.

Wette, Hermann. 1908. *Jost Knost—Der Herkules von Latop.* Leipzig: Fr. Wilhelm Grunow.

Wettengel, Michael. 1993. Staat und Naturschutz 1906–1945. *Historische Zeitschrift* 257:355–99.

Wey, Klaus-Georg. 1982. *Umweltpolitik in Deutschland: Kurze Geschichte des Umweltschutzes in Deutschland seit 1900.* Opladen: Westdeutscher Verlag.

Weyrauch, Robert. 1915. Ästhetische Gesichtspunkte bei Wasserbauten. *Schwäbisches Heimatbuch* 119–21.

Williams, Raymond. 1977. *Marxism and Literature.* Oxford: Oxford University Press.

———. 1980. Ideas of Nature. In *Problems in Materialism and Culture. Selected Essays.* London: Verso.

———. 1983. *Culture and Society 1780–1950.* 2d ed. New York: Columbia University Press.

———. 1989. Socialism and Ecology. In *Resources of Hope. Culture, Democracy, Socialism,* ed. R. Gable. London: Verso.

Williams, Rosalind. 1982. *Dream Worlds.* Berkeley: University of California Press.

Wilmowski, Tilo von. 1910. Amerikanische Reiseeindrücke. *Heimatschutz* 6:60–63.

Wilski. 1908. Der Waldbesitz der Stadt Frankfurt a. O. und seine Erschließung für das Publikum. *Helios—Abhandlungen und Mitteilungen aus dem Gesamtgebiete der Naturwissenschaft. Organ des Naturwissenschaftlichen Vereins des Regierungsbezirkes Frankfurt/Oder* 24/25:141–55.

Wilson, William H. 1989. *The City Beautiful Movement.* Baltimore: Johns Hopkins University Press.

Wirminghaus. 1913. Die soziale Bedeutung des Heimatschutzes. *Sauerländischer Gebirgsbote* 21 (11): 176–79.

"Wohnungsfürsorge." 1911–12. *Mitteilungen des Landesvereins Sächsischer Heimatschutz* 2:275–78.

Wolf, Gustav. 1918. Ein Reiseführer des Deutschen Bundes Heimatschutz. *Heimatschutz-Chronik* 2 (10–12): 19–23.

Wolff, Fritz. 1984. Der hessische Heimatbund in seiner Geschichte. *Hessische Heimat* n.s. 34: 119–40.

Wölfflin, Heinrich. 1907. Über das Erhalten von Altertümern. *Wanderers Freund* 12 (11): 83–84.

Wolschke-Bulmahn, Joachim. 1990. *Auf der Suche nach Arkadien. Zu Landschaftsidealen und Formen der Naturaneignung in der Jugendbewegung und ihrer Bedeutung für die Landespflege.* Munich: Minerva.

Worster, Donald. 1985. *Nature's Economy. A History of Ecological Ideas.* Cambridge: Cambridge University Press.

———. 1993a. *The Wealth of Nature. Environmental History and the Ecological Imagination.* New York: Oxford University Press.

———. 1993b. Paths across the Levee. In *The Wealth of Nature. Environmental History and the Ecological Imagination.* New York, Oxford: Oxford University Press.

———. 1993c. Restoring a Natural Order. In *The Wealth of Nature. Environmental History and the Ecological Imagination.* New York, Oxford: Oxford University Press.

Wrede, Stuart, and William Howard Adams, eds. 1991. *Denatured Visions. Landscape and Culture in the Twentieth Century.* New York: Museum of Modern Art.

Wurzer, Rudolf. 1992. Camillo Sittes Hauptwerk *Der Städtebau nach seinen künstlerischen Grundsätzen:* Anlaß, Vorbilder und Auswirkungen. *Alte Stadt* 19:1–15.

Wuttig. 1910. Bücherbesprechung—"Heimatschutz." *Raiffeisen-Bote. Mitteilungen über Genossenschaftswesen und Landwirtschaftsbetrieb von dem Verbande ländlicher Genossenschaften Raiffeisenscher Organisation für Rheinpfalz, Großherzogtum Hessen und Baden e. V.* 11:286.

"Zehnter Tag." 1909. Der zehnte Tag für Denkmalpflege und die Jahresversammlung des Deutschen Bundes Heimatschutz. *Mitteilungen des Württembergischen Bundes für Heimatschutz* 1 (2/3): 17–19.

Zentgraf. 1921. Dauerwaldwirtschaft. *Heimatschutz-Chronik* 5 (12) 4–5.

Zimmermann, Peter. 1975. *Der Bauernroman. Antifeudalismus—Konservatismus—Faschismus.* Stuttgart: J. B. Metzler.

Zimmermann, Rudolf. 1910–11. Deutsche Vogelfreistätten. *Heimat und Welt* 1:251–55.

———. 1917. Unsere Hecken und ihre Bedeutung für die Vogelwelt. *Mitteilungen des Landesvereins Sächsischer Heimatschutz* 6:224–26.

"Zur Moorkultur." 1912. *Blätter für Naturschutz* 2 (12): 22–23.

Index

William H. Rollins is Lecturer in the Department of German, University of Canterbury, Christchurch, New Zealand.